# Logical Methods

# Logical Methods

**Greg Restall and Shawn Standefer**

The MIT Press
Cambridge, Massachusetts
London, England

This book was set in Times New Roman by Shawn Standefer and Greg Restall. Printed and bound in the United States of America.

Library of Congress Cataloging-in-Publication Data

Names: Restall, Greg, 1969- author. | Standefer, Shawn, author.
Title: Logical methods / Greg Restall and Shawn Standefer.
Description: Cambridge, Massachusetts : The MIT Press, [2023] | Includes bibliographical references and index.
Identifiers: LCCN 2022006372 (print) | LCCN 2022006373 (ebook) | ISBN 9780262544849 | ISBN 9780262372701 (epub) | ISBN 9780262372695 (pdf)
Subjects: LCSH: Logic–Textbooks.
Classification: LCC BC71 .R4735 2023 (print) | LCC BC71 (ebook) | DDC 511.3–dc23/eng20220712
LC record available at https://lccn.loc.gov/2022006372
LC ebook record available at https://lccn.loc.gov/2022006373

10  9  8  7  6  5  4  3  2  1

To all our students at the University of Melbourne

# Contents

# How to Read This Book

Welcome to *Logical Methods*, an introduction to logic for philosophy students and for anyone who would like to learn how to use tools and techniques from formal logic to analyze arguments, to understand logical relationships between claims, to build models, and to understand the broad sweep of how formal logic has developed in the twentieth century and into the twenty-first.

## The Structure

There are many different ways to structure an introduction to logic. You have to make choices in how to introduce things: particularly, choices about what to emphasize, what to downplay, and what to leave out. For the curious, here are two guiding choices we have made in putting this book together. The *first* guiding choice was to put equal emphasis on the two major tools in modern formal logic, *proofs* and *models*. There are two significant traditions in modern formal logic, focused around these two different ways of analyzing and defining logical concepts. From the point of view of *proofs*, an argument is valid if you can construct a proof leading from the premises to the conclusion. An argument is invalid only if there is no such proof. From the point of view of *models*, an argument is invalid if there is some model according to which the premises are true and the conclusion is false. (We call such a model a *counterexample* to the argument.) An argument is valid only if it has no counterexample. It is one of the highlights of the twentieth-century development of formal logic to distinguish these two approaches and to show that (for certain ways of understanding proofs and certain ways of understanding models) these two ways of looking at validity come to the same judgment about validity. If there is a proof from some premises to a conclusion, then there is no counterexample rendering the premises true and the conclusion false. And conversely, if there is no counterexample, then there is some proof.

Many textbooks concentrate on models and treat proofs as secondary. Some textbooks concentrate on proofs, and models are an afterthought. If you have already done a course in logic, you may have already seen models described as giving a *semantics* of the concepts being modeled, since a model is *interpreting* our language, giving the *meanings* of terms, while proofs are described as *syntax*, having nothing to do with meaning. We think that this understanding is mistaken: if proofs and models can come to make the same judgments about validity, then proofs have as much to do with meaning—with *semantics*—as do models. So, we treat proofs and models with equal emphasis.

In the first part of the book, on Propositional Logic, we start by introducing *proofs* using the propositional connectives of *conjunction*, the *conditional*, *disjunction*, and *negation*, and we get familiar with how to construct these proofs and analyze their properties. Along the way, we explore the connections between proof rules for each different logical concept and the *meanings* of those concepts, and we show that there is a sense in which whenever we can construct a proof from premises to a conclusion in this vocabulary, we can convert it into a proof that is, in an important sense, *analytic*. Once we have these on hand, we introduce *models* for propositional logic, truth tables, and truth-value assignments. Then we look at the relationship between proofs and models and prove that (if we choose our proof rules and models appropriately), whenever we have a proof for an argument, we have no counterexample, and conversely, if we have no counterexample, there is some proof.

If you have done some logic before, you might think that the natural next step would be for us to introduce predicates, terms, and quantifiers and explore the language of first-order predicate logic. We don't do this. Our *second* guiding choice was to move from propositional logic to explore *modal* logic, the logic of possibility and necessity. We made this choice for two reasons. First, the modal concepts that we explore are of deep interest in their own right and arise naturally out of the considerations of the truth-table models we explore in the first part of the book. It is very natural to think of the different models of our propositional language as representing different *possibilities*, and so, we tackle the study of possibility and necessity head on, as a natural next step. Second, the logical tools we explore, both the proofs and the models, are less complex and intricate than proofs or models for first-order predicate logic. Moving from propositional logic, to modal logic, and only then, to first-order predicate logic, results in a smooth increase in levels of complexity, and each part builds on the tools developed in the previous parts.

So, in the second part of the book, devoted to Modal Logic, we use truth-table models as our launching pad, to define models for the concepts of *necessity* and *possibility*. So we introduce our concepts using models this time, unlike in part I, when we started with proofs. Only once we get to grips with these models, we then ask ourselves the question of how it is that we could *prove* things using modal concepts. So this part ends with a treatment of modal *proof*.

The third part of the book, devoted to quantifiers and Predicate Logic, then starts off where part II ended, looking at how the kinds of features distinctive to reasoning with *necessity* and *possibility* are shared with the kinds of reasoning we find when we try to prove things using the quantifiers *all* and *some* (the universal and existential quantifiers). So, we start the last part examining proofs for first-order predicate logic, with the quantifiers. Given those rules, it is natural to ask what models for languages with quantifiers look like, so we end on that note, defining and using models for first-order predicate logic.

Along the way, we will see many different examples of how to use these tools of proofs and models for propositional, modal, and predicate logic to construct and analyze arguments, to find their structure, to build counterexamples, and to gain a mastery of key concepts that are used again and again in philosophy and in any area where we take care to use proofs and counterexamples in clarifying our reasoning.

The final chapter of the book is a short Coda. We bring together strands from the second and third parts of the book to present simple models for quantified modal logic and then we close with some suggestions for the reader interested in learning more logic. We distinguish a few topics that came up in this book, and a few that we alluded to, and provide some references for the curious to follow up.

## The Language

This is a *textbook*. It is designed to give you the tools you need to learn these concepts and tools for yourself. As you read through the book, you'll see that each chapter involves a number of different kinds of text. Whenever you enter a new section, you will find us *introducing* and *motivating* the topic by describing our focus and giving some examples of what we are talking about. Whenever you start a new section, start at the beginning and read through this section, and go back to it and revisit it if you ever find yourself wondering *why* we're talking about this topic, how it connects to other topics, or what it's all about.

Once we've introduced a topic, you will probably find a *definition* (or two or three). The definitions are **key**. It is one thing to motivate an idea and to give an example, and you'll probably get "the idea" from that, but in formal logic, our job is to be precise and to be very *explicit* about what we mean. To do that, we will mark things out by highlighting a definition of a key term or a concept. Once we offer a definition of a concept, this is the standard that we will measure it by. Any question about that concept will be answered, ultimately, by going back to its definition. So, if you have a question about something, and it's one of the things we've given a definition for, that definition is the *first* place you should look to help answer that question.

We often show how definitions work by giving examples of things that *satisfy* the conditions we specified and things that *don't*. One of the things we start off with in chapter 1, for example, is the definition of the *formal language* we will use to describe the premises, conclusions, and intermediate steps for the proofs we study in the first few chapters. The point of such a definition is to help us focus on a particular kind of pattern or feature in what we're looking at. The point of having a *precise* definition is to have some clear focus on that pattern or structure. Whenever you see a definition, make sure to check your mastery of that definition by trying this for yourself. Where are the boundaries that this definition marks out? What is included, and what is excluded? The more you can do this, the better understanding you'll have.

In some sections, we will present *theorems*: general facts about the concepts and tools we have defined, which can be precisely stated and rigorously proved. Once we have defined things, we not only see what is ruled in and what is ruled out, but we can often gain a higher-level perspective on what we've defined by noticing patterns and proving more general facts about the things we've defined: to gain a deeper understanding of the patterns we have found. These theorems won't merely be *stated*: we will *prove* them, providing a *proof*, that is, some reasoning designed to explain why the theorem holds, based purely on the definitions of the concepts involved.

Notice that proofs are both a thing we will explicitly *study* in our tour through logic and a tool that we will *use* along the way. We will study formal proofs, which are precisely defined structures using a specific set of rules governing each of the different logical concepts we study in each part of this book. There are *also* the informal proofs we will use to help us understand the concepts we're defining. As you gain mastery of the formal proofs we are studying, you will see some of their features in the informal proofs we use along the way. (It will turn out, though, that some of the features we'll use in our *informal* proofs will go beyond what we can analyze in our formal theory in this introduction. This situation is rather like what you might get in an introductory linguistics textbook. It may *use* some features of the language in which it is written that it does not *explain*.) Nonetheless, it is a helpful exercise to use the insights you gain in your understanding of proof to help you see patterns and structures in the proofs we ourselves *do* along the way.

Finally, to help you keep all of these moving parts together in your own understanding, we will end each chapter with a summary of the key concepts involved. You can use this as a guide to make sure you've come to grips with the key ideas of each chapter before progressing to the next.

## Acknowledgments

We have learned so much from our own teachers, colleagues, friends, and students throughout the years. Thanks to all of you, and in particular, to Jeremy Avigad, Nuel Belnap, Jen Davoren, Catarina Dutilh Novaes, Allen Hazen, Lloyd Humberstone, Rod Girle, Anil Gupta, Ian Hinckfuss, Sheila Oates-Williams, Graham Priest, Stephen Read, John Slaney, and Neil Williams, who have inspired and encouraged us over many years. We hope to have something like the same impact on some of our own students in the coming years.

We also acknowledge the support of the Australian Research Council, which funded Shawn Standefer's postdoctoral position at the University of Melbourne, through research grant DP150103801.

We owe special thanks to our PhD student Timo Eckhardt, who read the text carefully and gave detailed feedback on each chapter. This helped us improve things considerably. We also owe a great deal to the first class for PHIL20030 in 2019, who went on the initial ride with us with this material while we were writing the first draft. Your enthusiasm, engagement, and encouragement sustained us over the semester, and we were delighted and surprised by how much you got out of the experience and how far you traveled with us.

Greg Restall and Shawn Standefer

St Andrews and Taipei
November 19, 2021

# PROPOSITIONAL LOGIC

# 1 Introduction

Consider these two different lines of reasoning:[1]

|     | *All footballers are bipeds.* |     | *All footballers are bipeds.* |
| --- | --- | --- | --- |
| (1) | *Sócrates is a footballer.* | (2) | *Sócrates is a biped.* |
|     | So, *Sócrates is a biped.* |     | So, *Sócrates is a footballer.* |

Argument (1) is *good* in a way that the argument (2) is *not*. The conclusion of argument (1)—to the effect that Sócrates is a biped—follows from the premises of argument (1). There is no way that all footballers could be bipeds and that Sócrates could be a footballer, without Sócrates also being a biped. Argument (2) doesn't fare so well. Its conclusion (that Sócrates is a footballer) might be true (it is!), but its truth needn't follow from the premises that argument (2) offers. There is a big difference in how the information hangs together in the premises and the conclusions of these two lines of reasoning.

\* \* \*

Why is that?

\* \* \*

There are really two questions we are asking when we ask that. First: for a good argument, like argument (1), why is it so good? What makes a conclusion follow from the premises in a good piece of reasoning? (Here are some ways to spell out this question: How can we construct good arguments? What shape might they have? What rules could we follow, or what patterns can we look for in making good reasoning like this?) Then, there is a second, related question: for a bad argument, like argument (2), what makes it so bad? What do we mean when we say a conclusion doesn't follow from the premises in a bad piece of reasoning? (Here are some ways to spell out this other question: What kinds of mistakes do we make in making bad arguments? What kind of evidence should we look for in finding these missteps? What could count as some grounds for conclusively showing that a conclusion doesn't follow from the premises of an argument?)

*Logic* is concerned with questions like these. We will be concerned with ways to understand and to work with the notion of logical consequence, as well as ways to both construct good arguments and detect and construct counterexamples to bad arguments.

---

1. Sócrates Brasileiro Sampaio de Souza Vieira de Oliveira (1954–2011) *was* a footballer. He was Brazil's captain in the 1982 World Cup. https://en.wikipedia.org/wiki/Sócrates

## 1.1  Arguments and Trees

We will start our exploration of these questions, in chapters 2, 3, and 4, with representing arguments and reasoning using what is called a *natural deduction* proof system. We will answer our first question by showing how to construct *proofs*, which step-by-step show how our premises lead to the conclusion of our argument. So, the first answer will be that we can show that an argument is good by constructing a *proof* that leads from the premises to the conclusion. (Then, starting from chapter 5, we will begin our answer to the second question by showing how to construct *models* to provide counterexamples to bad arguments.)

A proof, then, needn't just have premises and a conclusion. It can have internal steps that lead us from the premises to the conclusion. Here is an example of the kind of thing we mean. Consider this reasoning.

Melbourne is east of Adelaide, and Adelaide is east of Perth, so Melbourne is east of Perth. As Sydney is east of Melbourne, it follows that Sydney is east of Perth.

Here, the *conclusion* is *Sydney is east of Perth*. The *premises* are *Melbourne is east of Adelaide*, *Adelaide is east of Perth*, and *Sydney is east of Melbourne*. In the reasoning, we also got to the claim that *Melbourne is east of Perth*, but that was an intermediate step along the way. It is neither a premise nor a conclusion. One way to display how the claims are related to each other is to display them in a *tree*, like this:

$$\frac{\text{Sydney is east of Melbourne} \qquad \dfrac{\text{Melbourne is east of Adelaide} \quad \text{Adelaide is east of Perth}}{\text{Melbourne is east of Perth}}\ (1)}{\text{Sydney is east of Perth}}\ (2)$$

There are two steps to the argument, indicated by the lines and labeled with (1) and (2). These steps are called *inferences*. The *premises* of an inference go above the line, and the *conclusion* goes below the line.[2] The premises of the first inference, labeled (1), are *Melbourne is east of Adelaide* and *Adelaide is east of Perth*, and the conclusion is *Melbourne is east of Perth*.

The premises of the second inference, labeled (2), are *Melbourne is east of Perth* and *Sydney is east of Melbourne*, and the conclusion is *Sydney is east of Perth*. Notice that the conclusion of one inference serves as a premise of another. The sentences with no lines above them are the assumptions of the argument.[3] They are simply *given*, and we don't attempt to justify them in the argument. It turns out here that the assumptions of this argument *are* all true, but this need not be the case in general.

Now we can see how the argument maps onto the tree representation. The first sentence

Melbourne is east of Adelaide, and Adelaide is east of Perth, so Melbourne is east of Perth.

has two claims followed by a "so." These two claims are assumptions, stated flatly here, so they are premises of the inference indicated by the "so." The next sentence

As Sydney is east of Melbourne, it follows that Sydney is east of Perth.

---

2. Proofs have premises and inferences have premises. The premises of an inference in a proof may not be the premises of the proof. In the example, *Melbourne is east of Perth* is the premise of an inference but isn't a premise of the proof.

3. In the next chapter, we will introduce ways to *discharge* assumptions.

begins with a claim that serves as an assumption of the argument. The "it follows" indicates that the conclusion of an inference is going to come next. The question is whether the conclusion "Sydney is east of Perth" is supposed to follow from only "Sydney is east of Melbourne" or from something else in addition. In context in this argument, it is clear that a second premise is needed to infer that one place is east of another from a premise about being east of a place. Hence the use of "Melbourne is east of Perth" as another premise.

The tree representation can be used to give a notion of *dependence* on assumptions. The conclusion that Melbourne is east of Perth depends on the assumptions that Melbourne is east of Adelaide and that Adelaide is east of Perth. Those are the assumptions that are premises of the inference with the conclusion that Melbourne is east of Perth. It does not depend on the assumption that Sydney is east of Melbourne, since that is not one of the premises of inference (1). The conclusion of inference (2) depends on the assumption that Sydney is east of Melbourne, as well as the premise that Melbourne is east of Perth, but that premise wasn't assumed. It was the conclusion of an inference, so the conclusion of (2) also depends on whatever the premise that Melbourne is east of Perth depends on. So, the conclusion of the whole argument depends on all of the assumptions of the argument.

\* \* \*

When you display the reasoning in this way, you can see that both inferences follow a single pattern:

$$\frac{x \text{ is east of } y \qquad y \text{ is east of } z}{x \text{ is east of } z}$$

*Here is an exercise for you*: can you construct a *different* proof that leads from the same premises to the same conclusion (that *Sydney is east of Perth*) using inference steps that follow the same pattern but that combine them in some different way? Display your reasoning in a tree.

\* \* \*

In the first part of this course, we will look at proofs that have this sort of shape, but we won't focus on concepts like ... *is east of* ... that we used here. We will focus on basic patterns of inference for concepts like *and*, *if*, *or*, and *not*, which feature in so much of our reasoning and that can be used no matter what we are talking about.

The rest of this chapter will introduce the general structure of trees and then the formulas we will use to focus on patterns in reasoning involving *and*, *if*, *or*, and *not*. First, *trees*.

We have already seen that our tree made out of the statements *Sydney is east of Melbourne*, *Melbourne is east of Adelaide*, and so on, represented a notion of *dependence* between those statements. The assumptions depended on nothing else, other than themselves. The intermediate step depended on those two assumptions, and the final conclusion depended on everything that came before it. This dependence relation is an example of a *binary relation*, a relation that holds between two things from some *sets* of things. Sets are general collections of things, determined solely by their members.[4] Some binary relations that may be familiar are the "greater than" relation on the set of natural numbers, the "is

---

4. Some background on sets is included in the appendix, on page 13.

the same age as" relation on the set of people, and the "immediately precedes" relation on the set of days of the week.[5] We will encounter many binary relations and sets in this book. The dependence relation is an example of a special kind of binary relation, what we call an *order* on a collection, in this case, the collection of statements in the tree. This is a general phenomenon:

**Definition 1**    *A PARTIAL ORDER, $(X, \leq)$, is a set X and a binary relation $\leq$ on X such that*

- *for every element x of X, we have $x \leq x$,*
- *for every x and y in X, if $x \leq y$ and $y \leq x$, then $x = y$, and*
- *for every $x, y, z$ in X, if $x \leq y$ and $y \leq z$, then we also have $x \leq z$.*

Here, we think of $x \leq y$ as telling us that $x$ depends on $y$. (This stretches the notion of dependence a bit, in that we think of things depending on *themselves*, as well as the things they've been inferred from.)

The dependence relation in our reasoning was more than a strict order like this: a *tree* is a special kind of partial order.

**Definition 2**    *A TREE $\Pi$ is a partial order, $(T, \leq)$, such that*

- *there is a* root*, that is, an element t of T, such that for all x in T, $t \leq x$, and*
- *if x, y, and z are in T, $y \leq x$, and $z \leq x$, then either $y \leq z$ or $z \leq y$.*

Now we have added the idea that a tree has a single root: something that depends on everything in the tree, and the dependence relation branches outward. If $y$ and $z$ are items that both depend on $x$, then one of these depends on the other. Here is how you can see what is going on with this condition. Look at this tree:

$$\frac{\dfrac{a \quad b}{c} \quad d}{e}$$

Here the dependence relation goes like this. The bottom inference tells us that $e \leq c$ and $e \leq d$. The top inference tells us that $c \leq a$ and $c \leq b$. Chaining these two together, we have $e \leq a$ and $e \leq b$ too. And that's how dependence goes. Notice that $c \leq b$ and $e \leq b$. That is, both $c$ and $e$ depend on $b$. And here, $e$ depends on $c$, too, satisfying the second condition in definition 2. What would violate that condition would be adding the dependence relation $d \leq b$, for then, we'd have $c$ and $d$ both depending on $b$, while neither depends on the other. Drawing in a dependence link from $d$ to $b$ would make this not branch out like a tree. (We would have two different ways to get from the root $e$ up to the leaf $b$—to the left, through $c$, and to the right, via $d$.) If we wanted to add a dependence relation between $d$ and $b$, we would *copy* $b$ like this:

$$\frac{\dfrac{a \quad b^1}{c} \quad \dfrac{b^2}{d}}{e}$$

and now, we still have the structure of a tree. Now, $c \leq b^1$ while $d \leq b^2$.

---

5. All these relations relate things from one set to another thing in that same set. This is not required, as a relation can relate things from one set to things from a different set, such as the "is the birthday of" relation on the set of calendar dates and the set of people.

* * *

In general, finite trees can be pictured graphically, with the root at the bottom and the rest of the tree growing upward from it. The elements of $T$ are represented by nodes in the tree, and a line between two nodes indicates that the lower node is below the higher node in the ordering. Nodes that are not below any others are called *leaves*. We have seen cases in trees where we have *binary branching* dependence (where $e \leq c$ and $e \leq d$) and *unary "branching"* (where $d \leq b^2$ alone), and we will later see ternary, or three-way, branching and other structures in trees.

In later chapters, we will introduce inference rules that *discharge* assumptions. That is, we might make an assumption in the course of the reasoning and then let go of that assumption as the reasoning proceeds. These rules create new proofs from given proofs but with fewer assumptions. This will greatly broaden the scope of the kinds of arguments we can represent. For example, it permits us to represent *hypothetical reasoning*. Here is an example:

> I'm going to get a pet puppy. Suppose pets have to be registered with the council. Registration costs $50. So, I'll have to register my puppy with the council, and it follows that I'll have to pay $50. So, *if* pets have to be registered with the council, I'll have to pay $50.

We use this kind of reasoning all the time. In it, I made a *supposition*—I supposed that pets have to be registered with the council—and I used that supposition in a piece of my reasoning. This part of the reasoning has a structure like this:

$$\frac{\displaystyle \frac{\textit{I'm getting a puppy} \quad \textit{Pets have to be registered}}{\textit{I have to register my puppy}} \quad \textit{Registration costs \$50}}{\textit{I'll have to pay \$50}}$$

Now the final step of the reasoning *discharges* the assumption we made (that pets have to be registered), to conclude the *conditional* conclusion *if pets have to be registered with the council, I'll have to pay $50*. We'll represent this step like so:

$$\frac{\displaystyle \frac{\displaystyle \frac{\textit{I'm getting a puppy} \quad [\textit{Pets have to be registered}]^1}{\textit{I have to register my puppy}} \quad \textit{Registration costs \$50}}{\textit{I'll have to pay \$50}}}{\textit{If pets have to be registered, I'll have to pay \$50}}{}^1$$

The brackets around the assumption mean that the assumption has been *discharged* (it has done its work—we need not assume it any more) and the superscript 1 links it to the inference step labeled with 1, the point at which it was discharged in our reasoning.

Discharging assumptions also plays a role in representing *reasoning by cases*. Here's a simple example of reasoning like this:

> You're considering options for a pet. You can get either a cat or a dog. Suppose you get a cat. It would be very cute and cuddly. Cute and cuddly animals make you happy, so getting a cat would make you happy. On the other hand, suppose you get a dog. It would be very playful and need to go out for walks. Going on walks makes you happy. So, either way you would end up being happy.

The notion of dependence gets more complicated once we introduce discharging, so we will wait to formally define how dependence works in general until chapter 3, when our complete proof system has been defined. For now, we will rely on the following intuitive idea: assumptions depend on themselves, and the conclusion of an inference depends on anything a premise of the inference depends on.

## 1.2   Sentences and Formulas

We've begun looking at the structure of proofs. In this section, we'll begin to look at the items inside those proofs. In the examples we've seen so far, proofs involve sentences like *I'm going to get a pet puppy* or *Sydney is east of Perth*. These sentences are the kinds of things that we can use as the premises or conclusions in our reasoning. They are the kinds of things that can be true or false, that we can agree with or disagree with, and that are at the heart of our reasoning. Of course, not all sentences can do this work. A question, like *What is your puppy's name?*, or a request, like *Please open the window*, or a greeting, like *Hello!*, cannot be the premise or conclusion of a proof; they aren't the sorts of things that we look for evidence for or against, and they can't be true or false. The kinds of sentences that we find in our proofs are *declarative* sentences: those that declare how things are.[6] They are the kinds of things we can use to assert or deny; we can suppose they are the case and look for reasons for or against. Philosophers say that declarative sentences express *propositions*. They describe how things are; they propose that the world is a particular way.

In this subject, we'll be learning to work with the tools and methods of formal logic. That means we will focus on the forms and patterns we find in the declarative sentences in the languages that we speak, read, and write.[7] For most of this book, we will focus on what logicians call *propositional languages*, where there are declarative sentences and a small number of ways that sentences can be combined to form new declarative sentences. In English, we express these combinations using some simple "logical" words like *and*, *if*, *or*, and *not*. To help us focus on these concepts, we will introduce what we call a *formal language*—a language designed just to express the forms of propositions combined using these concepts. This will help us focus on the properties of just those logical concepts, and abstract away from the other details of the language.

---

6. Read Nuel Belnap's "Declaratives Are Not Enough" (1990) to rid yourself of the idea that declaratives are the only sentences that *really* matter to philosophy or to logic. Nonetheless, declaratives are where we will focus all of our time in this subject. Nuel Belnap (b. 1930) is an American logician who made many contributions to nonclassical logics, theories of truth, and modal logics. His 1959 PhD thesis, *The Formalization of Entailment*, was supervised by Alan Ross Anderson, with whom Belnap went on to do significant work on relevant logics, culminating in the two volumes of *Entailment* (Anderson and Belnap 1975; Anderson, Belnap, and Dunn 1992). He contributed to two approaches to the theory of truth, the prosentential theory, with Dorothy Grover and Joseph Camp (Grover, Camp, and Belnap 1975) and the revision theory, with Anil Gupta (Gupta and Belnap 1993). He made significant contributions to the logic of agency and branching time with his work on STIT logic (the logic of "seeing to it that"), including *Facing the Future* (Belnap, Perloff, and Xu 2001), with Michael Perloff and Ming Xu.

7. Focusing on form has the advantage of being applicable to languages that we *don't* know and can't speak, and that haven't yet arisen, as well as applying to nonnatural languages like digital representations in programming languages or the internal representations of databases.

**Definition 3**   *A* PROPOSITIONAL LANGUAGE *is given by defining the* FORMULAS *in that language. We start with a set* **Atom** *of* ATOMIC FORMULAS *(so each atomic formula is a formula). We reserve one special atomic formula* $\perp$.[8] *Then, to make new formulas out of old formulas, if A and B are formulas, then we can build new formulas like this:*

$$(A \wedge B) \qquad (A \vee B) \qquad (A \to B) \qquad \neg A$$

*These are also formulas, no matter what formulas A and B are. Finally, nothing else is a formula in this language, other than what you can build using these rules in some finite number of steps.*
    *The set of all formulas of the propositional language is called* **Form**.

The idea, of course, with our formulas is to help us focus on logical structure. Here is how that logical structure corresponds to some of the words we use in English.[9]

| symbol | words | concept |
|:---:|:---:|:---:|
| $\wedge$ | ... *and* ... | conjunction |
| $\vee$ | *either* ... *or* ... | disjunction |
| $\to$ | *if* ... *then* ... | conditional |
| $\neg$ | *it is not the case that* ... | negation |

The formation rules in our formal language correspond to the behavior of sentences in the languages we speak. If *A* and *B* are sentences in our language, then *A and B* is also a sentence. So is *either A or B*, and so is *if A then B* and *it is not the case that A*.[10,11]

* * *

So, that is how you build up a propositional language out of a set, **Atom**, of atoms. What are the atoms? We will use letters like

$$p \qquad q \qquad r \qquad s$$

in our set **Atom**, and if they run out, we'll add subscripted numbers, so we can have an unending supply of atoms.

$$p_1, p_2, p_3, \ldots \qquad q_1, q_2, q_3, \ldots \qquad r_1, r_2, r_3, \ldots \qquad s_1, s_2, s_3, \ldots$$

So, since *p* and *q* are formulas, so is $(p \vee q)$, and so is $(p \vee q)$ and so is $(q \to p)$ and so is $\neg p$. But since *p* is a formula *and p* is a formula (that's right, we've said the same thing twice), then so is $(p \wedge p)$, and so is $(p \vee p)$. There is no requirement in the definition that *A* and *B* be *different* formulas when you combine them.

---

8. We'll let you in on what $\perp$ means in chapter 3, but for now, you can think of it as a special false statement.

9. Disjunctions ("or") will always be *inclusive* in this text. This means that a disjunction *A* or *B* is true in three circumstances: (i) *A* is true and *B* is not, (ii) *B* is true and *A* is not, and (iii) both *A* and *B* are true. The contrast is *exclusive* disjunction, which is not true in case (iii). We will return to exclusive disjunction in an exercise in chapter 3.

10. Try this yourself, with sentences like "The cat sat on the mat," "the cow jumped over the moon," and "politics isn't like it used to be," or declarative sentences of your own choosing.

11. We will also see the connective "$\leftrightarrow$" in chapter 2, for "*if and only if*," the biconditional. However, we will not make this a separate part of our vocabulary, as we will show that it can be defined in terms of the conditional and conjunction.

We can combine complex formulas to make formulas, too, since $A$ and $B$ in our rules can be any formula you like. So, $(p \wedge q) \vee (q \to r)$ is a formula, and so is $\neg\neg(p \to p)$.

* * *

Why do we have parentheses in our formulas? The reason is simple. Saying something like

> *The cat sat on the mat and either the cow jumped over the moon or politics isn't like it used to be.*

which has the *form* $(p \wedge (q \vee r))$ has a different structure than

> *Either the cat sat on the mat and the cow jumped over the moon, or politics isn't like it used to be.*

which has the form $((p \wedge q) \vee r)$. If we didn't use our parentheses to distinguish the two forms, we'd just have

$$p \wedge q \vee r$$

which is *ambiguous* between the two different things you could mean. Helping us distinguish between different ways to read a sentence—*disambiguating* them—is one of the neat features of having a precisely defined formal language. The formal language is unbending and inflexible, which means we can use it to highlight the different ways that our natural languages can be flexible and ambiguous. Having an *unambiguous* grammar is why we have these parentheses whenever we combine two formulas into a new formula with an *and*, an *or*, or an *if*.

Having parentheses that come along with every conjunction, disjunction, or conditional means that you have a bit more writing than you really *need* to for the job of disambiguating. After all, $p \wedge (q \vee r)$ is no more ambiguous than $(p \wedge (q \vee r))$. So, from now on, we'll often leave out the *outermost* pair of parentheses if leaving them out would make no difference.

* * *

One neat feature of our unambiguous grammar is that each complex formula in our formal language has a unique *main connective*.

**Definition 4** *For every complex formula in our propositional language, its* MAIN CONNECTIVE *is defined like this:*

- *The main connective of* $\neg A$ *is* $\neg$, *and A is the* negand.
- *The main connective of* $A \wedge B$ *is* $\wedge$, *and A and B are the* conjuncts.
- *The main connective of* $A \vee B$ *is* $\vee$, *and A and B are the* disjuncts.
- *The main connective of* $A \to B$ *is* $\to$, *and A is the* antecedent *and B is the* consequent.

Since every complex formula (that is, every formula that isn't in Atom) is built up using the rules for $\wedge$, $\vee$, $\to$, and $\neg$, this definition does not leave any cases out.

Here is how you can identify the main connective of a formula in practice: to identify the main connective, you identify its atoms, and then you reconstruct the formula using the formation rules above. The *final rule* used is the rule for the main connective. Here is a simple example: Consider the formula

$$p \wedge (q \vee r).$$

The atoms in this formula are $p, q, r$. By the $\vee$ formation rule, $(q \vee r)$ is a formula. By the $\wedge$ formation rule, $p \wedge (q \vee r)$ is a formula, and this is the formula we are trying to construct, so its main connective is the $\wedge$.

Here are some examples of more complex formulas, where we have underlined the main connective.

- $p \underline{\vee} \neg q$
- $\underline{\neg}(p \vee q)$
- $\neg p \underline{\vee} q$
- $(p \wedge q) \underline{\rightarrow} r$
- $(p \rightarrow p) \underline{\rightarrow} ((q \rightarrow q) \rightarrow (r \rightarrow r))$
- $((p \rightarrow p) \rightarrow (q \rightarrow q)) \underline{\rightarrow} (r \rightarrow r)$

Notice that the last two formulas are *nearly* the same, but the parentheses make a big difference. $(p \rightarrow p) \rightarrow ((q \rightarrow q) \rightarrow (r \rightarrow r))$ is a conditional with $p \rightarrow p$ as its antecedent and $(q \rightarrow q) \rightarrow (r \rightarrow r)$ as its consequent, while $((p \rightarrow p) \rightarrow (q \rightarrow q)) \rightarrow (r \rightarrow r)$ has $r \rightarrow r$ as its consequent and $(p \rightarrow p) \rightarrow (q \rightarrow q)$ as its antecedent.

$$* * *$$

We are nearly at the end of our exploration of the structure of formulas. The last point we need to notice is the relationship between a formula and its parts, its *subformulas*. The subformulas of a formula are all of the formulas you make as you construct the formula out of atoms. So, for example, the subformulas of $((p \rightarrow p) \rightarrow (q \rightarrow q)) \rightarrow (r \rightarrow r)$ are $p$, $q$, and $r$ (its atoms) $p \rightarrow p$, $q \rightarrow q$ and $r \rightarrow r$, and $(p \rightarrow p) \rightarrow (q \rightarrow q)$, and then, for good measure, the formula itself. Here is the official definition:

**Definition 5 (Subformulas)**   *The set of* SUBFORMULAS *of a formula A*, sub($A$), *is defined like this:*

- sub($p$) = \{$p$\}, *where p is an atom,*
- sub($\neg A$) = sub($A$) $\cup$ \{$\neg A$\},[12]
- sub($A \wedge B$) = sub($A$) $\cup$ sub($B$) $\cup$ \{$A \wedge B$\},
- sub($A \vee B$) = sub($A$) $\cup$ sub($B$) $\cup$ \{$A \vee B$\}, *and*
- sub($A \rightarrow B$) = sub($A$) $\cup$ sub($B$) $\cup$ \{$A \rightarrow B$\}.

To work out the subformulas of a formula, such as $p \vee \neg q$, you proceed in stages according to the definition, which breaks a formula down into its component parts, collecting the formulas along the way. For the first stage,

$$\text{sub}(p \vee \neg q) = \text{sub}(p) \cup \text{sub}(\neg q) \cup \{p \vee \neg q\},$$

---

12. We use *set* notation here. If $S$ and $T$ are sets, then $S \cup T$ is a set, the *union* of the sets $S$ and $T$, which contains all of the elements of $S$ and all of the elements of $T$. Something is in $S \cup T$ if and only if it is either in $S$ or in $T$. We use braces to indicate small sets. \{$x$\} is the set containing only one thing, $x$. \{$x, y$\} is the set containing the two things $x$ and $y$. If you want to see more on sets and some of their properties, see the appendix to this chapter.

but $\text{sub}(\neg q) = \text{sub}(q) \cup \{\neg q\}$. As $\text{sub}(p) = \{p\}$ and $\text{sub}(q) = \{q\}$, we can put all these together to get

$$\text{sub}(p \vee \neg q) = \{p\} \cup \{q\} \cup \{\neg q\} \cup \{p \vee \neg q\} = \{p, q, \neg q, p \vee \neg q\}$$

So, $p \vee \neg q$ has *four* subformulas, $p$, $q$, $\neg q$, and the whole formula $p \vee \neg q$.

Here are some examples of subformulas.

- $\text{sub}(\neg(p \vee q)) = \{\neg(p \vee q), p \vee q, p, q\}$
- $\text{sub}(\neg p \vee q) = \{\neg p \vee q, \neg p, p, q\}$
- $\text{sub}((p \wedge q) \rightarrow r) = \{(p \wedge q) \rightarrow r, p \wedge q, p, q, r\}$
- $\text{sub}((p \rightarrow p) \rightarrow ((q \rightarrow q) \rightarrow (r \rightarrow r))) = \{(p \rightarrow p) \rightarrow ((q \rightarrow q) \rightarrow (r \rightarrow r)), p \rightarrow p, p,$
  $(q \rightarrow q) \rightarrow (r \rightarrow r), q \rightarrow q, q, r \rightarrow r, r\}$
- $\text{sub}(((p \rightarrow p) \rightarrow ((q \rightarrow q)) \rightarrow (r \rightarrow r)) = \{((p \rightarrow p) \rightarrow ((q \rightarrow q)) \rightarrow (r \rightarrow r), p \rightarrow p, p,$
  $((p \rightarrow p) \rightarrow (q \rightarrow q)), q \rightarrow q, q, r \rightarrow r, r\}$

Here is one reason that we might be interested in the subformulas of a formula. The subformulas of a formula are what you find when you analyze a formula into its component parts. One way to understand the program of *analytic philosophy* is in its focus on *analysis* of sentences in terms of their components and to understand the properties of a sentence as arising out of the features of its components. It is not our task here to argue for or against this approach, but it is enlightening to see what kinds of properties of formulas can arise out of properties of their subformulas and what kinds of properties do not have this feature. (We will return to this issue at a number of points throughout this book.)

As a general convention, we will use capital letters from early in the alphabet, "*A*," "*B*," "*C*," and so forth, to stand for arbitrary formulas. Lowercase letters from later in the alphabet, "*p*," "*q*," "*r*," and so forth, are particular atoms. The formula $p \wedge q$ has complexity 1, whereas $A \wedge B$ has complexity 1 plus the sum of the complexities of $A$ and $B$.

Definitions 4 and 5 provide the material for the useful concept of the *scope* of a connective. In the formula $\neg A$, $A$ is in the scope of the negation, as is any subformula of $A$. In the formula $A \wedge B$, $A$ and $B$ and their subformulas are in the scope of the conjunction. In the formula $A \vee B$, $A$ and $B$ and their subformulas are in the scope of the disjunction. In the formula $A \rightarrow B$, $A$, $B$ and their subformulas are in the scope of the conditional. Scope is useful for talking about the structure of formulas. In some contexts, which we will see later, it is useful to keep track of when connectives occur within the scope of other connectives, such as a disjunction occurring in the scope of a single negation, as in $\neg(p \vee q)$. The concept of scope will be especially useful when additional connectives are introduced later, in sections 7.3.1 and 10.1.

* * *

We will end this section returning to the issue that started this chapter. We are interested in ways of distinguishing good and bad arguments. For the first part of this book, we are focusing on the structure of the propositional connectives, the structure expressed in our formal language Form. So, the question becomes more specific. What arguments, whose premises and conclusions are taken from Form, are good? And how can we show that they are good?

**Definition 6 (Arguments)** *An argument in* Form *is defined by its premises and its conclusion. The premises of an argument consist in a set X of formulas, and its conclusion is a formula, A. We sometimes write the argument with premises X and conclusion A as*

$$X \succ A$$

Our task will be to give some account of when an argument is good. Our first approach will be to see how we can bridge the gap between the premises in $X$ to the conclusion $A$ into a step-by-step *proof*, which leads us from the premises to the conclusion, using principles determined by the meanings of the concepts involved.

## 1.3 Appendix: Some Useful Background

This section goes into three background topics in more depth: *sets, functions*, and *inductive arguments*. If you are already comfortable with these topics, you should skip this section and only come back here if you have questions about sets, functions, or reasoning by induction. If you don't know what these topics are yet, it's also best to skip this section until you come to a section in the next chapters that point you back here. On the other hand, if you are curious and want to be prepared for when we use more of these topics, feel free to read ahead.

Sets: We already saw the notion of sets when we defined the *subformulas* of a formula. Here is what we need to know about how sets work, in general. A set is a kind of *collection* of elements. If $S$ is a set and $t$ is an element, we can ask whether $t$ is a *member* of $S$ or not. We write "$t \in S$" to say that $t$ is a member of $S$ and "$t \notin S$" to say that $t$ is *not* a member of $S$. The crucial defining feature of sets is that they are *determined* by which members they have. In other words, if $S_1$ and $S_2$ are different sets, then there must either be some member of $S_1$ that isn't a member of $S_2$, or vice versa. That is, the only way to distinguish two sets is to find a member of one that is not a member of the other.

We can represent finite sets by writing out their members in a list between braces: $\{x, y, z\}$ is the set whose members are $x$, $y$, and $z$. The set $\{x\}$ is the set that contains only one member. We call it a *singleton* (it has a *single* member). If $S$ and $T$ are sets, then $S \cup T$—the *union* of $S$ and $T$—is the set whose members are the members of $S$ together with the members of $T$. In other words, something is a member of $S \cup T$ if and only if[13] it is *either* a member of $S$ *or* a member of $T$. Similarly, $S \cap T$—the *intersection* of $S$ and $T$—is the set whose members are those items that are *both* members of $S$ *and* members of $T$. So, $\{a, b, c\} \cup \{c, d\} = \{a, b, c, d\}$. The union of the three-element set $\{a, b, c\}$ and the two-element set $\{c, d\}$ has *four* elements. The intersection $\{a, b, c\} \cap \{c, d\}$ is the singleton set $\{c\}$, since $c$ is the only thing that is a member of both sets. Notice that if we have two sets, like $\{a, b\}$ and $\{c, d\}$ with no members in common, there is nothing in the intersection $\{a, b\} \cap \{c, d\}$. The intersection is the set $\{\ \}$ with no members. We call this the *empty set*, and we will write it as $\emptyset$. (There is only *one* empty set, since there can't be two different

---

13. Sometimes we will use the shorthand "iff" for "if and only if," especially when we are using a *lot* of "if and only ifs."

empty sets. If two sets are different, then there has to be something in one of the sets that is not in the other, and that cannot happen if both are empty.)

We say that the set $S$ is a *subset* of the set $T$ if every member of $S$ is a member of $T$. We write this "$S \subseteq T$." (Notice that according to this definition, a set is a subset of itself, since every member of $S$ is, of course, a member of $S$.) So, for example, $\{a, b\} \subseteq \{a, b, c\}$.

We say that the sets $S$ and $T$ are *disjoint* if no member of $S$ is a member of $T$ and no member of $T$ is a member of $S$. We write this "$S \cap T = \emptyset$." As an example, $\{a, b\}$ and $\{c, d\}$ are disjoint, in symbols, $\{a, b\} \cap \{c, d\} = \emptyset$.

To make sure you understand these concepts, you should confirm to your own satisfaction that for any sets $S$ and $T$, $S \cap T \subseteq S \subseteq S \cup T$ and that (therefore) $\emptyset$ is a subset of any set at all. (Why the "therefore"? Well, if $S \cap T \subseteq S$ for any sets $S$ and $T$, this includes the case where $S$ and $T$ are disjoint, and then, $S \cap T = \emptyset$, so $\emptyset \subseteq S$.)

If $S$ is a set, we can think of all of the subsets of $S$ as forming a set of its own. This is called the *power set* of $S$, or $\mathscr{P}S$. So, since the subsets of $\{a, b\}$ are $\emptyset$, $\{a\}$, $\{b\}$, and $\{a, b\}$, the power set $\mathscr{P}\{a, b\}$ is the set $\{\emptyset, \{a\}, \{b\}, \{a, b\}\}$.

\* \* \*

We will think of our propositional language as determining a set Form, whose members are all of the formulas in that formal language. So, if Atom is the set of *atomic* formulas, we have Atom $\subseteq$ Form, since every atomic formula is a formula. Another set that we will see from time to time is the set $\mathbb{N}$, of all natural numbers. It is the set defined as containing the number 0, and such that whenever it contains a number $n$, it also contains the next number $n + 1$. So, it contains 0, 1, 2, 3, and so on, but it does not contain the negative numbers, or fractions or infinite numbers. It contains all and only the finite counting numbers.

Functions:   Sometimes we are interested in operating on or transforming members of a set $S$. We have seen this already when we introduced the notion of the *subformulas* of a formula. For every $A$ in Form, we have defined sub($A$), which is a *set* of formulas. So, for every $A \in$ Form, we have defined sub($A$), which is a member of $\mathscr{P}$Form, the set of all *sets* of formulas. We say sub is a *function*, which sends members of Form to members of $\mathscr{P}$Form, which we write like this: "sub : Form $\mapsto \mathscr{P}$Form."

In general, "$f : S \mapsto T$" means that $f$ is a *function* from the set $S$ to the set $T$, so that for any element $s \in S$, the element $f(s)$ (the function $f$ *applied to* $s$) is a member of the set $T$.

Here is another example of a function, this time, a function from Form to $\mathbb{N}$, which associates with every formula a number, which we call the *complexity* of the formula. Here is the rule for defining the complexity of a formula:

- comp($A$) = 0 if $A \in$ Atom,
- comp($\neg A$) = comp($A$) + 1,
- comp($A \wedge B$) = comp($A$) + comp($B$) + 1,
- comp($A \vee B$) = comp($A$) + comp($B$) + 1,
- comp($A \rightarrow B$) = comp($A$) + comp($B$) + 1.

This defines the complexity value of all formulas in Form. Atoms have complexity zero, and each time we add a connective, we add 1 to the complexity value. In other words, the complexity of a formula is the number of connectives in it.

To check that this works as planned, here is comp($p \land (q \to \neg r)$). Spelling out the definition, step-by-step, we have

$$\text{comp}(p \land (q \to \neg r)) = \text{comp}(p) + \text{comp}(q \to \neg r) + 1$$

$$= 0 + \text{comp}(q) + \text{comp}(\neg r) + 1 + 1$$

$$= 0 + 0 + \text{comp}(r) + 1 + 1 + 1$$

$$= 0 + 0 + 0 + 1 + 1 + 1$$

$$= 3$$

Notice that the clauses of the definition of the complexity function (like the subformula function before it) parallel exactly the rules for constructing formulas, defining the function on atoms, and then showing how the function on *complex* formulas is defined in terms of the function already defined in terms of its components. We can think of the definition of the function as showing how you can compute the value of the function *alongside* the process of constructing the formula to which the function is applied. We will see this process again and again as we define different kinds of functions on the set of formulas.

This process is called an *inductive definition* of a function. We define the value of the function for the *base cases* (in this case, on the members of the set Atom) and then, for the objects that are constructed out of simpler objects (in this case, conjunctions, disjunctions, conditionals, or negations of formulas), we define the values of the function on the complex formulas *in terms of* the values of the function on the parts out of which the complex formula is made. This is called the *inductive step* in an inductive definition. Once the function has been defined for the base cases and the inductive steps, the function is defined for all its inputs.

Inductive arguments:   What goes for functions from Form to other sets also goes for *facts* about members of the set Form and other sets like them. We can prove a fact about all our formulas in exactly the same way, by proving that fact for the atomic formulas (this is the base case) and then showing that *if* the fact holds for some formulas, it also holds for the formulas built up out of those formulas (this is the inductive step). Then, in just the same way as we define a function on Form as the formulas are constructed, here we are verifying a truth about formulas, as they are constructed. Here is a simple example: we will verify that every formula contains an even number of parentheses.

[BASE CASE]  For the base case, every formula in Atom has zero parentheses, and zero is even.

[INDUCTIVE STEP]  For the inductive case, assume that $A$ and $B$ are formulas that each have an even number of parentheses, say $n$ and $m$, respectively. Then $(A \land B)$ has $n + m + 2$ parentheses, and $n + m + 2$ is an even number. Similarly, $(A \lor B)$ and $(A \to B)$ both have $n + m + 2$ parentheses. That leaves $\neg A$, which has $n$ parentheses, since the formation rule for negation does not add any parentheses, and $n$ is still an even number.

We can therefore conclude that *every* formula contains an even number of parentheses.

* * *

We can apply inductive arguments to anything that is built up in the same sort of way as formulas are. The simplest case is the numbers, $\mathbb{N}$, for these can be thought of as generated out of the starting number 0 using the +1 operation, that is, it is the set $\{0, 0+1, 0+1+1, 0+1+1+1, 0+1+1+1+1, \ldots\}$. A little more formally, 0 is a natural number and if $n$ is a number, then so is $n+1$. So, to show that every number has some feature, we need to show that 0 has that feature, and if a number $n$ has the feature, then so does $n+1$. Once we have done that, we can be sure that *all* numbers have the feature, as they are generated from 0 by closing the collection under the operation of adding one.

Here is a simple argument to illustrate this: we will argue that every number is either even or odd, in other words, every number is either divisible by 2 or is one more than a number divisible by 2.

[BASE CASE] For the base case, 0 is even, since it is divisible by 2 without remainder. So 0 is even or odd.

[INDUCTIVE STEP] For the inductive case, assume that $n$ is either even or odd. If it is even, then $n+1$ is odd, so it is even or odd. If $n$ is odd, then there is an even number $m$ such that $n = m+1$. Then $n+1 = m+1+1$, and $m+1+1$ is even. Therefore, $n+1$ is even, so it is even or odd. In both cases, $n+1$ is even or odd.

We can then conclude that *every* member of $\mathbb{N}$, that is, every natural number, is even or odd.

## 1.4 Key Concepts and Skills

Here are the key concepts you should make sure you've grasped before proceeding to the next chapter. Use this list as a checklist to see if you have mastered each skill and understood each concept.

- ☐ You can identify premises and conclusions in a course of reasoning presented in a natural language argument.
- ☐ You understand the definitions of the concepts *partial order* and *tree*. You know how to check whether a partial order is also a tree, and you can construct examples of partial orders that aren't trees. You can represent finite trees in tree diagrams.
- ☐ You can represent the structure of reasoning of simple arguments in the form of a tree, distinguishing premises and conclusions, as well as individual inference steps, and recognizing the ultimate conclusion of a proof.
- ☐ You can construct formulas in the formal propositional language Form. You know how to read formulas, recognizing conjunction ($\wedge$), disjunction ($\vee$), the conditional ($\rightarrow$), and negation ($\neg$), and you are able to detect whether something is actually a formula or if it is not formed using the formation rules of the formal language Form.
- ☐ You can identify the main connective of a complex formula and the subformulas of a formula.

A PARTIAL ORDER, $(X, \leq)$, is a set $X$ and a binary relation $\leq$ on $X$ such that

- for every element $x$ of $X$, we have $x \leq x$,
- for every $x$ and $y$ in $X$, if $x \leq y$ and $y \leq x$, then $x = y$, and
- for every $x, y, z$ in $X$, if $x \leq y$ and $y \leq z$, then we also have $x \leq z$.

A TREE $\Pi$ is a partial order, $(T, \leq)$, such that

- there is a *root*, that is, an element $t$ of $T$, such that for all $x$ in $T$, $t \leq x$, and
- if $x$, $y$, and $z$ are in $T$, $y \leq x$ and $z \leq x$, then either $y \leq z$ or $z \leq y$.

The set of formulas Form is defined as follows.

- Every atom in Atom is in Form.
- If $A$ and $B$ are in Form, then so are $(A \wedge B)$, $(A \vee B)$, $(A \rightarrow B)$, and $\neg A$.
- Nothing else is in Form.

The set of SUBFORMULAS of a formula $A$, sub$(A)$, is defined like this:

- sub$(p) = \{p\}$, where $p$ is an atom,
- sub$(\neg A) = $ sub$(A) \cup \{\neg A\}$,
- sub$(A \wedge B) = $ sub$(A) \cup $ sub$(B) \cup \{A \wedge B\}$,
- sub$(A \vee B) = $ sub$(A) \cup $ sub$(B) \cup \{A \vee B\}$, and
- sub$(A \rightarrow B) = $ sub$(A) \cup $ sub$(B) \cup \{A \rightarrow B\}$.

## 1.5  Questions for You

Use these questions to test your understanding of the concepts introduced in this chapter.

If you find that you can't answer these questions, go back to each section and make sure to work on the parts you don't yet understand.

### Basic Questions

1. Recall this reasoning:

> Melbourne is east of Adelaide, and Adelaide is east of Perth, so Melbourne is east of Perth. As Sydney is east of Melbourne, it follows that Sydney is east of Perth.

which we represented by this tree:

$$\cfrac{\text{Sydney is east of Melbourne} \qquad \cfrac{\text{Melbourne is east of Adelaide} \quad \text{Adelaide is east of Perth}}{\text{Melbourne is east of Perth}}\;{\scriptstyle(1)}}{\text{Sydney is east of Perth}}\;{\scriptstyle(2)}$$

Construct a *different* proof from the same premises to the same conclusion, using the same principles but combining them in a different way. How many ways are there to do this?

2. Consider this argument.[14]

> Hunger is caused by the stomach, by blood acting on the brain, or by all of the body's cells. If the stomach causes hunger, then removing stomach nerves in animals will interfere with normal eating. However, removing those nerves does not interfere with normal eating. So, the stomach doesn't cause hunger. Brain activity always starts with blood entering the brain. It follows that blood acting on the brain doesn't cause hunger. Thus, we can conclude that hunger is caused by all of the body's cells.

Represent this reasoning in the form of a tree, making sure to identify the ultimate conclusion and the intermediate conclusions we draw along the way and how they are related.

3. Which of these are formulas in our formal language **Form**, and which are not?

$$p \vee q \qquad p \vee q \rightarrow r \qquad \neg\neg p \qquad q \neg p \qquad p \wedge (q \vee r) \rightarrow \bot$$

$$(p \rightarrow q) \rightarrow ((p \rightarrow (q \rightarrow r)) \rightarrow (p \rightarrow r)) \qquad p \wedge q \wedge r$$

For those that aren't formulas, are they ambiguous? (Could they be made into correct formulas in different ways by adding parentheses?) If they are, disambiguate them by listing all of the different ways they can be made formulas, and consider for yourself the different things they could *mean*.

4. For the formulas you identified in the previous question, list all of their *subformulas*.

## Challenge Questions

1. Take some reasoning you've seen in other subjects (philosophy subjects or anything else) and try to map out that reasoning in the form of a tree. What do you notice? Do you see any logical concepts (conditionals, conjunction, disjunction, negation, or anything else) playing a role in this reasoning?

2. Why can't a formula be *infinitely long*, according to our definition of formulas? What would the problem be with a formula like

$$\cdots \neg\neg\neg\neg p$$

with an unending series of negations before the $p$, or

$$p_1 \wedge (p_2 \wedge (p_3 \wedge (p_4 \wedge \cdots)))$$

where the conjunction goes on forever?

---

14. This argument is adapted from Howard Pospesel and David Marans's text *Arguments: Deductive Logic Exercises* (1978, 57). It is a useful sourcebook for more arguments to analyze.

# 2 Connectives: *and* & *if*

In this chapter, we begin developing a NATURAL DEDUCTION proof system for statements in the language **Form**. The proof system will have a set of inference rules for constructing proofs, where the inference rules all arise out of the meanings of the logical concepts we use. In this chapter, we will start with two of our connectives: CONJUNCTION ($\wedge$) and the CONDITIONAL ($\rightarrow$).

Our aim in the next three chapters is to come to understand basic principles for constructing proofs and to understand some fundamental features of the logical concepts. Recall from the previous chapter that a proof is a tree, whose leaves are assumptions and whose root is the conclusion of the proof. As you read the tree from the top to the bottom, each inference leads you from one or more premises to a conclusion, which may in turn be a premise for another inference.

In our *natural deduction* system, our target is the *basic* inferences and the methods for combining them. We aim to construct proofs from the most fundamental principles. The first thing you can do in a proof is *assume* something. So, the most basic "proof" starts and ends in the same place. It is a single formula that we assume. This is the *rule of assumption*. The tree looks like this:

$$A$$

where $A$ is the assumption, and it's the conclusion. The proof starts and ends in the same place, so it doesn't go very far at all. For any longer proof, we need inference rules, and for these rules, we aim to only use the simplest rules possible. We break inference steps down into their most fundamental components, so we can see how complicated proofs are made from simple parts. The key idea that motivates the design of these basic components is that a basic inference involving a formula (let's say, the formula $A \wedge B$, where $A$ and $B$ are formulas) depends on the main connective of that formula (in this case, conjunction) and depends on *only* that connective.

## 2.1 Conjunction

We will start with perhaps the simplest connective, conjunction. In a conjunction, $A \wedge B$, the formulas on either side of the conjunction are called its conjuncts. When we think of inferences involving a conjunction, it seems obvious that the most fundamental way to prove a conclusion $A \wedge B$ is to prove it from its conjuncts, $A$ and $B$, like this:

$$\frac{A \quad B}{A \wedge B} \wedge I$$

This is what we call the *introduction rule* for a conjunction. It tells us that a conjunction follows from its conjuncts. We label the inference step with the name for its rule ($\wedge I$).

That is one thing we can do with a conjunction. The other simple thing we can do with a conjunction is to undo that step, by making an inference using a conjunction as a premise. This is what we call an *elimination* step, deconstructing the conjunction and using one of its components. There are two components of a conjunction, so we get two different instances of the elimination rule:

$$\frac{A \wedge B}{A} \wedge E \qquad \frac{A \wedge B}{B} \wedge E$$

Together, these rules say that a conjunction implies each of its conjuncts.

$$* * *$$

Now we can piece these rules together to make proofs. It seems like there is not very much we can do yet, because all we have to work with are assumptions and rules involving conjunction, but even these rules enable us to construct proofs that tell us something about how conjunction behaves. For example, using these rules, we can show that, in a sense, the order of the conjuncts in a conjunction does not matter. If we assume $p \wedge q$, we can prove $q \wedge p$. Here is how. First, we have two small proofs from $p \wedge q$ to the conjuncts $q$ and $p$, respectively:

$$\frac{p \wedge q}{q} \wedge E \qquad \frac{p \wedge q}{p} \wedge E$$

These are two proofs, which lead us to different conclusions from the same assumption $p \wedge q$. We can combine these two proofs, using $\wedge I$, like this:

$$\frac{\dfrac{p \wedge q}{q} \wedge E \qquad \dfrac{p \wedge q}{p} \wedge E}{q \wedge p} \wedge I$$

and now we have proved $q \wedge p$ from the assumption $p \wedge q$. The proof is simple: we have made the assumption $p \wedge q$ and decomposed it into its components, $q$ and $p$, and recombined them, in the other order, to get the conclusion $q \wedge p$. We have a proof *from $p \wedge q$ to $q \wedge p$*.

That was one example of a proof. Here is another, reasoning from $p \wedge (q \wedge r)$ to $(p \wedge q) \wedge r$.

$$\frac{\dfrac{p \wedge (q \wedge r)}{p} \wedge E \qquad \dfrac{\dfrac{p \wedge (q \wedge r)}{q \wedge r} \wedge E}{q} \wedge E}{p \wedge q} \wedge I \qquad \dfrac{\dfrac{p \wedge (q \wedge r)}{q \wedge r} \wedge E}{r} \wedge E}{(p \wedge q) \wedge r} \wedge I$$

Notice that this proceeds in the same sort of way: we make the assumption a number of times, decompose it into parts, and recompose them in a different order to get the desired conclusion. Here, there are three copies of the assumption $p \wedge (q \wedge r)$. We need all three because each time we make an assumption, it acts as a premise of only one inference.

* * *

So, there are two kinds of basic rules for an inference involving a conjunction: (1) an *introduction* rule, which shows how you can infer a conjunction $A \wedge B$ from premises, and (2) an *elimination* rule, which shows how you can use a conjunction $A \wedge B$ as a premise to conclude another formula.

What goes for conjunction will go for the other connectives, too. For each connective, we will have an introduction rule and an elimination rule. The introduction rules for a connective (let's call an arbitrary connective "∘" for the moment) say what *licenses* a formula with ∘ as its main connective. The elimination rules for a connective say what follows *from* a formula whose main connective is ∘.

* * *

Gerhard Gentzen[15] gave a compelling suggestion about how to think about the relation between the introduction and elimination rules, which has been taken up by many philosophers in the decades since his pioneering work. He said, "an introduction rule gives, so to say, a definition, of the constant in question" and "an elimination rule is only a consequence of the corresponding introduction rule, which may be expressed somewhat as follows: at an inference by an elimination rule, we are allowed to 'use' only what the principal sign of the major premise 'means' according to the introduction rule for this sign" (Prawitz 1965, 33).

Based on this suggestion from Gentzen, Dag Prawitz[16] formulated his *inversion principle* for rules. This principle says that if a proof uses the introduction rule for a connective and then immediately uses the elimination rule, with the introduced formula as the major premise, which is the formula displays the connective of the elimination rule, then the proof already contains a proof of the conclusion from a subset of the same premises that does not contain the pair of introduction and elimination steps. In a slogan, detours can always be eliminated. Prawitz's idea will be a guide for constructing our natural deduction rules.

* * *

PHILOSOPHICAL ASIDE: In at least two philosophical traditions in the twentieth century, people have taken the shape of rules like this to have significance for a general theory of meaning. For Michael Dummett,[17] the introduction rule for a concept gave you the *verification conditions* for a judgment involving that concept, and he takes this condition to place

---

15. Gerhard Gentzen, German logician: born in 1909, student of David Hilbert at Göttingen, died in 1945 in World War II. He was one of the first logicians to develop a natural deduction system, of the kind that we are using in this book. He also developed an alternative proof system, sequent calculus, and proved some important results in proof theory. http://www-groups.dcs.st-and.ac.uk/~history/Mathematicians/Gentzen.html

16. Dag Prawitz, Swedish logician: born in 1936 and still active at the University of Stockholm. His *Natural Deduction: A Proof-Theoretical Study* contained the first published proofs of the normalization theorem for natural deduction systems. He has made many contributions to structural proof theory. https://www.su.se/profiles/prawd-1.184063

17. Michael Dummett, English philosopher: born in 1925, Wykeham Professor of Logic at Oxford, and influential philosopher and campaigner for racial tolerance and equality in the second half of the twentieth century. Died in 2011. See his *Logical Basis of Metaphysics* (1991) for more on the connection between inference rules and verification conditions.

a significant constraint on the meaningfulness of our concepts. For Robert Brandom,[18] our grasp of inference rules is at the core of our understanding of a concept. Introduction rules for a concept give an account of what entitles us to the claim expressed, and the elimination rules make explicit to what that claim commits us. END ASIDE

* * *

An important feature of the rules is that they focus on the main connective of a formula. An *introduction rule* for a connective ○ has as its conclusion a formula whose main connective is ○. An *elimination rule* for a connective will have a premise with that connective as its main connective. Any rules that operate on connectives that are not the main connective of a formula cannot be truly *fundamental* and must somehow arise out of more basic rules that operate on main connectives.

* * *

The proofs we have seen so far start with complex formulas, break them down, and build up the components in new ways. Proofs don't have to have this direct structure. Consider this proof:

$$\cfrac{\cfrac{p \wedge (q \wedge r)}{p} \wedge E \qquad \cfrac{\cfrac{p \wedge (q \wedge r)}{q \wedge r} \wedge E}{r} \wedge E}{\cfrac{\boxed{p \wedge r}}{r} \wedge E} \wedge I$$

It takes what we can call a *detour*. The formula in the shaded box—$p \wedge r$—is not a subformula of the premise formula $(p \wedge (q \wedge r))$, and it is not a subformula of the conclusion formula $(r)$. We built this formula up in a $\wedge I$ step, only to break it down immediately in a $\wedge E$ step.

This proof is perfectly legitimate, but it is more complicated than it needs to be to do the job of deriving $r$ from $p \wedge (q \wedge r)$. We could cut out that detour and get a proof that omits that pair of steps:

$$\cfrac{\cfrac{p \wedge (q \wedge r)}{q \wedge r} \wedge E}{r} \wedge E$$

This is still a proof from the same premise to the same conclusion, but now it is more direct.

* * *

This is a general phenomenon. The general method for removing detours is to give *reduction steps* that take a proof with detours and yields a proof that eliminates that detour. The reduction step for conjunction detours replaces a proof (or a part of a proof) shaped like the left of the following diagram and replaces it (or that part of it) with the simpler

---

18. Robert Brandom, U.S. philosopher: born in 1950, Professor of Philosophy at Pittsburgh University. He is an influential contemporary proponent of a kind of *normative pragmatist* approach to philosophy, in the vein of the early American pragmatists. See his *Articulating Reasons* (2000) for an introduction to this approach to inference.

proof on the right. We will use "⇝" to indicate that one proof reduces to another according to the reduction steps we provide. (Here, $i$ is either 1 or 2, depending on whether we chose the first or the second conjunct in the elimination inference.)

$$\cfrac{\cfrac{\overset{\Pi_1}{A_1} \quad \overset{\Pi_2}{A_2}}{A_1 \wedge A_2} \; \wedge I}{A_i} \; \wedge E \qquad \leadsto \qquad \overset{\Pi_i}{A_i}$$

Here is how to read this diagram: if we ever have a proof where we introduced a conjunction $A_1 \wedge A_2$ from the conjuncts $A_1$ and $A_2$, these conjuncts came from somewhere. Maybe they are assumptions, or maybe they have their own proofs. Whatever that is, call $\Pi_1$ the proof leading up to $A_1$ (which may just be the assumption $A_1$ itself), and call $\Pi_2$ the proof leading up to $A_2$. Then, if we did an elimination step and chose (for example) $A_1$, then we could simplify the proof by cutting out the detour and choosing the proof $\Pi_1$, which ends in the formula $A_1$, and throwing away all of $\Pi_2$. On the other hand, if the elimination step chose $A_2$, we throw away $\Pi_1$ and keep $\Pi_2$ and its conclusion $A_2$, and we don't go through the step of introducing and eliminating the conjunction.

This principle shows us that whenever there is a detour through a conjunction introduction and elimination, it can be eliminated using a reduction like this, just like the example of eliminating a detour that we've already seen. Notice that with this reduction step, the resulting proof will be smaller because it throws away one of the branches of the proof. The fact that we can do this reduction means that there is a nice balance between the introduction and elimination rules for conjunction. If we introduce a conjunction from some assumptions, those assumptions themselves have enough in them to allow us to derive whatever else the elimination rules say we can infer *from* the conjunction. This feature is one that we should look for in the rules for the other concepts we will consider.

The next concept we will look at is the *conditional*. Its rules are a little more involved than the simple rules for conjunction.

## 2.2 Conditional

In a conditional, $A \rightarrow B$, $A$ is the *antecedent* and $B$ is the *consequent* of the conditional. The elimination rule for the conditional is sometimes called *modus ponens*.

$$\cfrac{A \rightarrow B \quad A}{B} \; \rightarrow E$$

This rule says that a conditional together with its antecedent implies its consequent. In this rule, we say that the conditional $A \rightarrow B$ is the *major premise* and the $A$ is the *minor premise*. For elimination rules with more than one premise, we will distinguish the *major premise* of the rule, the one containing the connective whose elimination rule is being applied, and the minor premise(s). The distinction will be used (soon) in specifying how to eliminate detours involving the connective and again (later) in specifying side conditions on rules.

The elimination rule for the conditional tells us that if we *have* a conditional, we can make the inference from the antecedent to the consequent. This can give us insight into what we should look for to introduce a conditional. We should expect that we can make an inference *to* a conditional $A \to B$ when we can already prove the consequent $B$ *from* the antecedent $A$. If this is the right idea, then to prove $A \to B$, one would assume the antecedent, $A$, and show that the consequent, $B$, follows. Then, one is licensed to conclude that if $A$, then $B$, dropping the assumption of $A$. Formally, the rule looks like this:

$$\frac{\begin{array}{c} [A]^1 \\ \Pi \\ B \end{array}}{A \to B} \ {\to}I^1$$

Before the inference to $A \to B$, we have a proof $\Pi$, which includes $A$ among its premises and has $B$ as a conclusion. Then, we make the inference step ${\to}I$ and conclude $A \to B$. The brackets around the assumption $A$ indicate that that assumption has been *discharged*. The inference discharging the assumption is marked with the same number (here, a 1) as the superscript on the brackets.[19]

The assumptions that the premise of ${\to}I$ depends on may differ from those that the conclusion depends on. The premise, here $B$, depends on some assumptions that include the copies of $A$ discharged by the rule ${\to}I$. The conclusion of the rule, here $A \to B$, does not depend on the discharged copies of $A$, but it does depend on the other assumptions that $B$ depends on. Here is an example to clarify how this works:

$$\frac{\dfrac{p \to (q \to r) \quad [p]^1}{q \to r} \ {\to}E \quad q}{\dfrac{r}{p \to r} \ {\to}I^1} \ {\to}E$$

This is a proof from $p \to (q \to r)$ and $q$ to $p \to r$. Let's break down how this proof is constructed. First, we have a proof from $p \to (q \to r)$ and $p$ to $q \to r$.

$$\frac{p \to (q \to r) \quad p}{q \to r} \ {\to}E$$

Here, the conclusion $q \to r$ depends on the two assumptions $p \to (q \to r)$ and $p$. Then, we add the assumption $q$, and we infer $r$ in another ${\to}E$ step.

$$\frac{\dfrac{p \to (q \to r) \quad p}{q \to r} \ {\to}E \quad q}{r} \ {\to}E$$

The conclusion $r$ depends on the assumption set $\{p \to (q \to r), p, q\}$. The next step is a ${\to}I$ inference, where we discharge the assumption $p$, to conclude $p \to r$:

$$\frac{\dfrac{p \to (q \to r) \quad [p]^1}{q \to r} \ {\to}E \quad q}{\dfrac{r}{p \to r} \ {\to}I^1} \ {\to}E$$

So here, $p$ is no longer among the assumptions: $p \to r$ depends on $\{p \to (q \to r), q\}$.

Let's construct another proof, this time using both the conjunction rules and the conditional rules. Let's show that we can prove $p \to (q \wedge r)$ from the assumption $(p \to q) \wedge (p \to$

---

19. The superscripts mark out what Leivant (1979) calls assumption classes.

*r*). To prove $p \rightarrow (q \wedge r)$ in the most direct way possible, we will assume $p$ and prove $q \wedge r$ and then use $\rightarrow I$. To prove $q \wedge r$, we prove $q$ and $r$, and we use $\wedge I$. To prove $q$ and prove $r$, we use the assumption of $(p \rightarrow q) \wedge (p \rightarrow r)$; break it down into $p \rightarrow q$ and $p \rightarrow r$, respectively, (using $\wedge E$); and then use the assumption $p$ to derive the $q$ and the $r$ that we want. Piecing these steps together, we get the following proof:

$$\cfrac{\cfrac{\cfrac{(p \rightarrow q) \wedge (p \rightarrow r)}{p \rightarrow q} \wedge E \quad [p]^1}{q} \rightarrow E \qquad \cfrac{\cfrac{(p \rightarrow q) \wedge (p \rightarrow r)}{p \rightarrow r} \wedge E \quad [p]^1}{r} \rightarrow E}{\cfrac{q \wedge r}{p \rightarrow (q \wedge r)} \rightarrow I^1} \wedge I$$

which takes us from the assumption $(p \rightarrow q) \wedge (p \rightarrow r)$ to the conclusion $p \rightarrow (q \wedge r)$.

* * *

Proofs using conditional rules could involve detours, just like proofs involving conjunctions. If we introduce the conditional $A \rightarrow B$ and then immediately eliminate it, we have a detour. These detours can be reduced, too. Here is how to do it:

$$\cfrac{\cfrac{\begin{array}{c}[A]^1 \\ \Pi_1 \\ B\end{array}}{A \rightarrow B} \rightarrow I^1 \qquad \begin{array}{c}\Pi_2 \\ A\end{array}}{B} \rightarrow E \qquad \leadsto \qquad \begin{array}{c}\Pi_2 \\ A \\ \Pi_1 \\ B\end{array}$$

This reduction step is more involved than the conjunction reduction step, which simply threw away part of the proof. This step constructs a new proof. The original proof has a subproof, $\Pi_2$, whose conclusion is $A$. In the new proof, rather than assuming $A$ for the $\rightarrow I$ step, $A$ is obtained *on the basis of the proof* $\Pi_2$, and then the rest of the steps in $\Pi_1$ are used to obtain $B$.

An example will clarify how this works in practice. This proof has a detour formula in a shaded box. This formula is introduced and then eliminated.

$$\cfrac{\cfrac{\cfrac{\cfrac{p \rightarrow (q \rightarrow r) \quad [p]^1}{q \rightarrow r} \rightarrow E \quad [q]^2}{\cfrac{r}{p \rightarrow r} \rightarrow I^1}}{q \rightarrow (p \rightarrow r)} \rightarrow I^2 \qquad \cfrac{p \wedge q}{q} \wedge E}{p \rightarrow r} \rightarrow E$$

The reduction step plugs the short proof of $q$ from $p \wedge q$ in for the discharged assumption of $q$.

$$\cfrac{\cfrac{p \rightarrow (q \rightarrow r) \quad [p]^1}{q \rightarrow r} \rightarrow E \quad \cfrac{p \wedge q}{q} \wedge E}{\cfrac{r}{p \rightarrow r} \rightarrow I^1} \rightarrow E$$

This new proof still concludes $p \to r$, and it operates from the same assumptions $\{p \to (q \to r), p \land q\}$ as the original proof, but now it proceeds directly, without making the detour through the formula $q \to (p \to r)$. Notice that the formulas used in this new proof are all subformulas of the premises $p \to (q \to r)$ or $p \land q$ or the conclusion formula $p \to r$, and no extraneous formulas are involved.

\* \* \*

The reduction step for the conditional *simplifies* the proof in the sense of cutting out a complex formula, but it may not make the proof *smaller*. This is due to a subtlety about how the introduction rule works. How many copies of the assumption $A$ are discharged by the rule? The example above discharges *one* copy per use of the rule. The rule, however, permits any number of copies to be discharged. (As a matter of fact, it allows even *zero* copies to be discharged, as we will see later.) Here is an example proof where two instances of $p$ are discharged. We prove $p \to q$ from $p \to (p \to q)$.

$$
\cfrac{\cfrac{p \to (p \to q) \quad [p]^1}{p \to q} \ {\to}E \quad [p]^1}{\cfrac{q}{p \to q} \ {\to}I^1} \ {\to}E
$$

Let us see how a proof may grow larger through a reduction step. Here, the detour formula is in a shaded box.

$$
\cfrac{\cfrac{\cfrac{\cfrac{p \to (p \to (p \to q)) \ [p]^1}{p \to (p \to q)} \ {\to}E \quad [p]^1}{p \to q} \ {\to}E \quad [p]^1}{\cfrac{q}{\boxed{p \to q}} \ {\to}I^1} \ {\to}E \quad \cfrac{\cfrac{r \land (p \land s)}{p \land s} \ \land E}{p} \ \land E}{q} \ {\to}E
$$

This reduces to the following:

$$
\cfrac{\cfrac{p \to (p \to (p \to q)) \quad \cfrac{\cfrac{r \land (p \land s)}{p \land s} \ \land E}{p} \ \land E}{p \to (p \to q)} \ {\to}E \quad \cfrac{\cfrac{r \land (p \land s)}{p \land s} \ \land E}{p} \ \land E}{\cfrac{p \to q \qquad \qquad \cfrac{\cfrac{r \land (p \land s)}{p \land s} \ \land E}{p} \ \land E}{q} \ {\to}E} \ {\to}E
$$

The proof with the detour has twelve nodes, and the proof without the detour has thirteen. The result is simpler, in the sense that there are fewer of the biggest detours, the detours containing the greatest number of logical connectives. In this case, there are simply fewer detours.

\* \* \*

Let's end this section with a short explanation of why the $\to I$ rule allows for some strange cases of discharging: first, the strange case where *no* instances of the discharged assumption are actually used in the proof and, then, the case where some instances of a formula are discharged and others are not.[20] Consider this proof that makes the assumption $p$ and proves the conclusion $q \to p$.

$$\dfrac{\dfrac{\dfrac{p \quad [q]^1}{p \wedge q} \ {\wedge}I}{p} \ {\wedge}E}{q \to p} \ {\to}I^1$$

It makes a detour through $p \wedge q$. You can see that the introduction of $p \wedge q$ allows us to involve the $q$ as an assumption upon which $p$ depends. It is then discharged, to get $q \to p$. When we apply the reduction step for conjunctions to this proof, the resulting simpler proof, without a detour, is this:

$$\dfrac{p}{q \to p} \ {\to}I$$

Here, the proof discharges *zero* copies of the assumption $q$, because in the original proof, the instance of $q$ was in the part of the proof that the reduction step threw away. So, the way we interpret ${\to}I$ steps is that if we have a proof of $B$, we can discharge *any* number (zero or more) of instances of the assumption $A$ in the proof of $B$, to conclude $A \to B$.

We can also see cases where *some* instances of a formula are discharged and others are not. Consider this proof, taking us from the assumption $p$ to the conclusion $q \to (p \wedge q)$.

$$\dfrac{\dfrac{p \quad [q]^1}{p \wedge q} \ {\wedge}I}{q \to (p \wedge q)} \ {\to}I^1$$

This is a perfectly acceptable proof. It discharges the assumption $q$ at the last step. Now, it is a reasonable constraint on our proofs that the rules are *general*. This should still be a proof if we replace $p$ and $q$ by any formulas we choose. In particular, it should still be a proof when we replace $q$ *by* $p$. In this case, we have

$$\dfrac{\dfrac{p \quad [p]^1}{p \wedge p} \ {\wedge}I}{p \to (p \wedge p)} \ {\to}I^1$$

But notice now, we discharged one of the $p$ assumptions and left the other. This counts as a proof, for us, even though it makes use of an assumption that is not strictly needed. We could well have discharged both instances of $p$ and got a different proof:

$$\dfrac{\dfrac{[p]^1 \quad [p]^1}{p \wedge p} \ {\wedge}I}{p \to (p \wedge p)} \ {\to}I^1$$

which would show that $p \to (p \wedge p)$ is logically valid and can be proved on the basis of no assumptions at all.

---

20. Feel free to skim or skip this discussion the first time you work through this material. Come back just before the end of the section, where we introduce the **definition** of proof trees starting on the next page.

* * *

Let's end this section with a definition, so we are as precise as we can be about the proofs we have been introducing. This is an interim definition of our natural deduction proof system, as we have not yet introduced the rules for $\neg$ or for $\vee$, but it is good practice to think of the rules we have given as defining a system of proofs, in just the same way as we defined a formal language in the previous chapter.

**Definition 7 (($\wedge$, $\rightarrow$) proofs)**   *The set of ($\wedge$, $\rightarrow$) proofs in a formal language* Form *is a set of trees of formulas, inductively defined in the following way.*

- *For any formula A, the tree whose only node is A is a ($\wedge$, $\rightarrow$) proof. Its conclusion is the formula A, and it depends on that formula A (and no other formulas).*
- *If $\Pi_1$ is a proof of A and $\Pi_2$ is a proof of B, then the tree*

$$\frac{\begin{array}{cc} \Pi_1 & \Pi_2 \\ A & B \end{array}}{A \wedge B} \wedge I$$

*is a proof with the conclusion $A \wedge B$, and in this proof, $A \wedge B$ depends on the assumptions in $\Pi_1$ on which A depends and the assumptions in $\Pi_2$ on which B depends.*

- *If $\Pi$ is a proof of $A \wedge B$, then the trees*

$$\frac{\begin{array}{c} \Pi \\ A \wedge B \end{array}}{A} \wedge E \qquad \frac{\begin{array}{c} \Pi \\ A \wedge B \end{array}}{B} \wedge E$$

*are proofs with the conclusions A and B, respectively, and in these proofs, the conclusions depend on the assumptions in $\Pi$ on which $A \wedge B$ depends.*

- *If $\Pi_1$ is a proof of $A \rightarrow B$ and $\Pi_2$ is a proof of A, then the tree*

$$\frac{\begin{array}{cc} \Pi_1 & \Pi_2 \\ A \rightarrow B & A \end{array}}{B} \rightarrow E$$

*is a proof with the conclusion B, and in this proof, B depends on the assumptions in $\Pi_1$ on which $A \rightarrow B$ depends and the assumptions in $\Pi_2$ on which B depends.*

- *If $\Pi$ is a proof of B, then*

$$\frac{\begin{array}{c} [A]^i \\ \Pi \\ B \end{array}}{A \rightarrow B} \rightarrow I^i$$

*(where i is some number not already used to flag discharged assumptions in the proof $\Pi$) is a proof with the conclusion $A \rightarrow B$, zero or more occurrences of the assumption A are discharged, and the conclusion $A \rightarrow B$ depends on the assumptions in $\Pi$ on which B depends, other than the occurrences of the assumption A that are discharged in this step. (We indicate this discharge by bracketing the discharged formulas and marking it with the number i.)*

*Nothing is a ($\wedge$, $\rightarrow$) proof in this language other than a tree that has been constructed in a finite number of steps according to the above clauses.*

This completes the definition of ($\wedge$, $\rightarrow$) proofs in our natural deduction system. Notice that these proofs are defined by induction in just the same way that formulas were defined in the previous chapter.

## 2.3   Biconditional

In this section, we will take a look at how we can combine inference rules to define inference rules for complex concepts defined in terms of simpler ones.

The biconditional, $A \leftrightarrow B$ (read: $A$ *if and only if* $B$), is sometimes understood as a defined connective, interpreted as a conjunction of two conditionals, as $(A \to B) \wedge (B \to A)$. However, we can think of the biconditional as another connective that has its own introduction and elimination rules. The appropriate rules for the biconditional are these:

$$\frac{A \leftrightarrow B \quad A}{B} \; {\leftrightarrow}E \qquad \frac{A \leftrightarrow B \quad B}{A} \; {\leftrightarrow}E \qquad \frac{\begin{array}{cc} [A]^i & [B]^j \\ \Pi_1 & \Pi_2 \\ B & A \end{array}}{A \leftrightarrow B} \; {\leftrightarrow}I^{i,j}$$

You can use a biconditional $A \leftrightarrow B$ to infer $B$ or to infer $A$ from $B$. (That is how to understand the elimination rules.) And the introduction rule is motivated from them: you can introduce a biconditional just when you can make both inferences, from $A$ to $B$ and from $B$ to $A$.

We could introduce the biconditional as a separate connective on our language, but we will not do so. Instead, we will show how, in a sense, these rules are already present in the language we have. If we think of $A \leftrightarrow B$ simply as a shorthand for $(A \to B) \wedge (B \to A)$, then these rules can be seen as composed out of the more basic rules for $\wedge$ and $\to$. Instead of the rules ${\leftrightarrow}E$, we use $\wedge E$ and $\to E$ in sequence, like this:

$$\frac{\dfrac{A \leftrightarrow B}{A \to B} \; {\wedge}E \quad A}{B} \; {\to}E \qquad \frac{\dfrac{A \leftrightarrow B}{B \to A} \; {\wedge}E \quad B}{A} \; {\to}E$$

The introduction rule for $\leftrightarrow$ can be composed out of the introduction rules for $\to$ and $\wedge$.

$$\frac{\dfrac{\begin{array}{c} [A]^1 \\ \Pi_1 \\ B \end{array}}{A \to B} \; {\to}I^1 \qquad \dfrac{\begin{array}{c} [B]^2 \\ \Pi_2 \\ A \end{array}}{B \to A} \; {\to}I^2}{A \leftrightarrow B} \; {\wedge}I$$

The upshot is that since our language has the conditional and conjunction already, we can express arguments that use the biconditional. Nothing essentially new is gained from adding it, except for allowing proofs to be a little shorter. Having fewer rules also makes some of the proofs by induction shorter, since there are fewer cases.

## 2.4   Derived Rules

In the previous section, we showed how we could define the biconditional using conjunction and the conditional, and we showed how we could recover the introduction and elimination rules for it using the other rules of our proof system. Adding a primitive biconditional with its own rules does not add anything essentially new to the proof system. We can show something similar for other rules as well.

Suppose we have a rule $R$

$$\frac{A_1 \cdots A_n}{C} \ (R)$$

and are considering adding this to the proof system. Will the addition of this rule result in a conclusion being provable from some assumptions that we did not have before? In many cases, we can show that the addition of the rule does not add anything essentially new. We need to define the crucial concept for showing this about rules.

**Definition 8 (Derived rule)**   *A rule $R$ is a* DERIVED RULE *in a proof system* $S$ *if, given proofs* $\Pi_1, \ldots, \Pi_n$ *of the premises $A_1, \ldots, A_n$ of the rule $R$, the conclusion $C$ of $R$ can be obtained by extending some of $\Pi_1, \ldots, \Pi_n$ with the primitive rules of $S$.*

The definition of derived rules is stated only for rules that do not discharge assumptions, but it can be extended to rules that discharge assumptions as well. For the moment, we will focus on rules that do not discharge assumptions.

An example will be helpful at this point. Consider the rule (Perm), which is short for "permutation."

$$\frac{A \to (B \to C)}{B \to (A \to C)} \ (Perm)$$

This is a derived rule of the proof system with the rules $\wedge E, \wedge I, \to E, \to I$. To show that a rule is a derived rule of a proof system, one gives a proof taking one from the premises to the conclusion.

$$\frac{\dfrac{\dfrac{A \to (B \to C) \quad [A]^i}{B \to C} \to E \qquad [B]^j}{\dfrac{\dfrac{C}{A \to C} \to I^i}{B \to (A \to C)} \to I^j}}{} \to E$$

Any proof using (Perm) can be converted into a proof that does not use (Perm) by replacing the (Perm) steps with the steps above, using new superscripts as appropriate. This shows that the addition of (Perm) does not add anything essentially new. It provides us with new *proofs*, namely, ones with (Perm) steps that we did not have before, but it does not provide new *provability facts*, what conclusions follow from a set of assumptions. We will say more about provability in the next chapter once more rules have been introduced.

When defining natural deduction proof systems, it is preferable, philosophically and technically, to provide a small set of rules that capture the meaning of the connectives. In the present case, this involves showing that detours can be eliminated using primitive rules that do not display any connectives apart from the one in the name of the rule. When using natural deduction proof systems, having to go back to primitive rules all the time can result in long proofs and make the systems unwieldy to use. Derived rules let us have our cake and eat it too. We can use only primitive rules that have the philosophical and technical features that we want and then show that some additional rules, which may lack those features, are derived rules. Once we have shown this, we can use the derived rules in giving proofs, which can make finding proofs easier.

It is important to note that the definition of detour is in terms of the primitive rules of the system. If we want to ask questions about detours in a proof, we need to replace all uses of derived rules with a series of steps using the primitive rules. While derived rules can make it easier to find proofs, they will often introduce detours when elaborated into primitive rules of the system. A challenge question further explores the concept of derived rules.

## 2.5    Key Concepts and Skills

☐ You should be able to *read* tree proofs using the rules $\wedge E$, $\wedge I$, $\to E$, and $\to I$. You should be able to check that a proof follows the rules, and you should be able to keep track of which assumptions are active at each stage of the proof.

☐ You should be able to *construct* simple tree proofs using the rules $\wedge E$, $\wedge I$, $\to E$, and $\to I$.

☐ You can perform *reductions* on tree proofs that involve detours, using the reduction steps.

Proof rules

$$\frac{A \quad B}{A \wedge B} \wedge I \qquad \frac{A \wedge B}{A} \wedge E \qquad \frac{A \wedge B}{B} \wedge E$$

$$\frac{\begin{array}{c} [A]^i \\ \Pi \\ B \end{array}}{A \to B} \to I^i \qquad \frac{A \to B \quad A}{B} \to E$$

## 2.6    Questions for You

### Basic Questions

1. Look at these proofs. Read them from top to bottom, and at every inference step, list which assumptions each formula depends on.

$$\frac{p \to q \quad \dfrac{\dfrac{[p \wedge r]^1}{p} \wedge E}{q} \to E \qquad \dfrac{[p \wedge r]^1}{r} \wedge E}{\dfrac{q \wedge r}{(p \wedge r) \to (q \wedge r)} \to I^1} \wedge I$$

$$\frac{\dfrac{\dfrac{[p \to (q \to r)]^3 \quad [p]^1}{q \to r} \to E \qquad \dfrac{[p \to q]^2 \quad [p]^1}{q} \to E}{\dfrac{\dfrac{r}{p \to r} \to I^1}{(p \to q) \to (p \to r)} \to I^2}}{(p \to (q \to r)) \to ((p \to q) \to (p \to r))} \to I^3$$

2. Construct proofs for the following arguments:

- $p \to q, r \to p, r \succ q$.
- $p \to q, r \to p \succ r \to q$.
- $p \to q \succ (r \to p) \to (r \to q)$.
- $p \succ q \to (p \wedge q)$.

- $p \wedge (q \rightarrow r) \succ q \rightarrow (p \wedge r)$.

3. Consider this proof.

$$
\cfrac{
\cfrac{
[q \rightarrow r]^2 \quad \cfrac{
\cfrac{p \rightarrow q \quad [p]^1}{q} \rightarrow E
}{r} \rightarrow E
}{
\cfrac{p \rightarrow r}{(q \rightarrow r) \rightarrow (p \rightarrow r)} \rightarrow I^1 \quad \rightarrow I^2
}
\quad [q \rightarrow r]^3
}{
\cfrac{
\cfrac{p \rightarrow r \qquad\qquad\qquad [q \rightarrow r]^3}{r} \rightarrow E \quad [p]^4
}{(q \rightarrow r) \rightarrow r} \rightarrow I^3
}
\rightarrow E
$$

It has a detour formula in a shaded box. Use the reduction step for detours to eliminate this detour formula. Does the proof you get have another detour formula (a conditional that is introduced and then immediately eliminated)? If so, reduce it, too, and keep reducing detour formulas until the result is a proof with no detours.

### Challenge Questions

1. What do you think of the requirement that a detour (a formula introduced and then eliminated) should always be able to be reduced in a proof? This requirement makes sense for $\wedge$ and for $\rightarrow$ and the rules we have for these concepts. Does it make sense for inference rules for *any* concept? Can you think of proof rules for a concept that might violate this condition?

2. Show that the following rules are derived rules of the proof system in this chapter.

   i. $\cfrac{A \wedge B}{B \wedge A}$ *(Comm $\wedge$)*

  ii. $\cfrac{A \rightarrow B \quad A \rightarrow C}{A \rightarrow (B \wedge C)}$ *($\rightarrow \wedge$)*

 iii. $\cfrac{A \rightarrow B \quad B \rightarrow C}{A \rightarrow C}$ *($\rightarrow$ Trans)*

 iv. $\cfrac{A \rightarrow (B \rightarrow C)}{(A \wedge B) \rightarrow C}$ *(Import)*

  v. $\cfrac{(A \wedge B) \rightarrow C}{A \rightarrow (B \rightarrow C)}$ *(Export)*

 vi. $\cfrac{A \rightarrow (A \rightarrow B)}{A \rightarrow B}$ *(Cont)*

("Comm" is short for "commutativity," "Trans" for "transitivity," and "Cont" for "contraction.")

3. (This is a question to provoke discussion. It does not have a straightforward answer.) What does $\rightarrow$ *mean*? If we can infer $q \rightarrow p$ from $p$ (and the rules we've given seem to have that consequence), does "$\rightarrow$" mean anything like "*if ... then*"? (Notice that we started with the rule $\rightarrow E$ [*modus ponens*], and this seems to be a basic feature of any

conditional, and $\to I$ was motivated by arguing that we can infer $A \to B$ whenever we can infer from $A$ to $B$, which is what we need for *modus ponens*.)

4. There are other ways to formulate the elimination rules in a manner that permits the elimination of detours. One way to do so is by using *general elimination rules*. These rules involve discharging assumptions, which are determined by the main connective of the major premise. In the next chapter, we will see a rule that does this, namely, $\vee E$. The general elimination rules for conjunction and implication are the following.[21]

$$\frac{A \wedge B \qquad \overset{\displaystyle [A]^i \quad [B]^j \quad X}{\underset{\displaystyle C}{\bigtriangledown \atop \Pi}}}{C} \wedge GE^{i,j} \qquad\qquad \frac{A \to B \quad A \quad \overset{[B]^i}{\underset{C}{\Pi}}}{C} \to GE^i$$

For $\wedge GE$, to work out consequences of a conjunction, we assume the conjuncts to reach some conclusion $C$. We can then conclude that $C$ follows from the conjunction and discharge the assumed conjuncts. For $\to GE$, what follows from a conditional and its antecedent is whatever can be reached from an assumption of the consequent.

For this question, supply derivations for the arguments in basic question 2 using the rules $\to I$, $\wedge I$, $\wedge GE$, $\to GE$, and the rule of assumption.

5. Show that the rules $\wedge E$ and $\to E$ are derived rules of the system using the rules $\to I$, $\wedge I$, $\wedge GE$, $\to GE$, and the rule of assumption. Next, show how to eliminate detours for the system using the general elimination rules.

6. The rule $\to I$ permits discharging zero occurrences of an assumption as well as discharging multiple occurrences of an assumption. Banning either of these options has an effect on what formulas are provable. For this question, only use $\to$ rules.

   • Find two provable formulas whose proofs require discharging zero occurrences of an assumption. Provide proofs to demonstrate this.

   • Find two provable formulas whose proofs require discharging multiple occurrences of an assumption. Provide proofs to demonstrate this.

   • Find one provable formula whose proof requires both discharging multiple occurrences of an assumption as well as discharging zero occurrences of an assumption. Provide a proof to demonstrate this.

   Informally explain why you think these formulas require the sort of discharging described in each part.

7. Suppose we add the following rule to our natural deduction system.[22]

$$\frac{\overset{[A \to B]^i}{\underset{A}{\Pi}}}{A} (PR)^i$$

---

21. The triangle notation in the rule $\wedge GE$ means that there is a proof $\Pi$ whose conclusion is $C$, whose assumptions may include those in the set $X$ as well as the formulas above the triangle, namely, $A$ and $B$, and $\Pi$ has no other undischarged assumptions.

22. "($PR$)" stands for "Peirce's rule," named after Charles Saunders Peirce (1839–1914). Peirce was an American philosopher and logician noted for his work on logical diagrams and his development of pragmatism.

Find a formula that is provable using this rule that is not provable without detours in the proof system of this chapter.

# 3 More Connectives: *not* & *or*

This chapter completes our introduction to natural deduction proofs by introducing the rules for NEGATION ($\neg$) and DISJUNCTION ($\vee$). The rules for disjunction are the more complex of the two, so we will start with negation. As we will see, it is for negation that we reserved a special Atom, the formula $\perp$, the *falsum*.

## 3.1 Negation and Falsum

Think about negation. How do we show that something is *not the case*? How do we prove $\neg A$? We could show that assuming $A$ leads to a conflict with the other things we take to be true. How could such a conflict arise in the first place? At least, if our assumptions ever lead to a contradiction—to conclude $B$, on the one hand, and $\neg B$, on the other. One way to keep track of this is to reserve the atomic formula $\perp$ for such a conflict. The negation elimination rule is then

$$\frac{\neg A \quad A}{\perp} \, \neg E$$

since if we have ever proved a negation and proved its negand (or assumed both, or assumed one and proved another), then we have found ourself in such a conflicted position.

With this in mind, we see what we must have in order to prove a negation. If I were to prove $\neg A$, I would need to be in such a position so as to be able to prove a conflict from the assumption of $A$. In other words, the introduction rule for negation looks rather like the introduction rule for the conditional, except instead of proving the consequent, we end up in the conflict situation, $\perp$:

$$\begin{array}{c} [A]^i \\ \Pi \\ \frac{\perp}{\neg A} \, \neg I^i \end{array}$$

To prove $\neg A$, we assume $A$ and show that this gives rise to a contradiction. With these rules, we can already demonstrate quite a lot of the power of negation. For example, we can prove (on the basis of no assumptions at all) the formula $\neg(p \wedge \neg p)$, since $p \wedge \neg p$ clearly leads to a contradiction:

$$\cfrac{\cfrac{\dfrac{[p \wedge \neg p]^1}{\neg p} \, \wedge E \quad \dfrac{[p \wedge \neg p]^1}{p} \, \wedge E}{\perp} \, \neg E}{\neg(p \wedge \neg p)} \, \neg I^1$$

Another simple principle is that we can argue from $p \wedge \neg q$ to $\neg(p \to q)$—if we have a conditional where the antecedent is true and the consequent is false, then the conditional is false:

$$\cfrac{\cfrac{\cfrac{p \wedge \neg q}{\neg q} \wedge E \qquad \cfrac{[p \to q]^1 \qquad \cfrac{\cfrac{p \wedge \neg q}{p} \wedge E}{q} \to E}{q}}{\cfrac{\bot}{q} \neg E}}{\cfrac{\bot}{\neg(p \to q)} \neg I^1}$$

The rules $\neg I$ and $\neg E$ only truly pin down the meaning of negation if we can distinguish $\bot$ from an arbitrary proposition. Here is one way we can do this. A contradiction is not like any arbitrary item of information. The falsum represents the absurd, or the *False*, to use Frege's term,[23] or "that than which nothing sillier can be conceived," to quote John Slaney (1989, 476). If a contradiction is sillier than anything else, it is *refuted* by anything else at all, so the argument from $\bot$ to $A$ is valid, due to the silliness of $\bot$.[24] (There are other ways of understanding $\bot$, but this is a prominent one, and we will proceed on this basis for the moment.[25])

$$\frac{\bot}{A} \bot E$$

One consequence of this rule is that $\neg p$ entails $p \to q$.

$$\cfrac{\cfrac{\cfrac{\neg p \qquad [p]^1}{\bot} \neg E}{q} \bot E}{p \to q} \to I^1$$

There are many other things we can do with the negation rules, in combination with the rules for conjunction and the conditional. See the questions at the end of this chapter for some pointers for where to start in constructing negation proofs for yourself.

\* \* \*

As with conjunction and the conditional, negation proofs can introduce detours. Here is a proof with a detour involving negation:

---

23. Gottlob Frege (1848–1925) was a German logician who worked at the University of Jena. He was an early pioneer of modern logic who was also made influential contributions to the philosophy of language. In his *Begriff-sschrift*, he developed a distinctive logical notation. In his two-volume work *Grundgesetze der Arithmetik*, Frege attempted to reconstruct much of mathematics using higher-order logical resources together with principles for defining new entities. Bertrand Russell demonstrated that Frege's theory was inconsistent using a version of what is known as Russell's paradox.

24. This uses the principle that if we have a proof from $A$ to $B$, then a refutation of $B$ can be made into a refutation of $A$. So being able to prove $A$ from $\bot$ is one way to respect the fact that evidence against $A$ is evidence against $\bot$. In other words, $\bot$ is *very easily* refuted.

25. For some of the other ways to understand $\bot$ and its role in logic and proofs, see Tennant (2017) or Kürbis (2015, 2019), particularly chapter 3.

$$\cfrac{\cfrac{\neg p \qquad \cfrac{[p \wedge q]^1}{p}\ \wedge E}{\cfrac{\bot}{\boxed{\neg(p \wedge q)}}\ \neg I^1} \qquad \qquad \cfrac{p \qquad q}{p \wedge q}\ \wedge I}{\cfrac{\bot}{r}\ \bot E}\ \neg E$$

Here, $\neg(p \wedge q)$ is introduced and immediately eliminated. Since the rules for negation look just like the rules for the conditional, the reduction step for negation looks just like the reduction step for the conditional, too:

$$\cfrac{\cfrac{\begin{array}{c}[A]^1 \\ \Pi_1 \\ \bot\end{array}}{\neg A}\ \neg I^1 \qquad \begin{array}{c}\Pi_2 \\ A\end{array}}{\bot}\ \neg E \qquad \rightsquigarrow \qquad \begin{array}{c}\Pi_2 \\ A \\ \Pi_1 \\ \bot\end{array}$$

Here, we cut out the assumption of $A$ and the inference to $\neg A$ from the derivation of the contradiction, and we use the *proof* $\Pi_2$ of $A$ instead, to supply the required proof of a contradiction without using the assumption of $A$. f

So, if you look at our proof of $r$ from $p$, $q$, and $\neg p$ involving a detour, our reduction step can simplify it. The reduction step plugs the derivation of $p \wedge q$ in for the assumption of $p \wedge q$ that is discharged by the $\neg I$ rule.

$$\cfrac{\neg p \qquad \cfrac{\cfrac{\boxed{p \qquad q}}{p \wedge q}\ \wedge I}{p}\ \wedge E}{\cfrac{\bot}{r}\ \bot E}\ \neg E$$

This reduction has produced a shorter derivation of the conclusion, $r$, from the assumptions $\neg p$, $p$, and $q$, but you can see that it has introduced (or revealed) a *new* detour, the conjunction $p \wedge q$ in a shaded box. You can use another reduction step to remove the detour and produce a simpler proof.

$$\cfrac{\cfrac{\neg p \qquad p}{\bot}\ \neg E}{r}\ \bot E$$

This proof has no detours, but it also has no assumptions of $q$. It reaches the desired conclusion from fewer assumptions.

There is no reduction step for falsum. Since there is no introduction rule for falsum, there is no way to have a detour using falsum. Without an introduction rule, there is no way it can be introduced and then eliminated. In the next chapter, however, we will introduce some further ways to manipulate proofs, some of which will involve falsum.

## 3.2   Disjunction

Now it is time to add the rules for disjunction, to complete our family of inference rules. Much like the elimination rules for conjunction, the *introduction* rules for disjunction are simple. They say that a disjunction follows from its disjuncts.

$$\frac{A}{A \vee B} \ \vee I \qquad \frac{B}{A \vee B} \ \vee I$$

It is important to keep in mind that in the $\vee I$ rule, the newly added disjunct can be anything at all.[26] If I have shown (or assumed) $A$, I can immediately weaken my claim to the claim $A \vee B$, where $B$ can be any formula.

The elimination rule for disjunction is more complex than any of the other rules we have seen. Here is why. Perhaps the most natural thing to do, when faced with a disjunction in a course of reasoning, would be to split your proof into two different *cases*, like this:

$$\frac{A \vee B}{A \qquad B} \ \vee E^{??}$$

where one case proceeds on the basis that $A$ holds and the other on the basis that $B$ holds. Unfortunately, that would give us proofs with a different shape. Now we have two different roots, corresponding to the two different cases, rather than a single conclusion to our proof. So, instead of totally rewriting everything we've done so far to allow for proofs with more than one conclusion, we will look for a different elimination rule for disjunction that still allows us to respect the idea that you deal with a disjunction by splitting your reasoning into different *cases*.[27]

The idea is the same. If I have assumed (or proved) $A \vee B$, then I can prove a conclusion $C$ if I can prove it from $A$ (in one case) and if I can prove it from $B$ (in the other). Something follows from a disjunction if it follows from each of the disjuncts separately. To show that $C$ follows from $A \vee B$, you show that $C$ follows from $A$ and that $C$ follows from $B$, in two different proofs—like this:

$$\frac{A \vee B \qquad \begin{matrix} [A]^1 \\ \Pi_1 \\ C \end{matrix} \qquad \begin{matrix} [B]^2 \\ \Pi_2 \\ C \end{matrix}}{C} \ \vee E^{1,2}$$

In this rule, we call $A \vee B$ the *major premise*, and the two premise copies of $C$ are the *minor premises*. It is worth emphasizing that every application of $\vee E$ requires two minor premise subproofs, one for each disjunct of the major premise. Let's see an example of how this works in practice, by proving $(p \vee q) \rightarrow r$ from the two premises $p \rightarrow r$ and $q \rightarrow r$. Since we want to prove $(p \vee q) \rightarrow r$, we assume $p \vee q$ and prove $r$. To do that, since we've assumed $p \vee q$, we can apply the rule $\vee E$ and assume $p$, on the one hand, in order to prove $r$, and assume $q$, on the other, *also* to prove $r$. Since we're allowed to assume $p \rightarrow r$ and $q \rightarrow r$,

---

26. This means that our disjunction is *inclusive*. It does not exclude the possibility that the disjunct $B$ is also true. $A \vee B$ could be true *both* because $A$ is true and because $B$ is true.

27. For a full development of a proof system with multiple conclusions along the lines indicated above, see Shoesmith and Smiley (1978).

this will be easy. Here is the shape of the proof to this stage:

$$p \lor q \quad \cfrac{p \to r \quad [p]^1}{r} \to E \quad \cfrac{q \to r \quad [q]^2}{r} \to E \\ \hline \phantom{xxxxxxxxxxxxxxxxxxxxx} r \phantom{xxxxxxxxxxxxxxxxxxxxx} \lor E^{1,2}$$

So here, we have a proof using $\lor E$; the major premise is the assumption of $p \lor q$, and the minor premises are the instances of $r$ we have proved twice, once under the assumption of $p$ (the first disjunct of $p \lor q$) and once under the assumption of $q$ (the second disjunct). We discharge both of those assumptions, appealing to $p \lor q$ and the rule $\lor E$, flagging the assumptions that are discharged with the labels 1 and 2. We can complete this proof by discharging the assumption $p \lor q$ in the next step:

$$[p \lor q]^3 \quad \cfrac{p \to r \quad [p]^1}{r} \to E \quad \cfrac{q \to r \quad [q]^2}{r} \to E \\ \cfrac{\phantom{xxxxxxxxxxxxxxxxxxxxx} r \phantom{xxxxxxxxxxxxxxxx} \lor E^{1,2}}{(p \lor q) \to r} \to I^3$$

and we have completed our proof: we have a proof from $p \to r$ and $q \to r$ to $(p \lor q) \to r$.

Here is another example, to illustrate how the disjunction rules can interact with the negation rules. Here is a proof from the assumption $\neg(p \lor q)$ to the conclusion $\neg p \land \neg q$.

$$\cfrac{\neg(p \lor q) \quad \cfrac{[p]^1}{p \lor q} \lor I}{\cfrac{\bot}{\neg p} \neg I^1} \neg E \qquad \cfrac{\neg(p \lor q) \quad \cfrac{[q]^2}{p \lor q} \lor I}{\cfrac{\bot}{\neg q} \neg I^2} \neg E \\ \hline \phantom{xxxxxxxxxxxxxxxxxxxxx} \neg p \land \neg q \phantom{xxxxxxxxxxxxxxxxxxxxx} \land I$$

As with the other rules, we can have detours involving the disjunction rules. The reduction for disjunction detours gives us another perspective on the $\lor E$ rule. According to the $\lor I$ rules, we can introduce a disjunction from either disjunct. If we eliminate that disjunction (using the disjunction to prove something), we need to be able to prove that thing *from either of the disjuncts*, since it could have been introduced from either. So, the reduction step for disjunction looks like this, where $i$ is either 1 or 2, depending on which case of the disjunction introduction rule is used.

$$\cfrac{\cfrac{\Pi}{A_i}}{A_1 \lor A_2} \lor I \quad \cfrac{[A_1]^1 \quad [A_2]^2}{\begin{array}{cc} \Pi_1 & \Pi_2 \\ C & C \end{array}} \\ \hline \phantom{xxxxxxxxxxxxxxxxxxxx} C \phantom{xxxxxxxxxxxxxxxxxxxx} \lor E^{1,2} \qquad \rightsquigarrow \qquad \begin{array}{c} \Pi \\ A_i \\ \Pi_i \\ C \end{array}$$

Here, if $A_1 \lor A_2$ came from $A_1$, then the proof $\Pi_1$ from $A_1$ is used in the reduction. If, on the other hand, $A_1 \lor A_2$ came from $A_2$, then the proof $\Pi_2$ from $A_2$ is used. In either case, we still have a proof of the formula $C$.

The following example shows how it is applied in practice. The detour formula is in a shaded box, as usual.

$$\cfrac{\cfrac{\dfrac{p \wedge q}{p} \, \wedge E}{\boxed{p \vee r}} \, \vee I \qquad \cfrac{\dfrac{[p]^1 \quad s}{p \wedge s} \, \wedge I}{(p \wedge s) \vee r} \, \vee I \qquad \cfrac{[r]^2}{(p \wedge s) \vee r} \, \vee I}{(p \wedge s) \vee r} \, \vee E^{1,2}$$

Carrying out the reduction steps yields this simpler proof:

$$\cfrac{\cfrac{\dfrac{p \wedge q}{p} \, \wedge E \qquad s}{p \wedge s} \, \wedge I}{(p \wedge s) \vee r} \, \vee I$$

$$* \, * \, *$$

The disjunction rule $\vee E$ lets us represent reasoning by cases. If we have a disjunction, $A \vee B$, then there are two cases, the case in which $A$ holds and the case in which $B$ does. One considers the $A$-case by supposing $A$ and reaching a conclusion, $C$. Then one considers the $B$-case by supposing $B$ and reaching the same conclusion, $C$. Either way, you get $C$, so you can conclude $C$.

The rule $\vee E$ requires reaching a conclusion from the supposition of each disjunct. What happens if you reach a conclusion from only one disjunct and try to use that as the conclusion, as in the following rule?

$$\cfrac{A \vee B \qquad \begin{array}{c} [A]^1 \\ \Pi_1 \\ C \end{array}}{C} \, \odot^1$$

The rule $\odot$ leads to disaster. As an example, every number is even or odd. Take my favorite number. It is either even or odd. Suppose it is even. Then it is divisible by 2. So, my favorite number is divisible by 2. But my favorite number is 3, which is not divisible by 2. The rule $\odot$, then, readily leads one to false conclusions.

$$* \, * \, *$$

Sometimes people use a rule rather like $\vee E$ but that allows for a little more flexibility. We don't require that the conclusions of the two subproofs are identical, but we merge them with a disjunction at the end.

$$\cfrac{A \vee B \qquad \begin{array}{c} [A]^1 \\ \Pi_1 \\ C \end{array} \qquad \begin{array}{c} [B]^2 \\ \Pi_2 \\ D \end{array}}{C \vee D} \, \vee E^{*1,2}$$

The rule $\vee E^*$ can be recovered from $\vee I$ and $\vee E$ in a straightforward way:

$$\frac{A \vee B \quad \dfrac{\begin{matrix}[A]^1\\ \Pi_1\\ C\end{matrix}}{C \vee D}\vee I \quad \dfrac{\begin{matrix}[B]^2\\ \Pi_2\\ D\end{matrix}}{C \vee D}\vee I}{C \vee D}\vee E^{1,2}$$

so adding the derived rule $\vee E^*$ to the system doesn't let us prove anything new, and since it doesn't shorten proofs by much, we won't use it.

### 3.3  Our System of Proofs

With all of the rules we have seen, we now have our complete definition of our system of proofs for the language of propositional logic. We take some care with the precise definition—repeating some of the definition given in the previous chapter—so that we have the full definition of our system of proofs in one place.

**Definition 9 (Propositional proofs)**   *The set of* PROPOSITIONAL PROOFS *in a formal language* Form *is a set of trees of formulas, inductively defined in the following way.*

- *For any formula A, the tree whose only node is A is a proof. Its conclusion is the formula A, and it depends on that formula A (and no other formulas).*
- *If $\Pi_1$ is a proof of A and $\Pi_2$ is a proof of B, then the tree*

$$\frac{\begin{matrix}\Pi_1 & \Pi_2\\ A & B\end{matrix}}{A \wedge B}\wedge I$$

*is a proof with the conclusion $A \wedge B$, and in this proof, $A \wedge B$ depends on the assumptions in $\Pi_1$ on which A depends and the assumptions in $\Pi_2$ on which B depends.*
- *If $\Pi$ is a proof of $A \wedge B$, then the trees*

$$\frac{\begin{matrix}\Pi\\ A \wedge B\end{matrix}}{A}\wedge E \qquad \frac{\begin{matrix}\Pi\\ A \wedge B\end{matrix}}{B}\wedge E$$

*are proofs with the conclusions A and B, respectively, and in these proofs, the conclusions depend on the assumptions in $\Pi$ on which $A \wedge B$ depends.*
- *If $\Pi_1$ is a proof of $A \rightarrow B$ and $\Pi_2$ is a proof of A, then the tree*

$$\frac{\begin{matrix}\Pi_1 & \Pi_2\\ A \rightarrow B & A\end{matrix}}{B}\rightarrow E$$

*is a proof with the conclusion B, and in this proof, B depends on the assumptions in $\Pi_1$ on which $A \rightarrow B$ depends and the assumptions in $\Pi_2$ on which B depends.*
- *If $\Pi$ is a proof of B, then*

$$\frac{\begin{matrix}[A]^i\\ \Pi\\ B\end{matrix}}{A \rightarrow B}\rightarrow I^i$$

*(where i is some number not already used to flag discharged assumptions in the proof $\Pi$) is a proof with the conclusion $A \rightarrow B$, zero or more occurrences of the assumption A are* discharged,

and the conclusion $A \to B$ depends on the assumptions in $\Pi$ on which $B$ depends, other than the occurrences of the assumption $A$ that are discharged in this step. (We indicate this discharge by bracketing the discharged formulas and marking them with the number $i$.)

- If $\Pi_1$ is a proof of $\neg A$ and $\Pi_2$ is a proof of $A$, then the tree

$$\frac{\begin{matrix} \Pi_1 & \Pi_2 \\ \neg A & A \end{matrix}}{\bot} \; \neg E$$

is a proof with the conclusion $\bot$, and in this proof, $\bot$ depends on the assumptions in $\Pi_1$ on which $\neg A$ depends and the assumptions in $\Pi_2$ on which $A$ depends.

- If $\Pi$ is a proof of $\bot$, then

$$\frac{\begin{matrix} [A]^i \\ \Pi \\ \bot \end{matrix}}{\neg A} \; \neg I^i$$

(where $i$ is some number not already used to flag discharged assumptions in the proof $\Pi$) is a proof with the conclusion $\neg A$, zero or more occurrences of the assumption $A$ are discharged, and the conclusion $\neg A$ depends on the assumptions in $\Pi$ on which $\bot$ depends, other than the occurrences of the assumption $A$ that are discharged in this step. (We indicate this discharge by bracketing the discharged formulas and marking them with the number $i$.)

- If $\Pi$ is a proof of $\bot$, then

$$\frac{\begin{matrix} \Pi \\ \bot \end{matrix}}{A} \; \bot E$$

is a proof with the conclusion $A$, and in this proof, $A$ depends on the assumptions in $\Pi$ on which $\bot$ depends.

- If $\Pi_i$ is a proof of $A_i$ ($i = 1, 2$), then the tree

$$\frac{\begin{matrix} \Pi_i \\ A_i \end{matrix}}{A_1 \lor A_2} \; \lor I$$

is a proof with conclusion $A_1 \lor A_2$, and in this proof, the conclusion depends on the assumptions in $\Pi_i$ on which $A_i$ depends.

- If $\Pi$ is a proof of $A_1 \lor A_2$, and if $\Pi_i$ is a proof of $C$, then

$$\frac{\begin{matrix} & [A_1]^j & [A_2]^k \\ \Pi & \Pi_1 & \Pi_2 \\ A_1 \lor A_2 & C & C \end{matrix}}{C} \; \lor E^{j,k}$$

(where $j$ and $k$ are numbers not already used to flag discharged assumptions in the proofs $\Pi$, $\Pi_1$, and $\Pi_2$) is a proof with the conclusion $C$, zero or more occurrences of $A_1$ in $\Pi_1$ are discharged, zero or more occurrences of $A_2$ in $\Pi_2$ are discharged, and in this proof, the conclusion $C$ depends on the assumptions in $\Pi$ on which $A_1 \lor A_2$ depends and (for $i = 1, 2$) the assumptions in $\Pi_i$ on which $C$ depends, other than the occurrences of $A_i$ that are discharged in this step. (We indicate this discharge by bracketing the discharged formulas and marking the discharged formulas in $\Pi_1$ with the number $j$ and in $\Pi_2$ with the number $k$.)

Nothing is a proof in our language other than a tree that has been constructed in some finite number of steps according to these rules.

We have completed our inductive definition of the set of all proofs in our propositional language and the dependence relation inside proofs. For each formula in a proof, we have an account of the formulas from among the assumptions above it, upon which it depends. Proofs, like formulas and like numbers, are inductively defined structures. We will start exploiting this fact in the next chapter, when we turn our attention to what we can do with proofs and what we can learn about what is provable, in general.

$$* * *$$

We introduced proofs as a way to give an account of what arguments are good and why they are good. Now we see that an argument $X \succ A$ is *valid* if the argument can be underwritten with a *proof* that leads from $X$ to $A$. That is, there is a proof $\Pi$ with conclusion $A$ and all of whose undischarged assumptions are found in the set $X$. In defining proofs and validity, we have defined a *logic* for our language. This logic has a name:[28]

**Definition 10 (Intuitionistic logic)** *Let $X$ be a set of formulas and $A$ a formula from* Form. *An argument $X \succ A$ is* VALID *according to* intuitionistic logic *if and only if there is a proof from premises in the set $X$ to the conclusion $A$. If $X \succ A$ is valid in intuitionistic logic, we write $X \vdash_{\mathsf{I}} A$.*

These statements $X \succ A$ and $X \vdash_{\mathsf{I}} A$ will sometimes be referred to as *sequents*.[29] We will often omit the set braces on the left of the turnstile, for example, by writing $B, C \vdash_{\mathsf{I}} A$, and even $\vdash_{\mathsf{I}} A$ when the premise set is empty, but you should keep in mind that the thing on the left of the turnstile is a *set* of formulas. The more explicit way to write these sequents would be $\{B, C\} \vdash_{\mathsf{I}} A$ and $\{\ \} \vdash_{\mathsf{I}} A$, or $\emptyset \vdash_{\mathsf{I}} A$.

Recall that according to our definition, it is not required that all the assumptions be used in a proof showing $X \vdash_{\mathsf{I}} A$. For example,

$$\frac{p \qquad q}{p \wedge q} \wedge I$$

is a short proof showing both: $p, q \vdash_{\mathsf{I}} p \wedge q$ and $p, q, r \vdash_{\mathsf{I}} p \wedge q$.

$$* * *$$

Intuitionistic logic differs from the classical two-valued logic you may have seen before in some ways, which will become clearer in the next chapter. We do not have $\neg\neg p \vdash_{\mathsf{I}} p$, and neither do we have $\vdash_{\mathsf{I}} p \vee \neg p$, to give just two examples. What to make of these facts—and, indeed, how to prove that you *cannot* prove something, that an argument is *not* valid—we will leave for the next two chapters.

### 3.4 Comments on Negation and Falsum

Earlier we noted that the negation rules are similar to the conditional rules and that the negation reduction step is similar to the conditional reduction step. It turns out that negation

---

28. The name "intuitionistic logic" goes back to the work of L. E. J. Brouwer, for whom reasoning and argumentation was to be understood as a process of *construction*, according to which the conclusion of a proof could be constructed (or verified) on the basis of the constructions or verifications of the premises. The particular formulation we follow goes back to the pioneering work of Gerhard Gentzen and Arend Heyting.

29. The terminology "sequent" goes back to Gentzen, for whom a sequent involves a *sequence* of formulas.

is definable in terms of the conditional and falsum: $\neg A$ can be defined as $A \rightarrow \bot$. It is not enough to give the definition; we also need to show that it works appropriately. In this case, that means showing that using the conditional and falsum rules, you can recover the negation rules.

Recovering the negation rules is immediate. The negation introduction rule is an instance of $\rightarrow I$.

$$\frac{\begin{array}{c} [B]^1 \\ \Pi \\ \bot \end{array}}{B \rightarrow \bot} \rightarrow I^1$$

The negation elimination rule is an instance of $\rightarrow E$.

$$\frac{\begin{array}{cc} \Pi_1 & \Pi_2 \\ B \rightarrow \bot & B \end{array}}{\bot} \rightarrow E$$

Any proof that contained instances of the negation rules can be rewritten using the defined negation and these rules to reach the same conclusion.

$$* \ * \ *$$

Notice that the recovery of the negation rules involved only the conditional rules. The falsum elimination rule did not play a part. The falsum elimination rule might strike you as odd, since there is not a separate falsum introduction rule. You might also be doubtful that an absurdity should imply *everything*. A natural question is what happens when $\bot E$ is dropped from our system of proofs. The answer is that the resulting system is quite similar to intuitionistic logic, but it differs in some interesting and important ways.

The system with all the rules of intuitionistic logic but $\bot E$ is called *minimal logic*. It was first formulated by Ingebrigt Johansson in 1936. We will use $X \vdash_M A$ to mean that there is a proof in *minimal logic* of $A$ from assumptions in the set $X$. (That is, there is a proof that makes no use of the $\bot E$ rule.) Minimal logic shares many features with intuitionistic logic, but one difference between the logics is how they treat contradictions.

In minimal logic, contradictions are not *explosive*, which is to say that they do not entail everything. In intuitionistic logic, contradictions are explosive, so for all formulas $A$ and $B$,

$$A, \neg A \vdash_I B.$$

There is a slightly weaker fact that holds in minimal logic, though: for all formulas $A$ and $B$,

$$A, \neg A \vdash_M \neg B.$$

Some logicians have thought that contradictions should not entail all formulas, nor should they entail all negated formulas. As they might say, it is hard to see why a contradiction, such as grass is green and is not green, should entail that Mount Taranaki is in Melbourne. One reason is that it does not seem like the contradiction concerning the color of grass has anything to do with the location of Mount Taranaki. Logics in which contradictions do not entail every formula are called *paraconsistent*. We will return to the idea of paraconsistency in section 6.5, once we have models on the table.

### 3.5 Key Concepts and Skills

☐ You should be able to *read* tree proofs using any or all of the rules ($\wedge E$, $\wedge I$, $\to E$, $\to I$, $\neg E$, $\neg I$, $\bot E$, $\vee E$, $\vee I$). You should be able to check that a proof follows these rules, and you should be able to keep track of which assumptions are active at each stage of the proof.

☐ You should be able to *construct* simple tree proofs using all the rules.

☐ You should understand the concept of a *derived rule* and be able to construct a proof to show how a given derived rule can be derived from the more fundamental rules of the calculus.

☐ You should be able to perform *reductions* on tree proofs that involve detours, using the reduction steps.

Proof rules

$$\frac{A}{A \vee B} \vee I \qquad \frac{B}{A \vee B} \vee I \qquad \frac{A \vee B \quad \overset{[A]^1}{\underset{C}{\Pi_1}} \quad \overset{[B]^2}{\underset{C}{\Pi_2}}}{C} \vee E^{1,2}$$

$$\frac{\overset{[A]^1}{\underset{\frac{\bot}{\neg A}}{\Pi}}}{} \neg I^1 \qquad \frac{\neg A \quad A}{\bot} \neg E \qquad \frac{\bot}{A} \bot E$$

### 3.6 Questions for You

**Basic Questions**

1. Read these two proofs from top to bottom, and at every step, list which assumptions each formula depends on.

$$\frac{[p \vee q]^3 \qquad \dfrac{\dfrac{\neg p \wedge \neg q}{\neg p} \wedge E \quad [p]^1}{\bot} \neg E \qquad \dfrac{\dfrac{\neg p \wedge \neg q}{\neg q} \wedge E \quad [q]^2}{\bot} \neg E}{\dfrac{\dfrac{\bot}{\neg(p \vee q)} \neg I^3}{}} \vee E^{1,2}$$

$$\frac{\dfrac{[p \vee q]^3 \quad \dfrac{\dfrac{(p \to \neg r) \wedge (q \to \neg r)}{p \to \neg r} \wedge E \, [p]^1}{\neg r} \to E \quad \dfrac{\dfrac{(p \to \neg r) \wedge (q \to \neg r)}{q \to \neg r} \wedge E [q]^2}{\neg r} \to E}{\neg r} \vee E^{1,2} \qquad [r]^4}{\dfrac{\dfrac{\bot}{\neg(p \vee q)} \neg I^3}{r \to \neg(p \vee q)} \to I^4} \neg E$$

2. Construct proofs for the following arguments:
   - $p \succ \neg\neg p$

- $\dot{p} \to r, q \to s \succ (p \land q) \to (r \land s)$
- $p \to r, q \to s \succ (p \lor q) \to (r \lor s)$
- $\neg p \lor \neg q \succ \neg (p \land q)$
- $\neg\neg\neg p \succ \neg p$

3. Show that the following are derived rules of the proof system.

   i. $\dfrac{\neg A \quad A}{B}$ (Expl)

   ii. $\dfrac{A \land (B \lor C)}{(A \land B) \lor (A \land C)}$ (Dist)

   iii. $\dfrac{A \to B \quad \neg B}{\neg A}$ (MT)

   iv. $\dfrac{\neg A}{\neg (A \land B)}$ ¬∧

   v. $\dfrac{A \to C \quad B \to C}{(A \lor B) \to C}$ (∨→)

   vi. $\dfrac{A \lor B \quad \neg A}{B}$ (DS)

   (Here, "Expl" stands for "explosion," "Dist" for "distribution," "MT" for "modus tollens," and "DS" for "disjunctive syllogism." These are relatively common names for these inferences that you may come across elsewhere.)

4. Here is a proof for the argument $(p \to r) \land (q \to r) \succ (p \land q) \to r$. It contains a detour formula, in a shaded box. Use reductions to eliminate the detour.

$$
\cfrac{
\cfrac{
\cfrac{\dfrac{(p \to r) \land (q \to r)}{p \to r} \land E}{\boxed{(p \to r) \lor (q \to r)}} \lor I
\quad
[p \to r]^1 \quad \dfrac{[p \land q]^3}{p} \land E \atop \cfrac{}{r} \to E
\quad
[q \to r]^2 \quad \dfrac{[p \land q]^3}{q} \land E \atop \cfrac{}{r} \to E
}{r} \lor E^{1,2}
}{(p \land q) \to r} \to I^3
$$

5. Suppose that $A, B \vdash_1 C \land D$. This tells us that there is a proof $\Pi$ whose set of undischarged assumptions may contain $A$ and $B$ and whose conclusion is $C \land D$. Which of the following statements about $\Pi$ are true, which are false, and which do you not have enough information to settle?

   i. $\Pi$ contains no detours.

   ii. $\Pi$ contains detours.

   iii. $\Pi$ ends in a $\land I$ step.

   iv. Both $A$ and $B$ appear in $\Pi$.

   v. $\Pi$ uses an elimination rule.

   vi. If $B$ is of the form $B_1 \to B_2$, then $\Pi$ contains a $\to E$ step.

   vii. $\Pi$ contains an assumption step.

   viii. There are no discharged assumptions in $\Pi$.

ix. $\Pi$ is the only proof of $A, B \succ C \wedge D$.

### Challenge Questions

1. Here is a proof, from $(p \rightarrow q) \vee r$ to $p \rightarrow ((q \vee r) \vee s)$. Does this proof contain any detours?

$$\cfrac{\cfrac{\cfrac{\cfrac{[p \rightarrow q]^3 \quad [p]^1}{q} \rightarrow E}{q \vee r} VI}{(p \rightarrow q) \vee r \quad p \rightarrow (q \vee r)} \rightarrow I^1 \quad \cfrac{\cfrac{[r]^4}{q \vee r} VI}{p \rightarrow (q \vee r)} \rightarrow I^2}{\cfrac{\cfrac{p \rightarrow (q \vee r)}{q \vee r} \quad [p]^5}{\cfrac{q \vee r}{(q \vee r) \vee s} VI} \rightarrow E} VE^{3,4}}{p \rightarrow ((q \vee r) \vee s)} \rightarrow I^5$$

If it does contain detours, what is the detour formula, and where is it introduced and eliminated? Are there any reduction steps to reduce the proof? If there are no detour formulas, how can you explain the presence of the formula $p \rightarrow (q \vee r)$, which is not a subformula of the undischarged assumption of the proof $((p \rightarrow q) \vee r)$ and is not a subformula of the conclusion $(p \rightarrow ((q \vee r) \vee s))$?

2. Describe in your own words a process for how to construct a proof for an argument. Imagine attempting to program a computer to construct a proof for $X \succ A$. How do you describe the process?

3. Let us say that a formula of the form $A \wedge \neg A$ is a contradiction. In intuitionistic logic, contradictions entail every formula because contradictions entail $\bot$ and by $\bot E$, $\bot$ entails everything. An influential alternative argument to the conclusion that contradictions entail everything was offered by C. I. Lewis, who will appear again in chapter 7. Lewis's argument uses the rule (DS), rather than $\neg E$ and $\bot E$. His argument is the following.

$$\cfrac{\cfrac{\cfrac{A \wedge \neg A}{A} \wedge E}{A \vee B} VI \quad \cfrac{A \wedge \neg A}{\neg A} \wedge E}{B} (DS)$$

Since the disjunct $B$ is arbitrary, contradictions entail everything. Which of the rules in the proof do you think is most plausible to give up in order to prevent this line of argument from working? Can you provide motivation for giving up your chosen rule? You can assume that $\bot$ is not in the language for purposes of this question.

4. Let us add the rule (PR)

$$\begin{array}{c} [A \rightarrow B]^i \\ \Pi \\ \cfrac{A}{A} (PR)^i \end{array}$$

to our proof system. Show that in the extended system, the rule *DNE*

$$\cfrac{\neg \neg A}{A} DNE$$

is derivable. Then, show that the rule *reductio*

$$
\begin{array}{c}
[\neg A]^i \\
\Pi \\
\dfrac{\bot}{A}\ \textit{reductio}^i
\end{array}
$$

is derivable in the system with (PR).

5. For this problem, we will use a second formal language, without $\bot$ but otherwise exactly like Form. The proof system on this language will use the following rules for negation, rather than $\neg I$ and $\neg E$, and it will not have $\bot E$.

$$
\begin{array}{cc}
\begin{array}{c}
[A]^i \\
\Pi \\
\dfrac{\neg A}{\neg A}\ (\neg I^*)^i
\end{array}
&
\begin{array}{c}
\dfrac{\neg A \quad A}{B}\ (\neg E^*)
\end{array}
\end{array}
$$

The proof system with these rules, along with $\to I$, $\to E$, $\lor I$, $\lor E$, $\land I$, and $\land E$, will be called J. First, provide reduction steps for the rules above to eliminate detours involving negation.

Next, show that, whenever $X$ and $A$ do not contain $\bot$, $X \vdash_J A$ iff $X \vdash_I A$. Note that the proofs in I may contain $\bot$, although it will not occur in $X$ or in $A$. Hint: You will need to show how to convert a proof in J to a proof in I, and conversely.

Can you provide any reasons for preferring the rules $\neg I$ and $\neg E$ over $(\neg I^*)$ and $(\neg E^*)$ or vice versa?

# 4    Facts about Proofs & Provability

In the last two chapters, we introduced proofs for propositional logic. In this chapter, we will look back and see what we can learn about logic from the point of view of proofs. In the first section, we will see what facts about validity follow from the analysis of validity in terms of proofs. In the rest of the chapter, we will look at the reduction process for proofs and see what this means about the connection between the premises and conclusion when we have a valid argument.

## 4.1    Facts about Provability

In this chapter, we will introduce some *metatheory*, proving general facts about the proof system we are working with. We will adopt some conventions that are fairly standard in proof theory, using $X, Y$ for $X \cup Y$, using $X, A$ or $A, X$ for $X \cup \{A\}$, and using $A, B$ for $\{A, B\}$. We will start with some general structural facts about provability in intuitionistic logic.

**Theorem 1**    *The following are true.*

1. $A \vdash_I A$.
2. *If* $X \vdash_I A$, *then* $X, B \vdash_I A$.
3. *If* $X \vdash_I A$ *and* $A, Y \vdash_I B$, *then* $X, Y \vdash_I B$.

*Proof.*    1. For $A \vdash_I A$, we can simply use a proof consisting of a single assumption of $A$.

2. Suppose that $X \vdash_I A$. Then there is a proof $\Pi$ of $A$ from some assumptions in $X$. These assumptions are also in the set $X \cup \{B\}$, so $X, B \vdash_I A$.

3. Suppose $X \vdash_I A$ and $A, Y \vdash_I B$. Then there is a proof $\Pi_1$ whose conclusion is $A$ and whose assumptions are from $X$, and there is a proof $\Pi_2$ whose conclusion is $B$ and whose assumptions are from $Y \cup \{A\}$. We can represent things pictorially like this:[30]

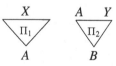

---

30. See the note on page 33 for an explanation of the triangle notation.

All of the assumptions of $A$ in $\Pi_2$ are replaced with copies of $\Pi_1$, like this:

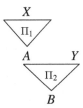

The result is a new proof, $\Pi_3$, whose conclusion is $B$ and whose assumptions are from $X \cup Y$, so we have $X, Y \vdash_\mid B$, as desired.

□

While the diagram for $\Pi_3$ in (3) displays one copy of $A$, you should keep in mind that there might be multiple assumptions of $A$ that are replaced by $\Pi_1$. For example, if there are two copies, the proof might look like this:

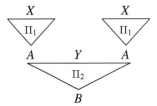

There may also be no assumptions of $A$ that are replaced, in which case, there are no copies of $\Pi_1$ that are inserted.

\* \* \*

These are some examples of structural features of the provability relation $\vdash_\mid$ that follow immediately from its definition in terms of *proofs*, given the structure of proofs—notice that when we verified these facts, we appealed to none of the specific connective rules. These follow from what are called the *structural rules* of the natural deduction system, the rules governing the structure of proofs as such, which do not apply specifically to only one sort of formula.

\* \* \*

Another important structural feature of our proof system is the following fact.

**Theorem 2 (Compactness)** *The provability relation $\vdash_\mid$ is compact in the following sense. If $X \vdash_\mid A$—even if the set $X$ is infinitely large—then there has to be some finite subset $X' \subseteq X$ where $X' \vdash_\mid A$.*

*Proof.* This fact follows immediately from the definition of $\vdash_\mid$ in terms of proofs, which are finite trees. If $X \vdash_\mid A$, then there is some proof for the argument $X \succ A$, and since proofs are, by definition, finite, there must have only been finitely many formulas in $X$ that were used in the proof as assumptions. Choose that collection of formulas to be $X'$, and we have $X' \vdash_\mid A$.                                                                       □

This fact, again, follows from the structural properties of the proof relation and requires no specific features of the connectives or rules in the logical system, other than the fact that proofs constructed by way of those rules are finite. It is not surprising that a propositional logic is compact. This result will have more significant consequences for us when we look at logic involving quantification in part III. There, compactness will have significant consequences for the kinds of concepts that will be expressible using logical resources.

$$* * *$$

The next collection of facts about provability gives us connections between the provability relation $\vdash_1$ and each of the connectives:

**Theorem 3** *For any set X of formulas, and any formulas A, B, and C, we have*

- $X \vdash_1 A \wedge B$ *if and only if* $X \vdash_1 A$ *and* $X \vdash_1 B$.
- $X \vdash_1 A \rightarrow B$ *if and only if* $X, A \vdash_1 B$.
- $X \vdash_1 \neg A$ *if and only if* $X, A \vdash_1 \bot$.
- $X, A \vee B \vdash_1 C$ *if and only if* $X, A \vdash_1 C$ *and* $X, B \vdash_1 C$.

These facts follow immediately from the inference rules for each connective. We write out the details of the conditional and disjunction cases and leave the other two as an exercise.

*Proof.* To show that $X \vdash_1 A \rightarrow B$ if and only if $X, A \vdash_1 B$, we first show that if $X \vdash_1 A \rightarrow B$, then $X, A \vdash_1 B$, and then we do the reverse: if $X, A \vdash_1 B$, then $X \vdash_1 A \rightarrow B$. So, let's suppose $X \vdash_1 A \rightarrow B$. That means we have some proof $\Pi$ for $X \succ A \rightarrow B$. We take that proof and extend it like this:

$$\frac{\overset{X}{\underset{\Pi}{\triangledown}}}{\dfrac{A \rightarrow B \quad A}{B}} \rightarrow E$$

which ensures that $X, A \vdash_1 B$, as we wanted to show. Conversely, if we have some proof $\Pi'$ for $X, A \succ B$, we can take that proof and extend it like this:

$$\frac{\overset{X \quad [A]^i}{\underset{\Pi'}{\triangledown}}}{\dfrac{B}{A \rightarrow B}} \rightarrow I^i$$

and this is a proof for $X \succ A \rightarrow B$, as we wanted to show. (The case for negation has exactly the same form as this one.)

For the disjunction case, suppose we have a proof $\Pi$ for $X, A \vee B \succ C$. We can extend it into proofs for $X, A \succ C$ and for $X, B \succ C$ with one step each, like this:

$$\begin{array}{cc} \dfrac{\underset{\phantom{X}}{X \quad \dfrac{A}{A \vee B}\,{\vee}I}}{\underset{\Pi}{\triangledown}} & \dfrac{\underset{\phantom{X}}{X \quad \dfrac{B}{A \vee B}\,{\vee}I}}{\underset{\Pi}{\triangledown}} \\ C & C \end{array}$$

Conversely, if we have proofs $\Pi_1$ and $\Pi_2$ for $X, A \succ C$ and for $X, B \succ C$, respectively, we can combine them into a proof for $X, A \vee B \succ C$ in the obvious way:

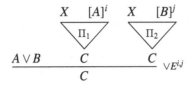

$\square$

(You should spend some time thinking about why this theorem does not have a fact of the form "$X \vdash_| A \vee B$ if and only if $X \vdash_| A$ or $X \vdash_| B$." One question for you at the end of the chapter asks you to explore this issue.)

As you gain familiarity with the logical connectives, with proofs (and with *models*, starting in the next chapter), you will gain confidence and an understanding of their behavior and what you can *do* with them. One thing you can do with them is show that some sets of formulas are inconsistent.

**Definition 11 (Inconsistency)**  *We say that a set $X$ of formulas is* INCONSISTENT *if and only if $X \vdash_| \perp$. That is, $X$ is inconsistent if and only if there is a proof of a contradiction from $X$.*

Given this definition, theorem 3 tells us that a negation $\neg A$ follows from $X$ if and only if when you add $A$ to $X$, the result is inconsistent. In other words, $\neg A$ follows from $X$ if and only if $A$ is *ruled out* by $X$. So, if you are attempting to evaluate whether a negation $\neg A$ is true (or whether it follows from the other commitments that you've made), then one thing to do is to *suppose A* and to see where it leads. If it leads to inconsistency, this is a good reason to infer $\neg A$ from your other commitments. If not, you have good reason to think that $A$ is not ruled out by what you've assumed so far. If there is reason to conclude $\neg A$, it must arise out of something other than the premises $X$ that you've granted so far.

\* \* \*

The results of this chapter so far have concerned *provability*. Now we turn from provability to *proofs*—to general features of *proofs themselves*. In particular, we will pay closer attention to the process of reduction and what we can learn from the process of eliminating detours in proofs. There are two reasons for this. First, it is interesting in its own right and will lead to a deeper understanding of how the inference rules work and the connection between proof and meaning and *analysis*. We will see that with the proof rules we have chosen, if there is a proof from $X$ to $A$, there is a proof that doesn't use anything *extra* outside the vocabulary involved in $X$ and $A$, in the sense that the only rules used will be introduction or elimination rules for the connectives occurring in $X$ and $A$. The connection between $X$ and $A$ arises out of the concepts in $X$ and $A$ themselves.

But there is a second reason to be interested in the process of reduction. One consequence of this analysis will be some new tools for thinking about provability. We will begin to get answers to questions about when something is *not* provable. If we can be sure that when there is a proof from $X$ to $A$, then there is a proof limited to the vocabulary given by $X$ and $A$, then there is a *much* smaller space to search when looking for proofs. This not only makes proof search easier. It gives us hope that we can *exhaust* the search space and be confident that there is no proof to be found. So, with these two motivations in hand, we will explore the process of cutting out detours in proofs—or *normalization*.

## 4.2 Normalization

You will have noticed when we introduced the introduction and elimination rules for each connective, we checked that they *matched* each other in an important sense: that we could perform a reduction on any detour. That is, whenever we introduced and then immediately eliminated a formula, we could have cut out that roundabout step and could have gone more directly to our conclusion.

A proof that has no detour formulas at all is said to be *normal*—not because proofs with detours are either odd or irregular—because normal proofs proceed from the premise to the conclusion in the most direct way possible.[31] Our aim is to give a general method for converting a proof into a normal proof (*normalization*) and then to see what we can do with normal proofs.

Well, that is *almost* the definition. Detours involving introduction and elimination steps are one way a proof can involve a detour. Another is a $\perp E$ step introducing a formula that is then a major premise in an elimination step, like this:

$$\cfrac{\cfrac{\cfrac{\Pi_1}{\perp}}{\boxed{A \to B}}\;\perp E \qquad \cfrac{\Pi_2}{A}}{B}\;\to E \qquad\rightsquigarrow\qquad \cfrac{\cfrac{\Pi_1}{\perp}}{B}\;\perp E$$

The rest of the family of $\perp E$ reductions are given here:

$$\cfrac{\cfrac{\cfrac{\Pi_1}{\perp}}{\boxed{\perp}}\;\perp E}{A}\;\perp E \qquad\rightsquigarrow\qquad \cfrac{\cfrac{\Pi_1}{\perp}}{A}\;\perp E$$

$$\cfrac{\cfrac{\cfrac{\Pi_1}{\perp}}{\boxed{A_1 \wedge A_2}}\;\perp E}{A_i}\;\wedge E \qquad\rightsquigarrow\qquad \cfrac{\cfrac{\Pi_1}{\perp}}{A_i}\;\perp E$$

$$\cfrac{\cfrac{\cfrac{\Pi_1}{\perp}}{\boxed{\neg A}}\;\perp E \qquad \cfrac{\Pi_2}{A}}{\perp}\;\neg E \qquad\rightsquigarrow\qquad \cfrac{\Pi_1}{\perp}$$

---

31. In fact, sometimes the direct route is not the *shortest*. Later we will see some examples where normal proofs are *longer* than nonnormal proofs, which take a shortcut through a more general principle.

$$\frac{\begin{array}{c}\Pi_1 \\ \bot\end{array}}{A \vee B} \bot E \quad \begin{array}{cc}[A]^i & [B]^j \\ \Pi_2 & \Pi_3 \\ C & C\end{array} \quad \rightsquigarrow \quad \begin{array}{c}\Pi_1 \\ \frac{\bot}{C} \bot E\end{array}$$
$$\frac{}{C} \vee E^{i,j}$$

A normal proof (without disjunction) is one that has no detour formulas.[32]

**Definition 12 (Detour formulas in proofs without disjunction)**  *A* DETOUR FORMULA *in a proof without disjunction is a formula instance that is introduced as either the conclusion of an introduction rule or the conclusion of a $\bot E$ rule and then is eliminated as the major premise of an elimination rule.*

**Definition 13 (Normal proofs for disjunction-free proofs)**  *A $\vee$-free proof $\Pi$ is a* NORMAL PROOF *iff $\Pi$ contains no detours.*

It was Dag Prawitz who first proved (Prawitz 1965) that there was a general process for normalizing proofs, and it is a central insight in the theory of natural deduction.[33] The procedure is not too hard to explain when our proof does not use the $\vee E$ rules. If we pick a detour formula to reduce, like this conditional $A \rightarrow B$:[34]

$$\frac{\begin{array}{c}[A]^1 \\ \Pi_1 \\ B\end{array}}{A \rightarrow B} \rightarrow I^1 \quad \begin{array}{c}\Pi_2 \\ A\end{array} \quad \rightsquigarrow \quad \begin{array}{c}\Pi_2 \\ A \\ \Pi_1 \\ B\end{array}$$
$$\frac{}{B} \rightarrow E$$

the resulting proof does not have *that* detour formula any more,[35] but the transformation might have introduced new detour formulas, where the pieces kept separate in the original proof are joined in the new one. Look at the site where the $A$ is proved *from* $\Pi_2$ and then used in $\Pi_1$ in the new proof. If the last step in $\Pi_2$ is an introduction, and the first step in $\Pi_1$ where that instance of $A$ was used is an elimination, this $A$ becomes a detour formula. And remember, the $A$ might have been discharged a number of times in the original proof in $\Pi_1$, so the substitution of the new proof $\Pi_2$ into $\Pi_1$ might occur a number of times—so we might have *many* new detour formulas. More than we started with. So this might not be much of a "reduction." Perhaps this process will make our proof *less normal*.

But wait. We have another piece of information that we can use. The original detour formula was $A \rightarrow B$ . The newly introduced detour formulas (if we have any) are copies of $A$. Notice that the new detour formula is *smaller* (less complex) than the old detour formula. In particular, it is a subformula of $A \rightarrow B$ . This gives us hope that the process

---

32. As you saw in exercise 1 on page 47 in chapter 3, the $\vee E$ rules complicate things quite a bit. We will get to this at the end of this section.

33. Independently, Andrés Raggio (1965) also proved the normalization theorem. Gentzen wrote a proof of the normalization theorem for intuitionistic natural deduction but never published it, as discussed by von Plato and Gentzen (2008).

34. Unlike the triangle notation for proofs used earlier in the chapter, when we write a discharged assumption above a "$\Pi$," there are no constraints on what other assumptions may be undischarged in $\Pi$.

35. Though it may contain the formula $A \rightarrow B$ elsewhere, of course—inside $\Pi_1$ or $\Pi_2$ somewhere.

doesn't go on forever, since no formula has infinitely many subformulas. If each reduction trades complex detour formulas for simpler detour formulas, we can be sure that this will eventually stop, because formula complexity is always finite, and no formula has any complexity below *zero*. But to be sure that we only trade in complex detour formulas for simpler detour formulas, we need to check one more thing. Notice that in the reduction step we are considering, the proof $\Pi_2$ is duplicated for each copy of $A$ that is discharged in $\Pi_1$. So, if $\Pi_2$ has detour formulas inside it—and detour formulas as complex as $A \rightarrow B$, then the resulting "reduced" proof may have *more* detour formulas of the same complexity as the one we eliminated. So, we need to avoid this if we want our process of reduction to lower the number of *most-complex* detour formulas. But notice that any of the formulas duplicated in this reduction are inside $\Pi_2$ and must occur "higher up"[36] in the whole proof than the detour formula $A \rightarrow B$. So to pick our detour formula to reduce, we pick a *highest* one in the proof (a detour formula where there are no detour formulas above it): there has to be a highest most-complex detour formula, as proofs don't go on forever, although there may be more than the highest most-complex detour formula; in that case, we can pick any one of the highest most-complex detour formulas to reduce. So, now we have all the pieces we need to describe the process and to verify that it works. We just need to make these ideas precise and spell out the details.

**Definition 14 (Complexity)**  *The* COMPLEXITY OF A DETOUR FORMULA *A in a proof* $\Pi$ *is the number* comp(A), *the number of connectives occurring in A.*

That is the complexity measure we're using. Now we need to make precise the notion of *height*.

**Definition 15 (Height)**  *The* HEIGHT *of a node a in a finite tree T is the number of steps it takes to get from the root of the tree to that occurrence of n. This can be defined inductively like this:*

- *The height of the root of a tree in that tree is* 0.
- *If the node a has height n in the tree T, then that occurrence has height n + 1 in the new tree:*

$$\frac{T \quad \cdots}{t}$$

*in which the node t is inserted as a new root, and T is inserted as one of the branches.*

*The height of the whole finite tree is the largest height of any node in that tree.*

To illustrate, here is a tree, together with a copy of the tree where we have replaced each node by its height.

$$\frac{\dfrac{\dfrac{b \quad c}{d} \quad e}{\quad f}}{g} \qquad \frac{\dfrac{\dfrac{3 \quad 3}{2} \quad 2}{1 \qquad 1}}{0}$$

Since 3 is the largest height, this is a tree of height 3—there are three steps from the root to the highest leaf.

* * *

---

36. "Higher up"? What does this mean? Good question. We define this precisely in definition 15.

There is one subtlety that deserves comment before we proceed to the normalization process, the distinction between formulas and occurrences of formulas. Proofs are trees whose nodes are labeled with formulas. A single formula may occur in multiple places in the proof. As an example, take the following proof.

$$\frac{\dfrac{p \wedge q}{p} \wedge E \quad \dfrac{p \wedge q}{q} \wedge E}{\dfrac{\boxed{p \wedge q}}{p} \wedge E} \wedge I$$

In this proof, the formula $p \wedge q$ occurs three times. It occurs twice as an assumption and it occurs once as a detour, in the shaded box. Each step in the process of normalizing a proof will reduce an occurrence of a detour formula. The reductions will only be carried out on the occurrences that are detours.

$$* * *$$

Now we get to describe the *process* for normalizing proofs. We leave out the disjunction rules, remember, because they complicate things somewhat. Wait for the end of the section to see how we can deal with normalization in general. It is more important to see the basic case first.

**Definition 16 (The normalization process for disjunction-free proofs)**   *Take a proof* $\Pi$. *If it contains* no *detour formulas, we are done, and it is normal. If* $\Pi$ *has some detour formulas to reduce, we pick among the detour formulas in* $\Pi$, *one with (a) the largest complexity of any detour formulas in* $\Pi$ *and (b) such that there is no detour formula with that same complexity at a* higher *position in* $\Pi$. *We reduce that detour, using the rules for reduction for* $\wedge$, $\rightarrow$, $\neg$ *from the previous chapter and the* $\perp E$ *reductions, from page 53, resulting in a proof* $\Pi'$. *We say* $\Pi \rightsquigarrow \Pi'$, *and* $\Pi'$ *is a* one-step *reduction of* $\Pi$. *Continue this process until the result is a normal proof.*

Our aim, then, is to verify this result.

**Theorem 4 (Detour reduction theorem for disjunction-free proofs)**   *For any proof* $\Pi$ (*not using the* $\vee$ *rules*), *the sequence*

$$\Pi \rightsquigarrow \Pi' \rightsquigarrow \Pi'' \rightsquigarrow \Pi'' \rightsquigarrow \cdots$$

*terminates at some stage, with a normal proof* $\Pi^*$, *which cannot be reduced.*

*Proof.*   We need to show that whenever $\Pi \rightsquigarrow \Pi'$, where we have picked a *most-complex* formula to reduce, at the *highest* point in $\Pi$ where a most-complex detour formula can be found, then $\Pi'$ has *fewer* most-complex detour formulas than $\Pi$. Once we prove this, we can be sure that the process won't go on forever, since each proof is finite. If, say, $\Pi$ has $n$ detour formulas of complexity $m$ (and none of any larger complexity), then in at most $n$ steps, we will have eliminated all of the detours of complexity $m$. Then we go on to formulas of complexity $m-1$. If our proof has $k$ detour formulas of that complexity, then in at most $k$ steps, we will have eliminated all of the detours of that size, and so on.

So, let's check that each of the reduction steps has the property of reducing the number of most-complex detour formulas if we applied them to a most-complex detour formula at the highest point in the tree. For conjunction, if we make this transformation:

$$
\begin{array}{c}
\dfrac{\begin{array}{cc} \Pi_1 & \Pi_2 \\ A_1 & A_2 \end{array}}{A_1 \wedge A_2} \wedge I \\[2pt]
\dfrac{\phantom{A_1 \wedge A_2}}{A_i} \wedge E
\end{array}
\quad \rightsquigarrow \quad
\begin{array}{c} \Pi_i \\ A_i \end{array}
$$

we notice that our new proof is strictly *smaller* than the original proof (we pruned one whole branch), and it eliminated the detour formula $A_1 \wedge A_2$ , so the new proof has strictly fewer detour formulas of that complexity.

We have already seen the conditional reduction:

$$
\begin{array}{c}
\begin{array}{c} [A]^1 \\ \Pi_1 \\ B \end{array} \\
\dfrac{}{A \to B} \to I^1 \quad \begin{array}{c} \Pi_2 \\ A \end{array} \\
\hline
B
\end{array} \to E
\quad \rightsquigarrow \quad
\begin{array}{c} \Pi_2 \\ A \\ \Pi_1 \\ B \end{array}
$$

Here, after the reduction, the new proof might have many copies of $\Pi_2$, one for each copy of $A$ that was discharged in $\Pi_1$, but if the indicated $A \to B$ is one of the *highest* detour formulas with that complexity, we can be sure that $\Pi_2$ has no detour formulas so complex, and as a result, the new proof has strictly fewer detour formulas of that complexity, since we have snipped out one instance of $A \to B$ .

Finally, the negation case has the same structure:

$$
\begin{array}{c}
\begin{array}{c} [A]^1 \\ \Pi_1 \\ \bot \end{array} \\
\dfrac{}{\neg A} \neg I^1 \quad \begin{array}{c} \Pi_2 \\ A \end{array} \\
\hline
\bot
\end{array} \neg E
\quad \rightsquigarrow \quad
\begin{array}{c} \Pi_2 \\ A \\ \Pi_1 \\ \bot \end{array}
$$

Here, after the reduction, the new proof might have many copies of $\Pi_2$, but if the indicated $\neg A$ is one of the *highest* detour formulas with that complexity, we can be sure that $\Pi_2$ has no detour formulas so complex, and as a result, the new proof has strictly fewer detour formulas of that complexity, since we have snipped out one instance of $\neg A$ .

Verifying that the $\bot E$ reductions also shrink the number of most-complex detour formulas is trivial. If you check the reductions (on page 53), you will see that there are *no* substitutions and the resulting proof deletes one of the most-complex detour formulas and properly shrinks the proof.

This completes our verification that Prawitz's normalization process always terminates—we end up with a normal proof.  □

This normalization strategy lets us prove the following theorem.

**Theorem 5 (Normalization theorem for disjunction-free proofs)** *If there is a proof $\Pi$ of $X \vdash_1 A$ (not using the $\vee$ rules), then there is a normal proof $\Pi'$ of $X \vdash_1 A$.*

The normalization theorem is useful. It means we can—if we like—focus on normal proofs when searching for proofs or thinking about proofs, since any proof that has detours can be converted into a normal proof. As we will see in the next section, this is sometimes very useful.

∗ ∗ ∗

The rest of this section comes with a  **WARNING LABEL** . Please only skim through the next few pages quickly on your first time through the text. The *important* content in the section is the material we have just covered and the explanation we give on this and the next page. Beyond that (once we start the definition of *repetition* and *detour sequences*), the material gets more complicated. If you are confident with proofs by induction, and you want to see all the details, go right ahead. If not, once you reach the definitions, feel free to skim lightly over the rest and start reading more slowly and deeply when you get to section 4.3.

∗ ∗ ∗

It's the *disjunction rules* that make all of this difficult. The main reason they complicate things is the structure of the disjunction elimination rule:

$$\frac{A \vee B \qquad \begin{matrix}[A]^1 \\ \Pi_1 \\ C\end{matrix} \qquad \begin{matrix}[B]^2 \\ \Pi_2 \\ C\end{matrix}}{C} \vee E^{1,2}$$

The complicating factor is the formula $C$. It is just sitting there *doing nothing* in the transition from premises to conclusion. No other rule has formulas that sit around idle like this. This can cause problems with normalization. Consider a proof that has a part looking like this:

$$\frac{\dfrac{\begin{matrix}\Pi_1 \\ A \vee B\end{matrix} \qquad \begin{matrix}[A]^1 \\ \Pi_2 \\ C \wedge D\end{matrix} \qquad \dfrac{\begin{matrix}[B]^2 \\ \Pi_3 \\ C\end{matrix} \quad \begin{matrix}[B]^2 \\ \Pi_4 \\ D\end{matrix}}{C \wedge D} \wedge I}{C \wedge D} \vee E^{1,2}}{C} \wedge E$$

Look closely at the structure of the proof. The conjunction $C \wedge D$ is introduced in one of the subproofs of the disjunction elimination step. $\Pi_3$ proves $C$ and $\Pi_4$ proves $D$. (We're not told how $C \wedge D$ got into $\Pi_2$. This proof might also end in a $\wedge I$ step, but it might not.) Is $C \wedge D$ a detour formula? It is not introduced and *immediately* eliminated. It is introduced in the $\wedge I$ step; it *hangs around* for one inference and is *then* eliminated. This is, in a sense, a detour, because we could have rewritten the proof like this, delaying the $\vee E$ step until *after* we eliminate the conjunction:

$$
\cfrac{\Pi_1}{A \vee B} \qquad \cfrac{\cfrac{[A]^1}{\quad} \\ \Pi_2 \\ \cfrac{C \wedge D}{C} \; \wedge E \qquad \cfrac{\cfrac{[B]^2 \quad [B]^2}{\Pi_3 \quad \Pi_4} \\ \cfrac{C \quad D}{\boxed{C \wedge D}} \; \wedge I}{C} \; \wedge E}{C} \; \vee E^{1,2}
$$

Now we can see the detour formula more clearly, and it can be eliminated in the usual way.

$$
\cfrac{\Pi_1}{A \vee B} \qquad \cfrac{\cfrac{[A]^1 \\ \Pi_2 \\ C \wedge D}{C} \; \wedge E \qquad \cfrac{[B]^2 \\ \Pi_3 \\ C}{}}{C} \; \vee E^{1,2}
$$

**Definition 17 (Repetition in a proof[37])** *An instance $C_{n+1}$ of a formula in a proof is said to be a* REPETITION *of another occurrence $C_n$ of that formula in the proof if and only if $C_{n+1}$ is the conclusion of a $\vee E$ inference, and $C_n$ is a minor premise of that inference.*

**Definition 18 (Detour sequences)** *A finite sequence of formula occurrences of the same formula in a proof is a* DETOUR *if the first element of the sequence is the conclusion of an introduction inference, or the conclusion of a $\perp E$ inference; each later element in the sequence is a repetition of the one before it, and the last element of the sequence is the major premise in an elimination inference.*

This is the generalization of the notion of a detour formula to cope with the complication of the disjunction rule. In a proof with this shape, there are *two* detour sequences: one of length 3, highlighted in shaded boxes,

$$
\cfrac{\Pi_1}{A \vee B} \qquad \cfrac{\cfrac{[A]^3 \\ \Pi_2 \\ A' \vee B'} \quad \cfrac{\cfrac{[A]^3[A']^1 \\ \Pi_3 \\ E \to (C \wedge D)} \quad \cfrac{[A]^3[A']^1 \\ \Pi_4 \\ E}{} }{C \wedge D} \; {\to}E}{\boxed{C \wedge D}} \quad \cfrac{\cfrac{[A]^3[B']^2 \quad [A]^3[B']^2}{\Pi_5 \quad \Pi_6} \\ \cfrac{C \quad D}{\boxed{C \wedge D}} \; \wedge I \qquad \cfrac{[B]^4 \; [B]^4}{\cfrac{\Pi_7 \; \Pi_8}{\cfrac{C \quad D}{C \wedge D} \; \wedge I}}}{} \; \vee E^{1,2}}{\cfrac{\boxed{C \wedge D}}{C} \; \wedge E} \; \vee E^{3,4}
$$

---

37. Not confident yet with proofs by induction? This is the point to start skimming and rejoin us at section 4.3 on page 62. The rest of this section is a detailed treatment of *how* to define the reduction process in the presence of the disjunction rules and verifying that it always terminates in a normal proof.

and the other of length 2, in the shaded boxes below:

$$
\cfrac{
\begin{array}{c}
\cfrac{
\begin{array}{c}
[A]^3 \\
\Pi_2 \\
\cfrac{A' \vee B'}{A \vee B}\ \Pi_1
\end{array}
\quad
\cfrac{
\cfrac{\begin{array}{cc}[A]^3[A']^1 & [A]^3[A']^1 \\ \Pi_3 & \Pi_4 \\ E\to(C\wedge D) & E\end{array}}{C\wedge D}\ {\to}E
\quad
\cfrac{\begin{array}{cc}[A]^3[B']^2 & [A]^3[B']^2 \\ \Pi_5 & \Pi_6 \\ C & D\end{array}}{C\wedge D}\ {\wedge}I
}{C\wedge D}\ {\vee}E^{1,2}
\end{array}
\quad
\cfrac{\begin{array}{cc}[B]^4 & [B]^4 \\ \Pi_7 & \Pi_8 \\ C & D\end{array}}{C\wedge D}\ {\wedge}I
}{C\wedge D}\ {\vee}E^{3,4}
$$

$$\cfrac{C\wedge D}{C}\ {\wedge}E$$

The instance of $C \wedge D$ in the conclusion of the $\to E$ step is not a part of a detour sequence as it was introduced in an elimination inference and not in an introduction.

Just as the concept of detour required generalization to that of detour sequence, the concept of detour requires generalization.

**Definition 19 (Normal proof)**  *A proof* $\Pi$ *is* NORMAL *iff* $\Pi$ *does not contain any detour sequences.*

A proof is not normal if it contains any detour sequences. Our reduction steps deal only with detour *formulas*, which are detour sequences of length 1. We have seen in one case, though, that you can swap the order of a $\vee E/\wedge E$ pair so the $\wedge E$ step goes first, and the sequence is shortened by 1. Because we swap the order of the rules, this is called a *permutation.*[38] The general phenomenon looks like this:

**Definition 20 (Permutation steps)**  *We permute a $\vee E$ inference with conclusion C in a proof below an elimination inference E*

$$\cfrac{C \quad \Delta}{D}\ E$$

*in which C is the major premise (and $\Delta$ represents the other components of the proof, if any—$\Delta$ is empty if this inference is a conjunction or falsum elimination, it is one proof if it is a negation or conditional elimination, and it is two proofs if E is a disjunction elimination), as follows:*

$$
\cfrac{
\cfrac{\begin{array}{ccc} & [A]^i & [B]^j \\ \Pi_1 & \Pi_2 & \Pi_3 \\ A\vee B & C & C\end{array}}{C}\ {\vee}E^{i,j} \quad \Delta
}{D}\ E
\qquad \hookrightarrow \qquad
\cfrac{
\begin{array}{ccc}
 & [A]^i & [B]^j \\
 & \Pi_2 & \Pi_3 \\
\Pi_1 & \cfrac{C\quad\Delta}{D}\ E & \cfrac{C\quad\Delta}{D}\ E \\
A\vee B
\end{array}
}{D}\ {\vee}E^{i,j}
$$

Notice that when we permute the $E$ step above the $\vee E$ step, it duplicates, so the proof can get bigger in this process. However, any detour sequence of instances of $C$ ending at this $E$ inference is shortened by 1 in the process.

The most complicated case of permutation is where $E$ is itself a disjunction elimination rule. However, there is no clash or threat of an infinite loop, because the requirement is that the formula $C$ is a *minor* premise of the $\vee E$ step and a *major* premise of the $E$ step. If we

---

38. The permutations defined here have nothing to do with the permutation rule discussed in section 2.4. These shift formulas and rules in a proof, while the rule swaps antecedents of conditionals.

have this configuration:

$$
\cfrac{\Pi_1 \quad \cfrac{\begin{array}{cc}[A]^i & [B]^j\\ \Pi_2 & \Pi_3\end{array}}{\quad}}{}
$$

$$
\cfrac{\begin{array}{ccc}
 & [A]^i & [B]^j\\
\Pi_1 & \Pi_2 & \Pi_3\\
A\vee B & C\vee C' & C\vee C'
\end{array}\ \vee E^{i,j}}{C\vee C'}\qquad
\cfrac{\begin{array}{cc}[C]^k & [C']^l\\ \Pi_4 & \Pi_5\\ D & D\end{array}}{D}\ \vee E^{k,l}
$$
$$
\overline{\qquad\qquad D \qquad\qquad}
$$

then the permutation transforms that configuration into this:

$$
\cfrac{\Pi_1}{A\vee B}\quad
\cfrac{\begin{array}{ccc}[A]^i & [C]^k & [C']^l\\ \Pi_2 & \Pi_4 & \Pi_5\\ C\vee C' & D & D\end{array}}{D}\ \vee E^{k,l}\quad
\cfrac{\begin{array}{ccc}[B]^j & [C]^k & [C']^l\\ \Pi_3 & \Pi_4 & \Pi_5\\ C\vee C' & D & D\end{array}}{D}\ \vee E^{k,l}
$$
$$
\overline{\qquad\qquad\qquad D \qquad\qquad\qquad}\ \vee E^{i,j}
$$

A couple of examples are in order to illustrate the permutations. We will start with an example in which the $E$ step has a single minor premise

$$
\cfrac{r\vee s\quad
\cfrac{\cfrac{[r]^1}{s\vee r}\,\vee I}{(p\wedge q)\to(r\vee s)}\,{\to}I^3\quad
\cfrac{\cfrac{[s]^2}{s\vee r}\,\vee I}{(p\wedge q)\to(s\vee r)}\,{\to}I^4}{(p\wedge q)\to(s\vee r)}\ \vee E^{1,2}\qquad
\cfrac{(p\wedge q)\wedge r}{p\wedge q}\,\wedge E}{s\vee r}\ {\to}E
$$

In this proof, the $\Delta$ portion is the minor premise subproof for the final $\to E$ step. The proof reduces to the following via the permutation reduction.

$$
\cfrac{r\vee s\quad
\cfrac{\cfrac{\cfrac{[r]^1}{s\vee r}\,\vee I}{(p\wedge q)\to(s\vee r)}\,{\to}I^3\quad \cfrac{(p\wedge q)\wedge r}{p\wedge q}\,\wedge E}{s\vee r}\,{\to}E\qquad
\cfrac{\cfrac{\cfrac{[s]^2}{s\vee r}\,\vee I}{(p\wedge q)\to(s\vee r)}\,{\to}I^4\quad \cfrac{(p\wedge q)\wedge r}{p\wedge q}\,\wedge E}{s\vee r}\,{\to}E}{s\vee r}\ \vee E^{1,2}}{s\vee r}
$$

Next, we will look at an example in which the $E$ step is $\vee E$.

$$
\cfrac{\cfrac{(p\wedge q)\vee(q\wedge r)\quad
\cfrac{\cfrac{[p\wedge q]^1}{p}\,\wedge E}{p\vee r}\,\vee I\quad
\cfrac{\cfrac{[q\wedge r]^2}{r}\,\wedge E}{p\vee r}\,\vee I}{p\vee r}\ \vee E^{1,2}\qquad
\cfrac{[p]^3}{r\vee p}\,\vee I\quad \cfrac{[r]^4}{r\vee p}\,\vee I}{r\vee p}}{r\vee p}\ \vee E^{3,4}
$$

The $\Delta$ portion of the proof is made up of the two minor premise subproofs, and it is these that have to be permuted upward. The permutation yields the following proof, where $A$ is the formula $(p\wedge q)\vee(q\wedge r)$.

$$
\cfrac{A\quad
\cfrac{\cfrac{\cfrac{[p\wedge q]^1}{p}\,\wedge E}{p\vee r}\,\vee I\quad \cfrac{[p]^3}{r\vee p}\,\vee I\quad \cfrac{[r]^4}{r\vee p}\,\vee I}{r\vee p}\ \vee E^{3,4}\qquad
\cfrac{\cfrac{\cfrac{[q\wedge r]^2}{r}\,\wedge E}{p\vee r}\,\vee I\quad \cfrac{[p]^5}{r\vee p}\,\vee I\quad \cfrac{[r]^6}{r\vee p}\,\vee I}{r\vee p}\ \vee E^{5,6}}{r\vee p}\ \vee E^{1,2}}{r\vee p}
$$

Now we are ready to describe Prawitz's process for normalizing proofs in its complete generality.

**Definition 21 (The normalization process for proofs)** *Take a proof* $\Pi$. *If it contains no detour sequences, we are done, and it is normal. If* $\Pi$ *has some detour sequence, consider the endpoint (that is, formula that is the major premise of an elimination step) of each most-complex detour sequence and select the sequences ending in the* highest *such endpoint. Permute all instances of* $\vee E$ *steps in those sequences below the endpoint of the sequences, using the permutation rules. Then the detour sequences are transformed into individual detour formulas, and in the process, we do not duplicate any other detour sequences of this same complexity. We reduce any resulting detours, using the rules for reduction for* $\wedge$, $\rightarrow$, $\neg$, *and* $\vee$ *from the previous chapter, resulting in a proof* $\Pi'$. *We say* $\Pi \rightsquigarrow \Pi'$, *and* $\Pi'$ *is a* one-step *reduction of* $\Pi$. *Continue this process until the result is a normal proof.*

**Theorem 6 (Normalization theorem)** *For any proof* $\Pi$, *the sequence*

$$\Pi \rightsquigarrow \Pi' \rightsquigarrow \Pi'' \rightsquigarrow \Pi'' \rightsquigarrow \cdots$$

*terminates at some stage, with a normal proof* $\Pi^*$, *which cannot be reduced.*

The proof has exactly the same form as the proof for theorem 4, except we are dealing with the more involved one-step reduction process $\rightsquigarrow$ where we permute $\vee E$ steps below an elimination step if we have a detour sequence of length longer than 1.

$$* \ * \ *$$

Phew! We're done.[39] Congratulations to you if you've made it through this far. Welcome back everyone else who skimmed over these three pages.

## 4.3 The Subformula Property

As we have hinted, a key property of normal proofs is their analyticity. The formulas inside the proof are found inside the premises or the conclusion.

**Definition 22 (Subformula property)** *A proof* $\Pi$ *for* $X \succ A$ *has the* SUBFORMULA PROPERTY *if and only if each formula in* $\Pi$ *is a subformula\* either of some formula in* $X$ *or of* $A$.

Can you see the small print proviso in the definition? We've put a little asterisk in the claim about subformulas. The reason is the behavior of negation and the contradiction formula $\perp$. Consider this proof:

$$\frac{\dfrac{\neg p \quad p}{\perp} \neg E}{q} \perp E$$

By our lights, this is a normal proof. Nothing is introduced. It does contain the formula $\perp$, but $\perp$ is a subformula neither of $\neg p$, nor of $p$, nor of $q$. There are two things we could do here. One is to say that $\perp$ is not really a *formula* but a punctuation mark or something that can't be put inside other formulas but simply says "*that's a contradiction!*" After all, if we never have $\perp$ occurring explicitly inside other formulas in the premises or conclusion

---

39. And **here** is where you can rejoin us.

of a normal proof, it always occurs only by itself and not in complex formulas like $\bot \rightarrow p$ or $q \vee \bot$. That's one option. The option we will take is less radical. We will introduce a slightly more liberal notion of a subformula, allowing $\bot$ to be a subformula of any negation formula.[40]

**Definition 23 (Liberal subformula)** *The set of* LIBERAL SUBFORMULAS *of a formula A,* $\text{sub}^*(A)$ *is defined like this:*

- $\text{sub}^*(p) = \{p\}$, *where p is an atom,*
- $\text{sub}^*(\neg A) = \text{sub}^*(A) \cup \{\neg A, \bot\}$,
- $\text{sub}^*(A \wedge B) = \text{sub}^*(A) \cup \text{sub}^*(B) \cup \{A \wedge B\}$,
- $\text{sub}^*(A \vee B) = \text{sub}^*(A) \cup \text{sub}^*(B) \cup \{A \vee B\}$, *and*
- $\text{sub}^*(A \rightarrow B) = \text{sub}^*(A) \cup \text{sub}^*(B) \cup \{A \rightarrow B\}$.

(Reading through this definition, you can see that if $A$ has no negations, $\text{sub}^*(A) = \text{sub}(A)$, and if $A$ contains a negation somewhere, $\text{sub}^*(A) = \text{sub}(A) \cup \{\bot\}$.)

With that little caveat, we can prove a powerful result concerning the structure of proofs. As with normalization, it will be easier to prove a restricted version of the result for the system without disjunction rules. For this result, we will need a definition and a lemma.

**Definition 24 (Track)** *Let* $\Pi$ *be a proof not using the* $\vee I$ *and* $\vee E$ *rules. A* TRACK *in* $\Pi$ *is a sequence of formula occurrences* $\langle A_1, \dots, A_n \rangle$, $n \geq 1$, *such that*

- $A_1$ *is an assumption of* $\Pi$ *(possibly discharged);*
- *for* $1 \leq i < n$, $A_i$ *is either the major premise of an elimination rule whose conclusion is* $A_{i+1}$ *or* $A_i$ *is a premise of an introduction rule whose conclusion is* $A_{i+1}$; *and*
- $A_n$ *is either the* conclusion *of the proof* $\Pi$ *or the* minor *premise of a* $\rightarrow E$ *or* $\neg E$ *rule.*

Consider this proof, in which we have labeled formula occurrences.

$$
\cfrac{\cfrac{1.\,p \wedge \neg q}{2.\,\neg q} \qquad \cfrac{5.\,[p \rightarrow q]^1 \qquad \cfrac{\cfrac{7.\,p \wedge \neg q}{8.\,p}\,{\wedge E}}{6.\,q}\,{\rightarrow E}}{\cfrac{3.\,\bot}{4.\,\neg(p \rightarrow q)}\,{\neg I^1}}}{}\,{\neg E}
$$

Here, one track consists of the formulas 1 to 4, consisting of the leftmost branch from leaf to root. The formulas 5 and 6 form another track, while the formulas 7 and 8 form another.

In all proofs, including this one, every formula is in *some* track or other.

**Lemma 1** *In any normal proof* $\Pi$ *not using the disjunction rules, every formula is in some track.*

*Proof.* We show this, as usual, by induction on the structure of the proof. If the proof is an assumption, the hypothesis holds immediately: the sole formula occurrence is a track on its own. Suppose, now that the proof is found by adding a single elimination inference, $\rightarrow E$, $\wedge E$, $\neg E$, or $\bot E$, at the end of a proof (or proofs) to which the assumption already holds. In this case, the new conclusion formula occurrence extends any track, including the conclusion of the major premise of this elimination step. If the inference has minor

---

40. Notice that this has exactly the same effect as thinking that $\neg A$ is shorthand for $A \rightarrow \bot$.

premises (that is, in the case of $\neg E$ and $\rightarrow E$), then the tracks in the proof of the *minor* premise terminate at this point.

On the other hand, if we have a proof (or two proofs) to which the induction hypothesis applies and we extend it (or them) with an introduction inference $\wedge I$, $\rightarrow I$, or $\neg I$, then we take any track in such a proof that terminates in the conclusion of that proof and extend it with the newly introduced conclusion formula occurrence at the end of the sequence. Again, this satisfies the condition for being a track, and our lemma is proved.               □

Notice that in the case of a proof containing an $\wedge I$ step, the conclusion formula of that inference will be present in *two* tracks, one including the left premise and one including the right.

Tracks are important because in *normal* proofs, they have an elegant structure:

**Lemma 2**   *Let $\Pi$ be a normal proof without the $\vee I$ and $\vee E$ rules. In any track $\langle A_1, \ldots, A_n \rangle$ of $\Pi$, if $A_i$ is the conclusion of an introduction rule, then for $j > i$, $A_j$ is not the conclusion of an elimination rule.*

*Proof.*   Let $\Pi$ be a normal proof and $\langle A_1, \ldots, A_n \rangle$ a track in $\Pi$. Suppose $A_i$ is obtained by an introduction rule and suppose, for a contradiction, that $k$ is the least number greater than $i$ such that $A_k$ is obtained by an elimination rule. Since $k = j + 1$, for some $j \geq i$, $A_j$ must be obtained by an introduction rule. There are then three cases, depending on which of $\wedge I$, $\rightarrow I$, and $\neg I$ was used to obtain $A_j$. If $A_j$ is the conclusion of $\wedge I$, then it is of the form $B \wedge C$. The elimination rule used to obtain $A_k$ must then be $\wedge E$, but then there is a detour in the proof, contradicting the assumption that $\Pi$ was normal. If $A_j$ is the conclusion of $\rightarrow I$, then it is of the form $B \rightarrow C$, so $A_k$ must be the conclusion of $\rightarrow E$. This is impossible, since that would mean that there is a detour in $\Pi$, which was assumed to be normal. The final case, in which $A_j$ is the conclusion of $\neg I$, is similar. Therefore, we conclude that $A_k$ is not the conclusion of an elimination rule, which contradicts the assumption that it was. Therefore, for all $k$ such that $i < k \leq n$, $A_k$ is not the conclusion of an elimination rule.               □

Lemma 2 tells us a lot about the structure of normal proofs in the proof system with the disjunction rules. A track in a normal proof split into two parts, the $E$-part and the $I$-part, either of which may be empty. The $E$-part, if it is nonempty, is the initial portion of the track, and the formulas in it are the conclusions of elimination rules breaking down the assumptions. Each formula $A_{i+1}$ in the $E$-part is *less* complex than $A_i$, unless $A_i$ is $\bot$, in which case $A_{i+1}$ may be more complex, but it will also be the final formula in the $E$-part. After the $E$-part, the $I$-part builds a formula, until the track ends, either in the conclusion of a proof or in the use of this formula as a minor premise of an elimination step. Each formula $A_{i+1}$ in the $I$-part is *more* complex than $A_i$. The lemma tells us that once we have started the $I$-part of a track, there will be no elimination rules used to obtain formulas farther down that track.

The lemma was proved for the system without disjunction rules. The wrinkle introduced by the disjunction rules is the potential for sequences of repetitions to occur via applications of $\vee E$. With a bit of care, sequences of repetitions can be dealt with, and later we will prove a version of the lemma for the full proof system, but that is more complex, so to get the simpler picture, let us turn now to our preliminary result.

**Theorem 7 (Normal proofs have the subformula property)**   *If* $\Pi$ *is a normal proof without the* $\vee I$ *and* $\vee E$ *rules, then it has the subformula property.*

*Proof.*   For this proof, we use the notion of the *order* of a track in a proof. We say a track is of order 0 if it ends in the conclusion of a proof. We say that a track has order $n + 1$ if it ends not in the conclusion but in the minor premise of an elimination step, whose *conclusion* (and *major premise*) is in a track of order $n$. It is clear that since proofs are finite, every track has a unique finite order.

Now it is straightforward to show that normal proofs have the subformula property, by first noticing that the subformula property holds for formulas in a track of order 0. Since $\Pi$ is normal, our track $\langle A_1, \ldots, A_n \rangle$ divides into an *I*-part and an *E*-part. If a formula in the track occurs in the *E*-part, it is a subformula of $A_1$, the first formula in the track, and if it is in the *I*-part, it is a subformula of the formula at the end, $A_n$. If the occurrence $A_1$ is not discharged in $\Pi$, we have the subformula property for formulas in this track. However, $A_1$ may be discharged in $\Pi$ by an application of $\rightarrow I$ or $\neg I$. In that case, the formula introduced by the application of this rule occurs (of the form $\neg A_1$ or $A_1 \rightarrow B$ for some $B$) in the *I*-part of the track, and so, the formulas in the *I* component of the track, all subformulas of $A_1$, are also subformulas of $A_n$, the final formula in our track.

Now we prove the rest of our result by induction on the order of tracks. Suppose the subformula property holds for each of the formulas in tracks of order $m$ in our proof $\Pi$, and consider some track $\langle A_1, \ldots, A_n \rangle$ of order $m + 1$. This starts at a leaf of $\Pi$ but now concludes in the minor premise of some inference step $\neg E$ or $\rightarrow E$ whose major premise ($\neg A_n$, or $A_n \rightarrow B$) and conclusion ($\bot$, or $B$) is present in some track of order $m$. As before, every formula in our track is either a subformula of $A_1$ or of $A_n$. Our final formula $A_n$ is the minor premise of the elimination rule, and so, is a subformula of the *major* premise, which occurs in the track of order $m$, to which the induction hypothesis applies. So subformulas of $A_n$ are among the subformulas of the premises and the conclusion of $\Pi$. The initial formula $A_1$, if undischarged, is one of the premises of $\Pi$, and if not, it is discharged either in an inference occurring in this track, or in the track of order $m$ to which this track is connected, or one of the tracks of lower order in $\Pi$, to which the hypothesis *also* applies. So, the subformula property holds for all of the formulas in our track of order $m + 1$. This completes the induction, and the subformula property is proved.                                          $\square$

To extend this proof to the system with disjunction rules will require generalizing the definition of a track, to allow tracks to *backtrack*. Without the disjunction rules, the intuition behind the definition is that a track in a normal proof proceeds from an assumption, down the major premises of elimination rules (if any) and then along the introduction rules (if any), as far as one can go. We either reach the conclusion (for a track of order 0) or the minor premise of an elimination rule (for tracks of higher orders). With disjunction rules, we need to complicate this idea, to account for the $\vee E$ rule.

$$
\begin{array}{ccc}
 & [A]^i & [B]^j \\
\Pi_1 & \Pi_2 & \Pi_3 \\
A \vee B & (1)\,C & (2)\,C \\
\hline
 & (3)\,C & 
\end{array} \ \vee E^{i,j}
$$

In this rule, there is a connection between the major premise of the rule, $A \vee B$, and the assumptions discharged by the rule, $A$ and $B$. (In an important sense, the $A$ and $B$ assumptions "arise from" the disjunction $A \vee B$.) So, the target concept of a track will take the occurrence (3) of $C$ to be the successor of occurrence (1) or (2), rather than the succcessor of $A \vee B$, and the successor of $A \vee B$ in a track will be one of the occurrences of the discharged $A$ or $B$, rather than the conclusion of the rule.

**Definition 25 (Track)**   *A* track *of a proof $\Pi$ is a sequence of formulas $\langle A_1, \ldots, A_n \rangle$, $n \geq 1$, such that*

- *$A_1$ is an assumption of $\Pi$ (possibly discharged) that is not discharged by an application of $\vee E$,*
- *for $1 \leq i < n$, either*
  - *$A_i$ is the major premise of $\rightarrow E$, $\wedge E$, $\perp E$, or $\neg E$ and $A_{i+1}$ is the conclusion,*
  - *$A_i$ is a minor premise of an application of $\vee E$ with $A_{i+1}$ as its conclusion,*
  - *$A_i$ is the major premise of an application of $\vee E$ and $A_{i+1}$ is an assumption discharged by that application, or*
  - *$A_i$ is a premise of an introduction rule whose conclusion is $A_{i+1}$.*
- *$A_n$ either is the conclusion of $\Pi$, or the minor premise of $\neg E$ or $\rightarrow E$, or the major premise of an application of $\vee E$ that does not discharge any assumptions.*

This definition of track extends the definition for the restricted system by incorporating sequences of repetitions introduced by $\vee E$ steps.

**Lemma 3**   *In any normal proof $\Pi$, every formula occurs in a track.*

*Proof.*   This proof works in exactly the same way as in the proof for lemma 1, by induction on the construction of the proof $\Pi$. We add the cases for $\vee I$ and $\vee E$ rules. If the proof $\Pi$ ends in a $\vee I$ step, then a track of $\Pi$ is found by extending any track of the proof leading up to $\Pi$, if that proof is not merely an assumption, or the assumption formula itself otherwise, with the newly introduced disjunction. The case for a $\vee E$ step is more complicated. In this case, $\Pi$ has the following form:

$$\frac{\begin{array}{ccc} & [A]^i & [B]^j \\ \Pi_1 & \Pi_2 & \Pi_3 \\ A \vee B & (1)\,C & (2)\,C \end{array}}{(3)\,C} \vee E^{i,j}$$

To find a track for this proof $\Pi$, notice that both $\Pi_2$ and $\Pi_3$ have a track leading to their conclusion formulas $C$ (occurrences (1) and (2), respectively). If either of these is a track that does not lead back to an assumption discharged in this $\vee E$ step (an $A$ in the case of $\Pi_2$ or a $B$ in the case of $\Pi_3$), then if we extend that track with the concluding instance $C$, at occurrence (3), the result is a track for $\Pi$. If, on the other hand, we have a track leading back to an $A$ instance in $\Pi_2$ or a $B$ instance in $\Pi_3$, we can extend the track with any track in $\Pi_1$, leading down to the conclusion $A \vee B$ of $\Pi_1$. The result, in either case, is also a track in $\Pi$.                                                                                       $\square$

Despite this expansion in our definition in light of the disjunction rules, tracks still split into introduction parts and elimination parts.

**Lemma 4**  *Let $\Pi$ be a normal proof. In a track $\langle A_1, \ldots, A_n \rangle$ of $\Pi$, if $A_i$ is the conclusion of an introduction rule, then for $j > i$, $A_j$ is not the conclusion of an elimination rule other than $\vee E$.*

*Proof.*  We apply the reasoning of our proof of lemma 2 almost unchanged. Let $\Pi$ be a normal proof and $\langle A_1, \ldots, A_n \rangle$ be one of its tracks. Suppose $A_i$ is obtained by an introduction rule and suppose, for a contradiction, that there is some $k > i$ where $A_k$ is the conclusion of an elimination rule, other than $\vee E$. We can assume that $k$ is the least such number greater than $i$. Since for some $j \geq i$, $k = j + 1$, the formula $A_j$ must either be an assumption discharged by an application of $\vee E$ or be obtained by some introduction rule.

The case where $A_j$ is obtained by an introduction rule is ruled out in just the same way as in the simpler case of this lemma, by the normality of $\Pi$: such a formula $A_j$ would be a detour. So consider the remaining case, where our $A_j$ is an assumption discharged by an application of $\vee E$. Then it is preceded in our track by $A_{j-1}$, which is of the form $B \vee A_j$ (or $A_j \vee B$) and occurs in $\Pi$ as the major premise of an application of $\vee E$. $A_{j-1}$ cannot arise as a result of some elimination step (other than another $\vee E$), which would violate the condition that $A_k$ is the first such occurrence in our track after $A_i$, so either it arises from an introduction ($\vee I$), which would again violate normality, since it is eliminated in the very next step, or it arises from yet another $\vee E$ inference. In that case, follow our track upward as before until the $\vee E$ inferences run out (the disjunction formula being repeated in the track all the while), to find the point at which an occurrence formula is at last introduced. In that case, we would still have a violation of normality, because a series of permutations of the intervening $\vee E$ inferences (in which our formula occurs as a minor premise and a conclusion) would have brought the initial inference (in which the disjunction is introduced) in contact with the final inference (in which that disjunction is eliminated), to normalize away this detour. Since our proof is normal, such a configuration cannot obtain, and so our lemma is proved.                                          $\square$

As with lemma 2, lemma 4 provides a lot of information about the structure of normal proofs. A track in a normal proof splits into two parts, an $E$-part and an $I$-part. Each part may be nonempty, and there may be subsequences of each part that make up a sequence of repetitions in the proof. Lemma 4 provides the resources needed to prove the desired theorem.

**Theorem 8 (Normal proofs have the subformula property)**  *If $\Pi$ is a normal proof, it has the subformula property.*

*Proof.*  The proof proceeds now in exactly the same way as in the proof for the restricted version of the subformula theorem, incorporating more comprehensive treatment of tracks, including the disjunction rules. The reasoning is largely unchanged, except for the fact that tracks of order 0 can terminate before reaching the conclusion of our proof, in the rare cases of a $\vee E$ inference in which no assumptions are discharged. In this case, notice that the track consists entirely of its $E$-part (since the inference at which it stops is $\vee E$), so in this case, every formula in that track is a subformula of the formula at its head. For the rest of the proof, the reasoning of the proof of theorem 7 applies, and the subformula property holds.                                          $\square$

The *subformula property* guarantees that the proofs are *analytic* in the sense that they do not need to appeal to anything that does not already appear in one of the premises or in the conclusion.[41] Analytic proofs simply unpack and reassemble what is given in the premises and the conclusion, to show the connection between them. This will become *very* useful when it comes to reasoning about proofs.

For example, if we are looking at a proof from no premises, of a conclusion that contains conditionals alone, then a normal proof of that theorem will not contain any rules for negation, conjunction, or disjunction. Proof search becomes much more targeted, and it is easier to tell when we have exhausted all of the options, and there is no proof to be found.

## 4.4  Consequences of Normalization

There are *many* interesting consequences of the normalization theorem. The first we will note is consistency of intuitionistic logic, which is to say that $\perp$ cannot be derived from the empty set of assumptions. This is the first result demonstrating *unprovability* in intuitionistic logic. When a formula $A$ is not provable from some assumptions $X$, we will write $X \nvdash A$.

**Theorem 9 (Consistency)**   *It is not the case that $\perp$ is provable in intuitionistic logic from the empty set of assumptions, which is to say $\nvdash_I \perp$.*

*Proof.*   Suppose that $\vdash_I \perp$. There is, then, a normal proof of $\perp$ from no assumptions. The only rules that have $\perp$ as a conclusion are the rule of assumption, $\neg E$, $\vee E$, and $\perp E$. Since there are no assumptions, the proof cannot consist solely of the rule of assumption. Since the proof is normal, $\perp$ cannot be the conclusion of $\perp E$. Suppose that $\perp$ is the conclusion of $\neg E$. Then, for some formula $A$, there are normal proofs $\Pi_1$ and $\Pi_2$ for $\vdash_I \neg A$ and $\vdash_I A$, respectively. The formula $\neg A$, however, is then not a subformula of the conclusion, $\perp$, or the assumptions, of which there are none. This contradicts theorem 8, and therefore, $\perp$ cannot come via $\neg E$. The argument for why it cannot come via $\vee E$ is similar. Since there are no other options, we conclude that, contrary to assumption, there is no proof of $\perp$ from the empty set of assumptions.                                                                           □

This is the first result showing unprovability using proof-theoretic methods. The normalization theorem implies that if we want to look for a proof for some argument, it is enough to try to find a normal proof. Normal proofs are constrained to have a certain sort of structure, and we can exploit this to limit the possible proofs that we must consider. If none of the possible proofs are proofs of the desired argument, then we can conclude that there is no proof for that argument. In chapter 6, we will see another way to demonstrate unprovability. Let us turn to the next consequence of the normalization theorem, the *disjunction property* for intuitionistic logic.

**Definition 26 (Disjunction property)**   *A consequence relation $\vdash$ has the* DISJUNCTION PROPERTY *if whenever $\vdash A \vee B$, then $\vdash A$ or $\vdash B$.*

---

41. You can come back now! It's all easy sailing through the rest of the chapter.

We can prove that intuitionistic logic has the disjunction property.

**Theorem 10** *Intuitionistic logic has the disjunction property.*

*Proof.* Suppose that $\vdash_I A \vee B$. Then, by the normalization theorem, there is a normal proof $\Pi$ of $\vdash_I A \vee B$. Since the main connective of $A \vee B$ is a disjunction, the final step of $\Pi$ must be $\vee I$. It is not an assumption (the proof has no assumptions), and it could not have entered the proof in an elimination step apart from $\bot E$. By the previous theorem, $\nvdash_I \bot$, so the final rule could not have been $\bot E$. Therefore, it enters the proof $\Pi$ by a $\vee I$ step, and the premise of this rule must be either $A$ or $B$. The proof obtained by omitting the final step then shows that either $\vdash A$ or $\vdash B$. $\qquad\square$

We will use the disjunction property for the next consequence, which is another unprovability result. The proof follows the general strategy described above, namely, to consider the structure of possible proofs and to show that none of those work.

The *law of excluded middle* is $A \vee \neg A$, or any formula of that form. An instance of the law of excluded middle is any formula obtained by replacing the "$A$" with a particular formula, such as $p \vee \neg p$. We cannot prove the law of excluded middle—in general—in intuitionistic logic. While there are instances of the law of excluded middle that are provable, we will show that there are instances that are not.

**Theorem 11** *The law of excluded middle is not provable in intuitionistic logic, $\nvdash_I p \vee \neg p$.*

The proof will be by reductio, which corresponds to the negation introduction rule in our formal system. This means that we will make an assumption and show that that leads to a contradiction. We can then discharge the assumption and conclude with the negation of the assumption.

*Proof.* Suppose that $\vdash_I p \vee \neg p$. By the disjunction property, $\vdash_I p$ or $\vdash_I \neg p$.

Suppose that $\vdash_I p$. There is a normal proof of $\vdash_I p$. There are no rules that introduce atoms apart from the assumption rule, but there are no assumptions in that proof. Therefore, there is no proof of $\vdash_I p$, which contradicts the assumption.

Suppose that $\vdash_I \neg p$. There is a normal proof of $\neg p$. Since the main connective is negation, it is the conclusion of a negation introduction rule. There is then a proof of $p \succ \bot$. As follows from question 2 on page 73, there is no proof of $p \vdash_I \bot$. There is then no proof of $\vdash_I \neg p$. This contradicts the assumption that there is such a proof.

In both cases, we reached a contradiction, so we have reached a contradiction under the assumption that $\vdash_I p \vee \neg p$. Therefore, we conclude that $\nvdash_I p \vee \neg p$. $\qquad\square$

We can use the unprovability of excluded middle to show that the double negation elimination (*DNE*) rule

$$\frac{\neg\neg A}{A} \; DNE$$

is not a derived rule in intuitionistic logic. If we were to add it to our system, we would be making more things provable, as we will now show.

*Proof.* Suppose that *DNE* is a derived rule in intuitionistic logic.

$$
\cfrac{
  \cfrac{
    [\neg(A \vee \neg A)]^2 \quad \cfrac{\cfrac{[A]^1}{A \vee \neg A}\;\vee I}{}
  }{
    \cfrac{\bot}{\neg A}\;\neg I^1
  }\;\neg E
}{}
$$

This means that with the addition of *DNE*, we can prove every instance of the law of the excluded middle. But we have just shown, on page 69, that the law of excluded middle is *not* derivable. Since the assumption that *DNE* is derivable results in the derivability of the law of excluded middle, we conclude that *DNE* is not derivable.                                       □

The rule *DNE* has a very different form than the other rules we have. It displays two copies of a connective, which none of the other rules do, and it provides a second way for negations to be eliminated. The addition of *DNE* results in a stronger natural deduction system, stronger in the sense that we can prove new things.

The system with *DNE* is called CLASSICAL LOGIC. We will write $X \vdash_C A$ to indicate that there is a proof of $A$ from the assumptions in $X$ in classical logic.

Classical logic differs from intuitionistic logic in many ways. Classical logic does not have the disjunction property (see definition 26). This follows from the fact that $p \vee \neg p$ is derivable in classical logic but neither $p$ nor $\neg p$ are. Classical logic does not have the subformula property. The failure of the disjunction property and the subformula property is related to the fact that not all proofs in classical logic can be converted into normal proofs of the conclusion from the same assumptions.

We can see this by looking at the above proof of $A \vee \neg A$. In this proof, there is a detour in the introduction of negation followed by *DNE*.[42] This detour cannot be eliminated, since the rest of the proof is an intuitionistic proof, and we have shown that $\nvdash_I A \vee \neg A$.

Since we do not have a normalization theorem for classical logic, we need another method for showing that a conclusion does not follow from some premises.[43] That method will be to use *models*, and we will look at these in the next chapter.

### 4.5   Key Concepts and Skills

☐ You should be able to reason about and verify simple general facts about provability ($\vdash_I$), such as the facts expressed in theorems 1, 2, and 3 and in questions 1 and 2 below.

---

42. The rule $\neg I$ introduces one negation and the rule *DNE* eliminates two, but the introduction/elimination pairing fits our definition of detour.

43. It is worth noting that there are proofs of normalization for classical logic, although they use a different concept of normal proof. As indicated by the examples above, the consequences of normalization for classical logic differ from those of normalization for intuitionistic logic. Proofs of normalization for classical logic were given by Prawitz (1965, chap. 3) and by Seldin (1989).

☐ You should be able to recognize detour formulas and detour sequences in nonnormal proofs and make a one-step reduction of the detour, including detours involving a formula being introduced in a $\bot E$ rule and eliminated in an elimination rule.

☐ You should be able to explain the significance of the normalization theorem and the subformula property.

☐ You can do simple proofs involving *DNE*.

---

**Reductions for $\bot$**

$$
\cfrac{\cfrac{\begin{array}{c}\Pi_1\\ \bot\end{array}}{\boxed{A \to B}}{\scriptstyle \bot E} \quad \begin{array}{c}\Pi_2\\ A\end{array}}{B}{\scriptstyle \to E} \quad \rightsquigarrow \quad \cfrac{\cfrac{\begin{array}{c}\Pi_1\\ \bot\end{array}}{B}{\scriptstyle \bot E}}{}
$$

$$
\cfrac{\cfrac{\cfrac{\begin{array}{c}\Pi_1\\ \bot\end{array}}{\boxed{\bot}}{\scriptstyle \bot E}}{A}{\scriptstyle \bot E}}{} \quad \rightsquigarrow \quad \cfrac{\begin{array}{c}\Pi_1\\ \bot\end{array}}{A}{\scriptstyle \bot E}
$$

$$
\cfrac{\cfrac{\begin{array}{c}\Pi_1\\ \bot\end{array}}{\boxed{\neg A}}{\scriptstyle \bot E} \quad \begin{array}{c}\Pi_2\\ A\end{array}}{\bot}{\scriptstyle \neg E} \quad \rightsquigarrow \quad \begin{array}{c}\Pi_1\\ \bot\end{array}
$$

$$
\cfrac{\cfrac{\begin{array}{c}\Pi_1\\ \bot\end{array}}{\boxed{A_1 \wedge A_2}}{\scriptstyle \bot E}}{A_i}{\scriptstyle \wedge E} \quad \rightsquigarrow \quad \cfrac{\begin{array}{c}\Pi_1\\ \bot\end{array}}{A_i}{\scriptstyle \bot E}
$$

$$
\cfrac{\cfrac{\begin{array}{c}\Pi_1\\ \bot\end{array}}{\boxed{A \vee B}}{\scriptstyle \bot E} \quad \begin{array}{c}[A]^i\\ \Pi_2\\ C\end{array} \quad \begin{array}{c}[B]^j\\ \Pi_3\\ C\end{array}}{C}{\scriptstyle \vee E^{i,j}} \quad \rightsquigarrow \quad \cfrac{\begin{array}{c}\Pi_1\\ \bot\end{array}}{C}{\scriptstyle \bot E}
$$

**Reductions for other connectives**

$$
\cfrac{\cfrac{\begin{array}{c}[A]^j\\ \Pi_1\\ B\end{array}}{\boxed{A \to B}}{\scriptstyle \to I^j} \quad \begin{array}{c}\Pi_2\\ A\end{array}}{B}{\scriptstyle \to E} \quad \rightsquigarrow \quad \begin{array}{c}\Pi_2\\ A\\ \Pi_1\\ B\end{array}
$$

$$
\cfrac{\cfrac{\begin{array}{c}\Pi_1 \quad \Pi_2\\ A_1 \quad A_2\end{array}}{\boxed{A_1 \wedge A_2}}{\scriptstyle \wedge I}}{A_i}{\scriptstyle \wedge E} \quad \rightsquigarrow \quad \begin{array}{c}\Pi_i\\ A_i\end{array}
$$

$$
\cfrac{\cfrac{\begin{array}{c}[A]^j\\ \Pi_1\\ \bot\end{array}}{\boxed{\neg A}}{\scriptstyle \neg I^j} \quad \begin{array}{c}\Pi_2\\ A\end{array}}{\bot}{\scriptstyle \neg E} \quad \rightsquigarrow \quad \begin{array}{c}\Pi_2\\ A\\ \Pi_1\\ \bot\end{array}
$$

$$
\cfrac{\cfrac{\begin{array}{c}\Pi\\ A_i\end{array}}{\boxed{A_1 \vee A_2}}{\scriptstyle \vee I} \quad \begin{array}{c}[A_1]^j\\ \Pi_1\\ C\end{array} \quad \begin{array}{c}[A_2]^k\\ \Pi_2\\ C\end{array}}{C}{\scriptstyle \vee E^{j,k}} \quad \rightsquigarrow \quad \begin{array}{c}\Pi\\ A_i\\ \Pi_i\\ C\end{array}
$$

| *Logics* | Assumption | $\to I/\to E$ | $\wedge I/\wedge E$ | $\vee I/\vee E$ | $\neg I/\neg E$ | $\bot E$ | DNE |
|---|---|---|---|---|---|---|---|
| Minimal ($\vdash_M$) | ✓ | ✓ | ✓ | ✓ | ✓ | | |
| Intuitionistic ($\vdash_I$) | ✓ | ✓ | ✓ | ✓ | ✓ | ✓ | |
| Classical ($\vdash_C$) | ✓ | ✓ | ✓ | ✓ | ✓ | ✓ | ✓ |

### 4.6  Questions for You

**Basic Questions**

1. Recall that $X$ is said to be inconsistent if and only if there is a proof of $\bot$ from $X$. Which of the following sets are inconsistent? For those that are inconsistent, prove $\bot$ from those premises. For those sets that aren't, try to explain why they aren't inconsistent.

   i. $p, q, \neg(p \wedge q)$

   ii. $p \vee q, \neg p \vee \neg q$

   iii. $\neg p, q, p \rightarrow q$

   iv. $p, \neg q, p \rightarrow q$

   v. $p, q, \neg(p \rightarrow q)$

   vi. $p \rightarrow q, \neg((q \rightarrow r) \rightarrow (p \rightarrow r))$

   vii. $\neg(p \rightarrow q), (q \rightarrow r) \rightarrow (p \rightarrow r)$

2. Which of these general "facts" about provability are really *facts*? For those that aren't, can you give any reasons why they aren't? For those that are, can you prove them?

   i. If $A \vdash_I B$, then $\neg B \vdash_I \neg A$.

   ii. If $A \vdash_I B$, then it's not true that $B \vdash_I A$.

   iii. Either $A \vdash_I B$ or $B \vdash_I A$.

   iv. $X, A \vdash_I \neg A$ if and only if $X \vdash_I \neg A$.

   v. $X \vdash_I A \vee B$ if and only if $X \vdash_I A$ or $X \vdash_I B$.

3. Complete the proof of theorem 3 by showing that

   • $X \vdash_I A \wedge B$ if and only if $X \vdash_I A$ and $X \vdash_I B$.

   • $X \vdash_I \neg A$ if and only if $X, A \vdash_I \bot$.

   (You can follow the reasoning in the proof of theorem 3 pretty closely. In particular, $\neg A$ is rather like $A \rightarrow \bot$.)

4. Use *DNE* to find arguments to show that the following *classical validities* hold:

   i. $\neg(p \rightarrow q) \vdash_C p$,

   ii. $\neg(p \wedge q) \vdash_C \neg p \vee \neg q$,

   iii. $\vdash_C ((p \rightarrow q) \rightarrow p) \rightarrow p$.

   In the first two cases, first try proving the double negation of the conclusion and then appeal to *DNE* at the end. For example, first find an argument for $\neg(p \rightarrow q) \succ \neg\neg p$ and do *this* by constructing a proof for $\neg(p \rightarrow q), \neg p \succ \bot$. This is the same sort of strategy we used for proving $A \vee \neg A$ using *DNE*.

   Beware, the argument for the last one is rather tricky. Don't be worried if it takes you a while. The hint for this is not to try to prove the double negation of $((p \rightarrow q) \rightarrow p) \rightarrow p$ but to prove $\neg\neg p$ from the assumption $(p \rightarrow q) \rightarrow p$ first.

5. Consider the following proof, from the premise $(p \rightarrow q) \vee r$ to the conclusion $p \rightarrow ((q \vee r) \vee s)$. Mark out all of the detour sequences in this proof.

$$
\cfrac{
(p \rightarrow q) \vee r \quad
\cfrac{
\cfrac{
\cfrac{\cfrac{[p \rightarrow q]^3 \quad [p]^1}{q} \rightarrow E}{q \vee r} \vee I
}{p \rightarrow (q \vee r)} \rightarrow I^1 \quad
\cfrac{
\cfrac{[r]^4}{q \vee r} \vee I
}{p \rightarrow (q \vee r)} \rightarrow I^2
}{p \rightarrow (q \vee r)} \vee E^{3,4} \quad [p]^5
}{
\cfrac{\cfrac{\cfrac{q \vee r}{(q \vee r) \vee s} \vee I}{p \rightarrow ((q \vee r) \vee s)} \rightarrow I^5}{} 
} \rightarrow E
$$

You will notice that there are sequences involving the minor premises of the $\vee E$ inference. Permute the $\rightarrow E$ inference above the $\vee E$ inference, so the detour sequences are reduced to length 1.

Then eliminate those detours in the proof, using reduction steps. Is the resulting proof normal? If so, verify that it has the subformula property. If not, reduce it, and keep reducing it, until you have a normal proof, verifying that this proof indeed has the subformula property.

## Challenge Questions

1. We say that $A$ and $A'$ are *logically equivalent* if $A \vdash_I A'$ and $A' \vdash_I A$. That is, there is a proof from $A$ to $A'$ and a proof from $A$ to $A'$. Show the following general facts about provability, assuming that $A$ and $A'$ are logically equivalent formulas.
   i. If $X \vdash_I A$, then $X \vdash_I A'$.
   ii. If $X, A \vdash_I B$, then $X, A' \vdash_I B$.
   iii. $A \wedge B$ is logically equivalent to $A' \wedge B$.
   iv. $A \rightarrow B$ is logically equivalent to $A' \rightarrow B$.
   v. $B \rightarrow A$ is logically equivalent to $B \rightarrow A'$.
   vi. $\neg A$ is logically equivalent to $\neg A'$.
   vii. $A \vee B$ is logically equivalent to $A' \vee B$.
   viii. Explain why it follows for any complex formula $C(A)$ with $A$ as subformula, $C(A)$ is logically equivalent to $C(A')$, where $C(A')$ is found by replacing the $A$ in $C$ by $A'$.
2. A set $X$ of formulas is *purely positive* if and only if it does not contain $\neg$ or $\bot$ as a subformula. This question will help you show that *no* set of purely positive formulas is inconsistent. If $X$ is purely positive, we cannot have $X \vdash_I \bot$.
   i. First explain why, if there is a proof for $X \succ \bot$, then there is also a proof for $X^p \succ \bot$, where $X^p$ is found by replacing all of the atoms in every formula in $X$ by the one atom $p$ that is not $\bot$.
   ii. Then explain why every purely positive formula made up from the atom $p$ is either equivalent to $p$ or equivalent to $p \rightarrow p$. (Recall from the previous question that $A$ is equivalent to $B$ if $A \vdash_I B$ and $B \vdash_I A$.)
   iii. Then explain why there is no normal proof for $p \succ \bot$, or $p \rightarrow p \succ \bot$ or for $p, p \rightarrow p \succ \bot$, without appealing to the subformula property in normal proofs.

iv. Then put all of this together to conclude that if $X$ is purely positive, then we never have $X \vdash_I \bot$.

3. Provide a detailed proof of theorem 8 on page 67.

4. We defined a set $X$ of formulas as inconsistent iff there is a proof of $\bot$ from $X$ in our intuitionistic natural deduction system. This might be termed "inconsistent in I." We can define inconsistency in M as follows: A set $X$ of formulas is inconsistent in M iff $X \vdash_M \bot$. Is it the case that $X$ is inconsistent in I iff $X$ is inconsistent in M? If so, provide a proof. If not, provide a counterexample.

5. The definition of inconsistency (in I) in this chapter is in terms of proving $\bot$. We can call this $\bot$-inconsistency (in I), or inconsistency (in I) in the sense of entailing $\bot$. An alternative notion is absolute inconsistency (in I), or inconsistency (in I) in the sense of entailing everything. Say that a set $X$ of formulas is absolutely inconsistent (in I) iff for all formulas $A$, $X \vdash_I A$. Prove that a set of formulas $X$ is $\bot$-inconsistent (in I) iff $X$ is absolutely inconsistent (in I).

   We can define these two notions in M by replacing $\vdash_I$ with $\vdash_M$. Show that in M, a set $X$ of formulas can be $\bot$-inconsistent without being absolutely inconsistent. Are any sets of formulas absolutely inconsistent (in M)? If so, provide one, and if not, provide a proof that there are none.

6. The quantum disjunction connective, $\bar{\nabla}$, has the following rules.

$$\frac{A}{A\bar{\nabla}B} \bar{\nabla}I \qquad \frac{B}{A\bar{\nabla}B} \bar{\nabla}I \qquad \frac{A\bar{\nabla}B \quad \overset{\displaystyle [A]^i}{\underset{\displaystyle C}{\Pi_2}} \quad \overset{\displaystyle [B]^j}{\underset{\displaystyle C}{\Pi_3}}}{C} \bar{\nabla}E^{i,j}$$

The $\bar{\nabla}E$ rule has a restriction, that the only undischarged assumptions in $\Pi_2$ are the indicated occurrences of $A$ and the only undischarged assumptions in $\Pi_3$ are the indicated occurrences of $B$, and each of these occurrences is discharged by $\bar{\nabla}E$. In short, $\bar{\nabla}E$ does not allow auxiliary assumptions in the minor premise proofs, unlike $\vee E$. Show that the system for $\{\wedge, \bar{\nabla}\}$ normalizes. That system uses the rule of assumption, $\wedge I$, $\wedge E$, $\bar{\nabla}I$, and $\bar{\nabla}E$.

7. The proof system for $\{\wedge, \bar{\nabla}\}$ does not have a proof of $A \wedge (B\bar{\nabla}C) \succ (A \wedge B)\bar{\nabla}(A \wedge C)$. What goes wrong when you attempt to give a proof of this sequent without detours? (As an extra challenge, use the normalization theorem of the previous question to prove that this sequent is not derivable.) Show that this sequent becomes derivable when $\vee$ is added with the rules $\vee I$ and $\vee E$.

   When we add regular disjunction to a system with quantum disjunction, quantum disjunction collapses into regular disjunction. Is there anything we can conclude about the meanings of disjunction and quantum disjunction from this collapse?[44]

44. Quantum disjunction is discussed by Dummett (1991), and it is important in the subsequent discussions of harmony and stability, for which see Steinberger (2011), Francez and Dyckhoff (2012), Francez (2017), Jacinto and Read (2017), or Tranchini (2018), among others.

# 5 Models & Counterexamples

In the last three chapters, our focus has been constructing proofs and understanding their properties. We have begun to explore one half of the answer to the question we set ourselves in chapter 1. An argument is good when you can break it down into its most basic steps, and each of those steps arises out of the meanings of the concepts involved. A proof to that effect shows that an argument is *valid*.

A close analysis of the structure of proofs—and the process of *normalization*—can give us some insight into when an argument is *not* so good. The normalization theorem entails that every argument that has a proof has a *normal* proof. To show that some conclusion does not follow from some premises, it is enough to search through all the possible normal proofs. This technique, which is called *proof search*, works but is labor intensive.[45] Furthermore, in systems, such as our proof system for classical logic, where we do not *have* a normalization theorem, we need another way to show that an argument is invalid. Either way, we want an answer to the other half of our first question. We want to understand what makes an argument *invalid*. We can understand this in terms merely of its failure to be valid, its failure to be underwritten by a *proof*, but this is not the only way to understand invalidity. There is another way to look at logical concepts, not by way of *proof* but by way of *models*. So, models will be our subject for the next two chapters.

The motivating idea is simple. Think back to the bad argument from chapter 1:

> *All footballers are bipeds.*
> *Sócrates is a biped.*
> So, *Sócrates is a footballer.*

Why can we be sure that there is no way to prove that conclusion from *those* premises? One reason might have occurred to you. Perhaps Sócrates is one of the bipeds who doesn't play football. If the world were like this—if every footballer happened to be a biped, but some bipeds weren't footballers, and Sócrates just happened to be one of the non-football-playing bipeds—then the premises would be true, and the conclusion wouldn't be.

This way of thinking involves coming up with a situation—a *hypothetical* situation—and spelling that situation out specifically enough to show that, *relative to that situation*, the premises come out true, and the conclusion *doesn't*. Then we can conclude that the argument can't be valid, because it doesn't apply in *that* circumstance. We have found a *counterexample* to the argument. It is invalid.

---

45. Whether that labor is your labor or the labor of a computer program, proof search is a lot of work.

## 5.1 Models and Truth Tables

In general, a *model* is a way of specifying a situation. It represents the important features of a situation in a way that shows how the components of that situation relate to each other.[46] A *valuation* is one simple kind of model for sentences in a language, where we specify one way things could be, by making choices for the *semantic values* for each component of the language. In the case of our simple propositional language, in which the atoms are declarative sentences, we will work with *Boolean models*.[47] These models assign the formulas in the language one of two *truth values*, which are the semantic values we will use here. The way a Boolean model specifies a situation is by assigning 1 (true) or 0 (false) to each atomic formula, and these values are used to calculate the value of all of the complex formulas in the language. So a valuation *models* a situation by selecting which things are *true* in that situation and which things are *false*.

**Definition 27 (Boolean valuation, model)** *A* BOOLEAN VALUATION $v$ *is a function* $v$: Atom $\mapsto \{0, 1\}$ *such that* $v(\bot) = 0$.

*A given Boolean valuation $v$ is extended to a* BOOLEAN MODEL *on the whole language, which is to say that it is a function* Form $\mapsto \{0, 1\}$, *inductively as follows:*[48,49]

$$
\begin{aligned}
v(\neg A) = 1 &\quad \textit{iff} \quad v(A) = 0 \\
v(A \land B) = 1 &\quad \textit{iff} \quad v(A) = 1 \textit{ and } v(B) = 1 \\
v(A \lor B) = 1 &\quad \textit{iff} \quad v(A) = 1 \textit{ or } v(B) = 1 \\
v(A \to B) = 1 &\quad \textit{iff} \quad v(A) = 0 \textit{ or } v(B) = 1
\end{aligned}
$$

A valuation $v$ is specified by giving a truth value to every atom in Atom, with $v(\bot) = 0$, and this determines a Boolean model for the whole language that determines the value of a complex formula in terms of the value of its constituents, in an inductive process, of building up from the parts to the whole. In fact, every valuation $v$ that has $v(\bot) = 0$ uniquely determines a Boolean model on the whole language.

**Lemma 5** *Given a Boolean valuation $v$, there is exactly one Boolean model $u$ such that for all $p \in$ Atom, $u(p) = 1$ iff $v(p) = 1$.*

*Proof.* Let $v$ be a Boolean valuation. Suppose that $u_1$ and $u_2$ are distinct Boolean models such that $u_1(p) = 1$ iff $v(p) = 1$ and $u_2(p) = 1$ iff $v(p) = 1$. We will prove by induction on the structure of the formula $A$ that $u_1(A) = u_2(A)$ for all formulas $A$. There are two base cases. Suppose $A$ is an atom, $p$. Then $u_1(p) = 1$ iff $v(p) = 1$ and $u_2(p) = 1$ iff $v(p) = 1$ by definition,

---

46. Think of a model car, which represents, in miniature, many of the physical components of the car, their shapes, the way they fit together, and their colors and textures. It doesn't represent *all* of the features of the car—not the weight or what they cost, or their chemical composition—but it represents some of those features.

47. Boolean models are named after George Boole (1815–1864), an English logician who is famous for his work on algebraic logic. His most famous book is *The Laws of Thought* (1854).

48. Remember, "iff" is shorthand for "if and only if."

49. You might wonder *why* we have made these choices for how $v$ works. For $\land$ and $\lor$, $\neg$ and $\bot$, things are relatively straightforward. A conjunction is true iff both conjuncts are true. A disjunction is true iff either one disjunct is true or the other is (including the case where both are). A negation is true iff the negand is not. For the conditional, we have fewer nice things to say, except that it's clear that a conditional is *false* when the antecedent is true and the consequent is false. On the other hand, we do have $B \vdash_I A \to B$, so it seems plausible that if $B$ is true, $A \to B$ should be true. Similarly, we have $\neg A \vdash_I A \to B$, so when $A$ is false, $A \to B$ should be true, too.

so $u_1(p) = 1$ iff $u_2(p) = 1$. The other base case is where $A$ is $\bot$, in which case $u_1(\bot) = 0$ and $u_2(\bot) = 0$, by definition.

The inductive hypothesis is that $u_1(B) = 1$ iff $u_2(B) = 1$, for formulas $B$ less complex than $A$.

Suppose $A$ is of the form $\neg B$. Then, $u_1(\neg B) = 1$ iff $u_1(B) = 0$. By the inductive hypothesis, $u_1(B) = 0$ iff $u_2(B) = 0$. Since $u_2(B) = 0$ iff $u_2(\neg B) = 1$, we can chain the biconditionals together to get $u_1(\neg B) = 1$ iff $u_2(\neg B) = 1$.

Suppose $A$ is of the form $B \wedge C$. Then, $u_1(B \wedge C) = 1$ iff $u_1(B) = 1$ and $u_1(C) = 1$. By the inductive hypothesis, $u_1(B) = 1$ iff $u_2(B) = 1$ and $u_1(C) = 1$ iff $u_2(C) = 1$. Therefore, $u_1(B \wedge C) = 1$ iff $u_2(B) = 1$ and $u_2(C) = 1$, and $u_2(B) = 1$ and $u_2(C) = 1$ iff $u_2(B \wedge C) = 1$. Chaining the biconditionals together yields that $u_1(B \wedge C) = 1$ iff $u_2(B \wedge C) = 1$.

Suppose $A$ is of the form $B \wedge C$. Then, $u_1(B \vee C) = 1$ iff $u_1(B) = 1$ or $u_1(C) = 1$. By the inductive hypothesis, $u_1(B) = 1$ iff $u_2(B) = 1$ and $u_1(C) = 1$ iff $u_2(C) = 1$. Therefore, $u_1(B \vee C) = 1$ iff $u_2(B) = 1$ or $u_2(C) = 1$. As $u_2(B) = 1$ or $u_2(C) = 1$ iff $u_2(B \vee C) = 1$, we can obtain that $u_1(B \vee C) = 1$ iff $u_2(B \vee C) = 1$.

The case where $A$ is of the form $B \rightarrow C$ is similar to the preceding and is left to the reader. $\square$

The preceding lemma establishes that any Boolean valuation $v$ can be uniquely extended to a Boolean model. This means that we can specify a Boolean valuation just on Atom (subject to the constraint on $\bot$) and thereby specify a Boolean model. Having noted that, we will let the distinction between valuations and models lapse. In later chapters, such as chapter 7, we will introduce new kinds of models, which will also have the same feature that specifying values on atoms, possibly with an additional parameter, will be sufficient to determine values for every formula of the language.

One way to represent the rules in definition 27 is to represent them in a table, like so:

| $A$ | $B$ | $A \wedge B$ | $A \vee B$ | $A \rightarrow B$ | $\neg A$ | $\bot$ |
|---|---|---|---|---|---|---|
| 0 | 0 | 0 | 0 | 1 | 1 | 0 |
| 0 | 1 | 0 | 1 | 1 | 1 | 0 |
| 1 | 0 | 0 | 1 | 0 | 0 | 0 |
| 1 | 1 | 1 | 1 | 1 | 0 | 0 |

Here, each row of the table represents the four different possibilities for the values $v(A)$ and $v(B)$, and in each row, the value in the column for a complex formula (or for $\bot$) is the value that $v$ assigns to the complex formula, given that $v$ has made that particular choice for $v(A)$ and for $v(B)$. In the third row of this table, for example, we assume that $v$ has assigned 1 to $A$ and 0 to $B$. Then, $v(A \wedge B) = 0$ and $v(A \rightarrow B) = 0$, but $v(A \vee B) = 1$. $v(\neg A) = 0$ (here, the value depends only on the value of $A$) and $v(\bot) = 0$ as it is always.

A Boolean model assigns a truth value for *every* atom in our language and giving $\bot$ the value 0. It is maximally opinionated. In practice, we don't need to specify the value of *every* atom if we are interested in analyzing how the value of a complex formula depends on the values of its parts.

**Example 1** *We will start with some simple examples. First, let's suppose $v(p) = 1$ and $v(q) = 0$, and we want to evaluate $v(p \wedge q)$. We don't need to know the values $v$ assigns to other atoms to evaluate this. We can write the reasoning like this: $v(p \wedge q) = v(p) \wedge v(q)$ (where now we think of "$\wedge$"*

*as applying to the* values 0 *and* 1) *and this is* $1 \wedge 0$, *and we can look up our table to see that* $1 \wedge 0 = 0$, *so* $v(p \wedge q) = 0$.

*Here's another example: Let's suppose* $v'$ *is a different model, where* $v'(p) = 0$ *and* $v'(q) = 1$. *Let's evaluate* $(p \vee q) \to p$ *in that model. We have*

$$v'((p \vee q) \to p) = (v'(p) \vee v'(q)) \to v'(p),$$

*and we can substitute the values into the right-hand side to continue the evaluation*

$$(0 \vee 1) \to 0 = 1 \to 0 = 0.$$

*So,* $v'((p \vee q) \to p) = 0$.

*In our last example, we'll evaluate* $(p \to (q \vee \neg r))$ *in a model* $v''$ *such that* $v''(p) = 1$, $v''(q) = 0$, *and* $v''(r) = 1$. *We can evaluate* $v''(p \to (q \vee \neg r))$ *using the equations*

$$= 1 \to (0 \vee \neg 1) = 1 \to (0 \vee 0) = 1 \to 0 = 0.$$

This method is fine for evaluating a complex formula in terms of the values of its parts. We can do this in a more efficient way, relative to a whole set of valuations, to gain a lot more information about a formula. Take our second example $(p \vee q) \to p$. We found that $v'((p \vee q) \to p) = 0$. We might wonder if this is the only way to assign values to $p$ and to $q$ that makes this formula false. We could check all the possibilities in a "truth table" like this:

| $p$ | $q$ | $(p$ | $\vee$ | $q)$ | $\to$ | $p$ |
|-----|-----|------|--------|------|-------|-----|
| 0   | 0   | 0    | 0      | 0    | 1     | 0   |
| 0   | 1   | 0    | 1      | 1    | 0     | 0   |
| 1   | 0   | 1    | 1      | 0    | 1     | 1   |
| 1   | 1   | 1    | 1      | 1    | 1     | 1   |

We have written out a row for each of the different possible combinations for $v(p)$ and $v(q)$, and in each row, under the formula $(p \vee q) \to p$, we have written under each atom, or each main connective of a subformula, the value of that formula in that row. So, the column under the main connective of the whole formula—the conditional—shows the value of the whole formula in each valuation. And here we see that indeed the valuation $v'(p) = 0$ and $v'(q) = 1$ is the only way to assign values to $p$ and to $q$ that makes this formula false. The other three possibilities assign the value 1 instead.

This truth table visually represents all the different ways the formula $(p \vee q) \to p$ can be modeled by Boolean valuations. There are more than four different Boolean valuations for this formula, but these different valuations differ in the values they give to atoms *other* than $p$ and $q$. *All* of the infinite variety of Boolean valuations neatly arrange themselves into one of these four rows, depending on the choices they make concerning the truth value of $p$ and of $q$. For example, no matter what a valuation $v$ assigns to $r$, it will fit into one of the rows of the truth table above.

Truth tables give us an easy way to represent another distinction between formulas. Some formulas are true in *every* valuation whatsoever. Some formulas are true in *none*. And the remainder of our formulas (like the example $(p \vee q) \to p$) are true in some valuations and not in others. These distinct types are worth giving a name:

**Definition 28 (Boolean tautologies, contradictions, and contingencies)**  *A formula A is said to be a* BOOLEAN TAUTOLOGY (*a* TAUTOLOGY *for short*) *iff v(A) = 1 for every model v. It is said to be a* BOOLEAN CONTRADICTION (*a* CONTRADICTION *for short*) *iff v(A) = 0 for every model v. If A is neither a tautology nor a contradiction, A is said to be a* BOOLEAN CONTINGENCY (*a* CONTINGENCY *for short*). *That is, if for some model v and v', v(A) = 1, and v'(A) = 0.*

If you do a truth table for a formula, with a row for each different combination of truth values for the atoms, a tautology is a formula that has the value 1 in every row under its main connective, while a contradiction has the value 0 in every row under its main connective.

| $p$ | $q$ | $(p$ | $\land$ | $q)$ | $\to$ | $p$ | $(p$ | $\land$ | $q)$ | $\land$ | $\neg$ | $q$ |
|---|---|---|---|---|---|---|---|---|---|---|---|---|
| 0 | 0 | 0 | 0 | 0 | 1 | 0 | 0 | 0 | 0 | 0 | 1 | 0 |
| 0 | 1 | 0 | 0 | 1 | 1 | 0 | 0 | 0 | 1 | 0 | 0 | 1 |
| 1 | 0 | 1 | 0 | 0 | 1 | 1 | 1 | 0 | 0 | 0 | 1 | 0 |
| 1 | 1 | 1 | 1 | 1 | 1 | 1 | 1 | 1 | 1 | 0 | 0 | 1 |

In the Questions for You section at the end of the chapter, there will be other examples for you to work through to hone your skills on Boolean valuations.

## 5.2  Counterexamples and Validity

So, let's put Boolean models to work in answering our question about arguments. A *counterexample* to an argument from premises $X$ to a conclusion $A$ tells us that the argument is invalid. What is a counterexample? It is a model that makes the premises true and the conclusion false.

**Definition 29 (Counterexample)**  *A* Boolean counterexample *to the argument $X \succ A$ is a model v that assigns 1 to all formulas in the set X and assigns 0 to A.*

Because we will see this notion—of assigning 1 to *all* the formulas in a set $X$—we will expand our notation accordingly. We will write $v(X) = 1$ (where $X$ is a set of formulas) to mean $v(B) = 1$ for each formula $B$ in $X$, or equivalently, $v(X) = 0$ if and only if there is some $B$ in $X$ where $v(B) = 0$.[50] We will also write $v(X, A) = 1$ to mean $v(X) = 1$ and $v(A) = 1$, and we will use $v(X, Y) = 1$ to mean $v(X) = 1$ and $v(Y) = 1$.

Absence of counterexamples provides another way to understand *validity*. Let's give this an official definition:

**Definition 30 (Validity)**  *An argument from $X$ to $A$ is* VALID, *according to Boolean models, iff there is no valuation v where $v(X) = 1$ and $v(A) = 0$, that is, there are no counterexamples to the argument.*
   *We will use the notation $X \models_{CL} A$ to mean that the argument $X \succ A$ is valid, according to Boolean models. The notation $X \not\models_{CL} A$ will mean that there is a counterexample to the argument $X \succ A$.*

Throughout the book, we will use $\models$ for validity in terms of models and $\vdash$ for validity in terms of proofs. Subscripts on the turnstiles indicate what kinds of models or proofs are under consideration.

---

50. What about when $X$ is the empty set, we hear you ask? What is $v(\{\ \})$? Well, $v(\{\ \})$ can't be 0 because there is nothing in $\{\ \}$ to get the value 0, so, by default, $v(\{\ \}) = 1$, for every valuation $v$.

\* \* \*

Let's give some examples.

**Example 2**  *Take the argument $p \lor q \succ p \land q$. This has a counterexample, setting $v(p) = 1$ and $v(q) = 0$. Then we have $v(p \lor q) = 1$, but $v(p \land q) = 0$, so $p \lor q \nvDash_{CL} p \land q$.*

**Example 3**  *Consider the argument $q \land (r \land s) \succ q \land p$. For a counterexample, take a valuation $v$ with $v(q) = 1, v(r) = 1, v(s) = 1, v(p) = 0$. Then, $v(r \land s) = 1$, so $v(q \land (r \land s)) = 1$, but $v(q \land p) = 0$.*

Next we will look at two examples of *valid* arguments. The approach to establishing validity is to assume there is a counterexample and show that, however the counterexample makes the premises true and the conclusion false, this assumption leads to a contradiction. We can then conclude that there is, in fact, no counterexample.

**Example 4**  *Consider $\neg(p \to q) \succ p \land \neg q$. Suppose we have a counterexample, $v$, so $v(\neg(p \to q)) = 1$ but $v(p \land \neg q) = 0$. The former implies $v(p \to q) = 0$, which implies that $v(p) = 1$ and $v(q) = 0$. Then, $v(\neg q) = 1$, so $v(p \land \neg q) = 1$, which contradicts the assumption that $v$ was a counterexample. Thus, $\neg(p \to q) \vDash_{CL} p \land \neg q$.*

**Example 5**  *Suppose that the argument $p, \neg p \succ q$ has a counterexample, $v$. Then $v(p) = 1, v(\neg p) = 1$, and $v(q) = 0$. If $v(\neg p) = 1$, then $v(p) = 0$, which contradicts the assumption that $v(p) = 1$. Therefore, there is no counterexample, so $p, \neg p \vDash_{CL} q$.*

That last example strikes some people as very surprising. No model makes the premises both true, so no model can serve as a counterexample to the argument.

The reasoning in these examples can also be represented in a truth table, where we systematically work through all of the possibilities for the valuations. The example for $\neg(p \to q) \succ p \land \neg q$ goes like this:

| $p$ | $q$ | $\neg$ | $(p$ | $\to$ | $q)$ | $p$ | $\land$ | $\neg$ | $q$ |
|---|---|---|---|---|---|---|---|---|---|
| 0 | 0 | 0 | 0 | 1 | 0 | 0 | 0 | 1 | 0 |
| 0 | 1 | 0 | 0 | 1 | 1 | 0 | 0 | 0 | 1 |
| 1 | 0 | 1 | 1 | 0 | 0 | 1 | 1 | 1 | 0 |
| 1 | 1 | 0 | 1 | 1 | 1 | 1 | 0 | 0 | 1 |

Here, we have a row for each different kind of valuation, for the different possible combinations of values for $p$ and $q$, and in each row, we write the values for the premise and the conclusion of the argument and each of their subformulas. We see that there is no valuation where the premise is true and the conclusion is false, so the argument has no counterexample.

When using truth tables to assess whether a formula is a tautology or to assess whether an argument is valid, it is important that every possible assignment of truth values to the atoms is considered. You must make sure that you do not overlook any possibilities. One way to ensure that no possibilities are missed is to set up the truth table in a systematic way. If the formula, or argument, under consideration has, apart from $\bot$, $n$ different atoms, where $n \geq 0$, then there will be $2^n$ many rows in the complete truth table for that formula, or argument. The truth table for an argument whose only atom is $\bot$ has only one row. The truth table for an argument with one non-$\bot$ atom will have two rows, for an argument that has two non-$\perp$ atoms four rows, for an argument with three non-$\perp$ atoms eight rows, and for an argument with four non-$\perp$ atoms sixteen rows.

We will describe one systematic way to fill out the truth table for a formula, or argument, with $n$ different non-$\perp$ atoms. After writing out the truth table with the non-$\perp$ atoms on the left and the formula(s) to be evaluated on the right, fill in the columns under any $\perp$s with 0s. Then, fill in the column under the first atom with the first $\frac{1}{2}(2^n)$ rows as 0 and the next $\frac{1}{2}(2^n)$ rows as 1, creating two blocks of truth values. We will illustrate with the table for the argument $p \wedge (q \wedge r) \succ p \wedge r$.

| $p$ | $q$ | $r$ | $p$ | $\wedge$ | $(q$ | $\wedge$ | $r)$ | $p$ | $\wedge$ | $r$ |
|---|---|---|---|---|---|---|---|---|---|---|
| 0 | | | | | | | | | | |
| 0 | | | | | | | | | | |
| 0 | | | | | | | | | | |
| 0 | | | | | | | | | | |
| 1 | | | | | | | | | | |
| 1 | | | | | | | | | | |
| 1 | | | | | | | | | | |
| 1 | | | | | | | | | | |

For the next atom's column, the two blocks are split in half, with the first half of each filled in with 0s and the second half of each filled in with 1s. More carefully, there are $2^{n-2}$ rows of 0s, followed by $2^{n-2}$ rows of 1s, and then $2^{n-2}$ rows of 0s, followed by $2^{n-2}$ rows of 1s.

| $p$ | $q$ | $r$ | $p$ | $\wedge$ | $(q$ | $\wedge$ | $r)$ | $p$ | $\wedge$ | $r$ |
|---|---|---|---|---|---|---|---|---|---|---|
| 0 | 0 | | | | | | | | | |
| 0 | 0 | | | | | | | | | |
| 0 | 1 | | | | | | | | | |
| 0 | 1 | | | | | | | | | |
| 1 | 0 | | | | | | | | | |
| 1 | 0 | | | | | | | | | |
| 1 | 1 | | | | | | | | | |
| 1 | 1 | | | | | | | | | |

For the second atom, there are twice as many blocks of truth values as the first, and for the next atom, each of *these* blocks is split in half, with the top half filled in with 0s and the bottom half filled in with 1s. Each block of 0s is $2^{n-3}$ rows and is followed by a block of $2^{n-3}$ rows of 1s. There are four such pairs in total for that column.

| $p$ | $q$ | $r$ | $p$ | $\wedge$ | $(q$ | $\wedge$ | $r)$ | $p$ | $\wedge$ | $r$ |
|---|---|---|---|---|---|---|---|---|---|---|
| 0 | 0 | 0 | | | | | | | | |
| 0 | 0 | 1 | | | | | | | | |
| 0 | 1 | 0 | | | | | | | | |
| 0 | 1 | 1 | | | | | | | | |
| 1 | 0 | 0 | | | | | | | | |
| 1 | 0 | 1 | | | | | | | | |
| 1 | 1 | 0 | | | | | | | | |
| 1 | 1 | 1 | | | | | | | | |

If there were more atoms, we would continue this procedure of breaking the rows into blocks of 0s and 1s. For atom number $k$, $1 \leq k \leq n$, each block of 0s will be $2^{n-k}$ rows and each block of 1s will be $2^{n-k}$ rows. The column under atom $k$ is filled in with alternating blocks of 0s and 1s. In the column under the $n$th non-$\perp$ atom, each block will be only a single row. Since the argument in the example above only had three atoms, we have finished the initial setup, having listed all the possible assignments of truth values to these atoms, and we can fill in the rest of the truth table.

We will begin by copying the truth values for the atoms from the left under the atoms in the argument. This is not required, but it can make it easier to fill out the table.

| $p$ | $q$ | $r$ | $p$ | $\wedge$ | $(q$ | $\wedge$ | $r)$ | $p$ | $\wedge$ | $r$ |
|---|---|---|---|---|---|---|---|---|---|---|
| 0 | 0 | 0 | 0 | | 0 | | 0 | 0 | | 0 |
| 0 | 0 | 1 | 0 | | 0 | | 1 | 0 | | 1 |
| 0 | 1 | 0 | 0 | | 1 | | 0 | 0 | | 0 |
| 0 | 1 | 1 | 0 | | 1 | | 1 | 0 | | 1 |
| 1 | 0 | 0 | 1 | | 0 | | 0 | 1 | | 0 |
| 1 | 0 | 1 | 1 | | 0 | | 1 | 1 | | 1 |
| 1 | 1 | 0 | 1 | | 1 | | 0 | 1 | | 0 |
| 1 | 1 | 1 | 1 | | 1 | | 1 | 1 | | 1 |

Next, we evaluate the subformulas whose components have been assigned truth values. In this case, we can evaluate the conjunctions that have truth values listed under both of their conjuncts, namely, $q \wedge r$ and $p \wedge r$.

| $p$ | $q$ | $r$ | $p$ | $\wedge$ | $(q$ | $\wedge$ | $r)$ | $p$ | $\wedge$ | $r$ |
|---|---|---|---|---|---|---|---|---|---|---|
| 0 | 0 | 0 | 0 | | 0 | 0 | 0 | 0 | 0 | 0 |
| 0 | 0 | 1 | 0 | | 0 | 0 | 1 | 0 | 0 | 1 |
| 0 | 1 | 0 | 0 | | 1 | 0 | 0 | 0 | 0 | 0 |
| 0 | 1 | 1 | 0 | | 1 | 1 | 1 | 0 | 0 | 1 |
| 1 | 0 | 0 | 1 | | 0 | 0 | 0 | 1 | 0 | 0 |
| 1 | 0 | 1 | 1 | | 0 | 0 | 1 | 1 | 1 | 1 |
| 1 | 1 | 0 | 1 | | 1 | 0 | 0 | 1 | 0 | 0 |
| 1 | 1 | 1 | 1 | | 1 | 1 | 1 | 1 | 1 | 1 |

If the table is not complete, we proceed to evaluate a subformula whose components have been assigned truth values. In this case, an additional conjunction now has truth values under both conjuncts, $p \wedge (q \wedge r)$, so we can fill that column in next.

| $p$ | $q$ | $r$ | $p$ | $\wedge$ | $(q$ | $\wedge$ | $r)$ | $p$ | $\wedge$ | $r$ |
|---|---|---|---|---|---|---|---|---|---|---|
| 0 | 0 | 0 | 0 | 0 | 0 | 0 | 0 | 0 | 0 | 0 |
| 0 | 0 | 1 | 0 | 0 | 0 | 0 | 1 | 0 | 0 | 1 |
| 0 | 1 | 0 | 0 | 0 | 1 | 0 | 0 | 0 | 0 | 0 |
| 0 | 1 | 1 | 0 | 0 | 1 | 1 | 1 | 0 | 0 | 1 |
| 1 | 0 | 0 | 1 | 0 | 0 | 0 | 0 | 1 | 0 | 0 |
| 1 | 0 | 1 | 1 | 0 | 0 | 0 | 1 | 1 | 1 | 1 |
| 1 | 1 | 0 | 1 | 0 | 1 | 0 | 0 | 1 | 0 | 0 |
| 1 | 1 | 1 | 1 | 1 | 1 | 1 | 1 | 1 | 1 | 1 |

Finally, as the truth table has been fully filled out, we can check whether there are any counterexamples to the argument. The only row in which the premise, $p \wedge (q \wedge r)$, is true is the final row, but the conclusion, $p \wedge r$, is also true there. Therefore, there are no Boolean counterexamples to this argument, so it is valid.

## 5.3  Model-Theoretic Validity

In this section, we will look at some facts about validity defined using Boolean models. We'll be looking at the same kinds of facts that we saw in section 4.1 in the previous chapter, but instead of talking about *provability* (defined by way of proofs), we are defining validity using *models*.

For the proofs, it will be useful to state this obvious fact, which we'll codify in a lemma.

**Lemma 6 (Validity as preservation of truth)**   *The argument from X to A is valid iff for all models v, if $v(X) = 1$, then $v(A) = 1$.*

*Proof.*   This is immediate consequence of the fact that in Boolean models, the only values are 0 and 1, and no model assigns *both* 0 and 1 to a single formula. We reason as follows. If $X \models_{CL} A$, then there is no model where $v(X) = 1$ and $v(A) = 0$. So, in any model $v$ where $v(X) = 1$, we have $v(A) = 1$, too. Conversely, if for every model $v$, if $v(X) = 1$ implies that $v(A) = 1$, we can have no counterexample, as there is no model where $v(X) = 1$ and $v(A) = 0$, and so $X \models_{CL} A$.                                                                                       $\square$

So, sometimes we will think of Boolean validity as an absence of counterexamples, and sometimes we will think of it as preservation of truth. The two notions are equivalent. Now we can take a look at the facts we proved concerning provability (in intuitionistic logic). The corresponding facts hold for Boolean validity, too.

**Theorem 12**   *The following are all true.*

1. $A \models_{CL} A$.
2. *If $X \models_{CL} A$, then $X, B \models_{CL} A$.*
3. *If $X \models_{CL} A$ and $A, Y \models_{CL} B$, then $X, Y \models_{CL} B$.*

*Proof.*   1. We need to show that every model $v$ is such that if $v(A) = 1$, then $v(A) = 1$. But, that is a tautology, so we are done.

2. Suppose that $X \models_{CL} A$. To show that $X, B \models_{CL} A$, suppose that $v$ is a counterexample, so $v(B) = 1$ and $v(X) = 1$ but $v(A) = 0$. From the initial assumption and the assumption that $v(X) = 1$, we obtain $v(A) = 1$, which contradicts the assumption that there is a counterexample. Therefore, $X, B \models_{CL} A$.

3. Suppose that $X \models_{CL} A$ and $A, Y \models_{CL} B$. Let $v$ be a model such that $v(X) = 1$ and $v(Y) = 1$. We need to show that $v(B) = 1$. From the assumption that $X \models_{CL} A$ and $v(X) = 1$, we obtain $v(A) = 1$. As $v(Y) = 1$, $v(A, Y) = 1$, so from the assumption that $A, Y \models_{CL} B$, we conclude $v(B) = 1$. Therefore, $X, Y \models_{CL} B$.

                                                                                                $\square$

Notice that the verifications of these facts are *similar* to the verifications of the corresponding facts in theorem 1, but instead of combining formal proofs, these verifications simply talked about the presence or absence of counterexamples. We cannot combine two counterexamples to make another counterexample. This is unlike the case for provability, where we used the operation of chaining proofs together to make larger proofs.

Let's look at the other facts we proved in section 4.1. If validity acts like provability, we should be able to prove this theorem.

**Theorem 13 (Boolean compactness)**   *The validity relation* $\models_{CL}$ *is* COMPACT *in the following sense. If* $X \models_{CL} A$—*even if the set* $X$ *is infinitely large*—*then there is some* finite *subset* $X' \subseteq X$ *where* $X' \models_{CL} A$.

Think for a moment about how we could prove this. Using the contrapositive, we would want to show that if there is a counterexample to each argument $X' \succ A$, where $X'$ is a finite subset of $X$, then there is also a counterexample to $X \succ A$. There is no guarantee that we will be able to combine each of these counterexamples together to make a *single* model that makes *everything* in $X$ true. (The model that satisfies one finite subset *might* conflict with a model that satisfies another.) It turns out that this theorem is true, but the techniques used for proving it directly (operating on models) are rather mathematically sophisticated and beyond the scope of this text.[51]

$$* * *$$

Let's end this section looking at the connection between Boolean validity and the connectives.

**Theorem 14**   *For any set* $X$ *of formulas, and any formulas* $A$, $B$, *and* $C$, *we have*

- $X \models_{CL} A \wedge B$ *if and only if* $X \models_{CL} A$ *and* $X \models_{CL} B$.
- $X \models_{CL} A \rightarrow B$ *if and only if* $X, A \models_{CL} B$.
- $X \models_{CL} \neg A$ *if and only if* $X, A \models_{CL} \perp$.
- $X, A \vee B \models_{CL} C$ *if and only if* $X, A \models_{CL} C$ *and* $X, B \models_{CL} C$.

*Proof.*   Let's take these items one by one. First, if $X \models_{CL} A \wedge B$, then for any model $v$, if $v(X) = 1$, we have $v(A \wedge B) = 1$ too. Since $v(A \wedge B) = 1$, we have $v(A) = 1$ and $v(B) = 1$, and so, $X \models_{CL} A$ and $X \models_{CL} B$. Conversely, suppose $X \models_{CL} A$ and $X \models_{CL} B$, then if we have a valuation $v$ where $v(X) = 1$, since $X \models_{CL} A$, we have $v(A) = 1$, and since $X \models_{CL} B$, we have $v(B) = 1$. So, $v(A \wedge B) = 1$. It follows, then, that $X \models_{CL} A \wedge B$.

To show that $X \models_{CL} A \rightarrow B$ if and only if $X, A \models_{CL} B$, it is straightforward to prove this by showing that $X \not\models_{CL} A \rightarrow B$ iff $X, A \not\models_{CL} B$.[52] $X \not\models_{CL} A \rightarrow B$ iff there is some $v$ where $v(X) = 1$ and $v(A \rightarrow B) = 0$, and $v(A \rightarrow B) = 0$ iff $v(A) = 1$ and $v(B) = 0$. So, $v(X) = 1$ and $v(A \rightarrow B) = 0$ iff $v(X) = 1$ and $v(A) = 1$ and $v(B) = 0$, which holds iff $v(X, A) = 1$ and $v(B) = 0$, that is, $X, A \not\models_{CL} B$.

Similarly, $X \not\models_{CL} \neg A$ iff there is some $v$ where $v(X) = 1$ and $v(\neg A) = 0$, which holds iff $v(X) = 1$ and $v(A) = 1$, that is, $v(X, A) = 1$, which holds iff $v(X, A) = 1$ and $v(\perp) = 0$ (since $v(\perp) = 0$ for *any* $v$), and this holds iff $X, A \not\models_{CL} \perp$. So, $X \not\models_{CL} \neg A$ iff $X, A \not\models_{CL} \perp$, and hence, $X \models_{CL} \neg A$ iff $X, A \models_{CL} \perp$ too.

---

51. A challenge question in the next chapter asks you to prove this theorem using ideas presented there.
52. Note that we are proving the contrapositive of the biconditional, which is classically equivalent to proving the biconditional.

For disjunction, let's suppose $X, A \vee B \models_{CL} C$. In order to show that $X, A \models_{CL} C$, we will suppose $v(X, A) = 1$, and we'll try to show $v(C) = 1$. Now, since $v(A) = 1$, we have $v(A \vee B) = 1$, and so, $v(X, A \vee B) = 1$, and since $X, A \vee B \models_{CL} C$, we have $v(C) = 1$. The case to show $X, B \models_{CL} C$ is shown in the same way. To show the converse, let's suppose $X, A \models_{CL} C$ and $X, B \models_{CL} C$. We wish to show $X, A \vee B \models_{CL} C$. To do this, suppose $v(X, A \vee B) = 1$, in order to show that $v(C) = 1$. Since $v(A \vee B) = 1$, we either have $v(A) = 1$ or $v(B) = 1$, and hence, $v(X, A) = 1$ or $v(X, B) = 1$. Since we have both $X, A \models_{CL} C$ and $X, B \models_{CL} C$, in either case, we can conclude $v(C) = 1$, as desired. □

We said that a set of sentences $X$ is *inconsistent* iff $X \vdash_I \bot$, if a contradiction can be proved from $X$. An analogous definition, unsatisfiability, is appropriate for validity defined in terms of models.

**Definition 31 (Boolean unsatisfiability)** *A set $X$ of formulas is said to be* UNSATISFIABLE *if and only if $X \models_{CL} \bot$. That is, there is no model $v$ where $v(X) = 1$.*

In the previous chapter, we saw that the corresponding facts hold for intuitionistic provability. A natural question is how similar these two notions of logical consequence turn out to be. What is the relationship between validity in terms of provability in intuitionistic natural deduction ($X \vdash_I A$) and validity according to models ($X \models_{CL} A$)? We will look at this relationship in the next chapter.

### 5.4 Key Concepts and Skills

☐ You need to be familiar with the definitions of Boolean valuations and models, and given a valuation, you can calculate the value of a complex formula, in terms of the values of its atoms.

☐ You should be able to complete truth tables for formulas.

☐ You need to know what it means for a formula to be a tautology, a contradiction, or a contingency, and you can use truth tables to test for whether a formula is a tautology or a contradiction or contingent.

☐ You can test arguments using Boolean models.

☐ You should be able to reason about and verify simple general facts about validity ($\models_{CL}$), such as the facts expressed in theorems 12 and 14, as well as in questions 1 and 3 below.

A BOOLEAN VALUATION $v$ is a function $v: \mathsf{Atom} \mapsto \{0, 1\}$ such that $v(\bot) = 0$.
Given a Boolean valuation $v$, we extend $v$ to all formulas inductively using the
following truth tables.

| $A$ | $B$ | $A \wedge B$ | $A \vee B$ | $A \rightarrow B$ | $\neg A$ | $\bot$ |
|-----|-----|--------------|------------|-------------------|----------|--------|
| 0   | 0   | 0            | 0          | 1                 | 1        | 0      |
| 0   | 1   | 0            | 1          | 1                 | 1        | 0      |
| 1   | 0   | 0            | 1          | 0                 | 0        | 0      |
| 1   | 1   | 1            | 1          | 1                 | 0        | 0      |

A valuation $v$ extended to the whole language will be called a BOOLEAN MODEL.
A BOOLEAN COUNTEREXAMPLE to the argument $X \succ A$ is a model $v$ that assigns
1 to all formulas in the set $X$ and assigns 0 to $A$.
An argument is BOOLEAN VALID iff it has no Boolean counterexamples.

## 5.5  Questions for You

**Basic Questions**

1. Complete truth tables for these formulas, and decide whether they are tautologies,
   contradictions, or contingencies:
    i.   $\neg(p \wedge \neg p)$
    ii.  $p \wedge (p \rightarrow \bot)$
    iii. $(p \wedge (p \rightarrow \bot)) \rightarrow \bot$
    iv.  $p \rightarrow \neg q$
    v.   $(p \rightarrow \neg q) \rightarrow (q \rightarrow \neg p)$
    vi.  $q \rightarrow (p \wedge (p \rightarrow q))$

2. Which of these arguments are valid? For those that are, explain why (using valuations),
   and for those that aren't, provide a counterexample.
    i.   $\neg(p \wedge q) \succ \neg p \wedge \neg q$
    ii.  $p \rightarrow q, q \rightarrow r \succ p \rightarrow r$
    iii. $p \rightarrow q, q \rightarrow r \succ \neg p \rightarrow \neg r$
    iv.  $\neg(p \rightarrow q) \succ p$
    v.   $\neg(p \wedge q) \succ \neg p \vee \neg q$
    vi.  $\succ ((p \rightarrow q) \rightarrow p) \rightarrow p$

3. Which of these general "facts" about validity are really *facts*? For those that aren't, can
   you give any reasons why they aren't? For those that are, can you prove them?
    i.   If $A \vDash_{\mathsf{CL}} B$, then $\neg B \vDash_{\mathsf{CL}} \neg A$.
    ii.  If $A \vDash_{\mathsf{CL}} B$, then it's not true that $B \vDash_{\mathsf{CL}} A$.
    iii. Either $A \vDash_{\mathsf{CL}} B$ or $B \vDash_{\mathsf{CL}} A$.
    iv.  $X, A \vDash_{\mathsf{CL}} \neg A$ if and only if $X \vDash_{\mathsf{CL}} \neg A$.
    v.   $X \vDash_{\mathsf{CL}} A \vee B$ if and only if $X \vDash_{\mathsf{CL}} A$ or $X \vDash_{\mathsf{CL}} B$.

4. Suppose that $A, B \models_{CL} C \wedge D$. Let $v$ be a Boolean valuation. Which of the following statements are true, which are false, and which do you not have enough information to evaluate?

   i. If $v(A) = 1$ and $v(B) = 1$, then $v(C \wedge D)$ can be 0.

   ii. If $v(C) = 0$, then $v(A) = 0$.

   iii. If $v(C) = 0$, then $v(A) = 0$ or $v(B) = 0$.

   iv. If $v(C \wedge D) = 0$ and $v(A) = 1$, then $v(B) = 0$.

   v. If $v(A) = 1$ and $v(B) = 0$, then $v(C \wedge D) = 0$.

   vi. If $v(C \wedge D) = 1$, then $v(A) = 1$.

   vii. If $v(A) = 1$ and $v(B) = 1$, then $v(\neg(C \wedge D)) = 0$.

## Challenge Questions

1. We say that $A$ and $A'$ are *Boolean equivalent* if $A \models_{CL} A'$ and $A' \models_{CL} A$. That is, if $v(A) = v(A')$ for every valuation $v$. Show the following general facts about provability, assuming that $A$ and $A'$ are Boolean equivalent formulas.

   i. If $X \models_{CL} A$, then $X \models_{CL} A'$.

   ii. If $X, A \models_{CL} B$, then $X, A' \models_{CL} B$.

   iii. $A \wedge B$ is Boolean equivalent to $A' \wedge B$.

   iv. $A \to B$ is Boolean equivalent to $A' \to B$.

   v. $B \to A$ is Boolean equivalent to $B \to A'$.

   vi. $\neg A$ is Boolean equivalent to $\neg A'$.

   vii. $A \vee B$ is Boolean equivalent to $A' \vee B$.

   viii. Explain why it follows for any complex formula $C(A)$ with $A$ as a subformula that $C(A)$ is Boolean equivalent to $C(A')$, where $C(A')$ is found by replacing the $A$ in $C$ by $A'$.

2. A formula is in *disjunctive normal form* (DNF) if it is constructed from some number of *disjunctions* of *conjunctions* of *literals*, where a literal is an atom (not including $\perp$) or a negation of an atom (not including $\perp$). We define these notions formally as follows:

- $A$ is a *literal* iff $A$ is a member of **Atom**, other than $\perp$, or $A$ is the *negation* of some member of **Atom**, other than $\perp$.

- If $A$ is a literal, then we say $A$ is also a *conjunction of literals* (of conjunction length 1). If $B$ and $C$ are both *conjunctions of literals* (of conjunction lengths $n$ and $m$, respectively), then $(B \wedge C)$ is also a *conjunction of literals* (of conjunction length $n + m$).

- If $C$ is some conjunction of literals, then we say that $C$ is also a *disjunction of conjunctions of literals* (of disjunction length 1). If $D$ and $E$ are disjunctions of conjunctions of literals (with disjunction lengths $k$ and $l$, respectively), then $(D \vee E)$ is a disjunction of conjunctions of literals with disjunction length $k + l$.

So, for example, $((p \wedge \neg q) \vee \neg r) \vee (\neg p \wedge (r \wedge s))$ is a disjunction of conjunctions of literals, with disjunction length 3. Its disjuncts have conjunction lengths 2, 1, and 3, respectively.

In this question, we will show how to find, for any formula not in disjunctive normal form, another formula equivalent to it, which does not violate the constraints of DNF in that way.

- $\bot$ is equivalent to $p \wedge \neg p$.
- $A \rightarrow B$ is equivalent to $\neg A \vee B$.
- $\neg\neg A$ is equivalent to $A$.
- $\neg(A \wedge B)$ is equivalent to $\neg A \vee \neg B$.
- $\neg(A \vee B)$ is equivalent to $\neg A \wedge \neg B$.
- $A \wedge (B \vee C)$ is equivalent to $(A \wedge B) \vee (A \wedge C)$.
- $(B \vee C) \wedge A$ is equivalent to $(B \wedge A) \vee (C \wedge A)$.

   i. For the first part of this question, show that if a formula in Form is *not* a disjunction of conjunction of literals, then it has a *subformula* of one of the following forms:

$$\bot \quad A \rightarrow B \quad \neg\neg A \quad \neg(A \wedge B)$$

$$\neg(A \vee B) \quad A \wedge (B \vee C) \quad (B \vee C) \wedge A$$

   ii. Take the formula $p \rightarrow (r \wedge \neg(p \vee \neg q))$. Choose a subformula of this formula of one of the forms shown in part (i) and replace it with an equivalent formula, using the equivalences listed above. Is the result in DNF? If not, find another subformula to convert, and continue, until the result is in DNF.

  iii. Do you think this process could work for *any* formula? Could the process of simplification go on forever, or will it always terminate in a formula in DNF? Why or why not?

3. In truth tables, we can think of the connectives as representing functions on the set $\{0, 1\}$ of truth values. Negation represents a unary function, sending the input 1 to the output 0, and vice versa. There are sixteen binary truth functions, sending two truth values to some truth value as output. Three of these functions are represented by the binary connectives we have adopted. An example of one of the remaining binary truth functions is the $\star$ connective, whose truth table is below.

| $A$ | $B$ | $A \star B$ |
|-----|-----|-------------|
| 0 | 0 | 0 |
| 0 | 1 | 0 |
| 1 | 0 | 1 |
| 1 | 1 | 0 |

We can define $\star$ using the connectives we have by producing a formula with the same truth table as $\star$. Verify that $A \star B$ has the same truth table as $\neg(A \rightarrow B)$. Adding the connective $\star$ to the language does not give us anything essentially new, because we can define it using the connectives we have. Show that all sixteen binary truth functions can be defined using the connectives of our language.

4. A set of connectives is TRUTH FUNCTIONALLY COMPLETE if, for $n \geq 0$, every $n$-ary truth function can be defined by a formula whose only connectives are from that set. Show that the set $\{\neg, \rightarrow, \wedge, \vee, \bot\}$ is truth functionally complete by providing a procedure for defining every $n$-ary truth function.

Then, prove that $\{\vee, \neg\}$ is truth functionally complete. Show that $\{\rightarrow, \perp\}$ is also truth functionally complete.

5. Show that the set of connectives $\{\vee, \wedge, \rightarrow\}$ is not truth functionally complete. Show that the set of connectives $\{\vee, \perp\}$ is not truth functionally complete.

6. Show that the binary connective $\downarrow$, with the truth table

| $A$ | $B$ | $A \downarrow B$ |
|-----|-----|------------------|
| 0 | 0 | 1 |
| 0 | 1 | 0 |
| 1 | 0 | 0 |
| 1 | 1 | 0 |

is truth functionally complete on its own, which is to say that $\{\downarrow\}$ is truth functionally complete.[53] Find at least one other binary connective that is also truth functionally complete.

---

[53]. The connective $\downarrow$ is sometimes called "NOR": $A \downarrow B$ is *neither A nor B*, and the notation "$\downarrow$" is due to Charles Sanders Peirce. Peirce's student, Christine Ladd-Franklin (1847–1930), showed that a generalization of $\downarrow$ could be used to represent all of the inferences of Aristotle's syllogistic logic (Uckelman 2021). Ladd-Franklin studied with the mathematician James J. Sylvester and the logician Charles Sanders Peirce at Johns Hopkins University. She was admitted when applying under the name C. Ladd. When it was revealed that she was a woman, the board of the university attempted to revoke the offer, but with Sylvester's encouragement and support, she continued her study, writing a dissertation "On the Algebra of Logic" with Peirce as her supervisor. She was the first American woman to formally receive graduate instruction in mathematics and symbolic logic. Since women were not allowed to graduate from Johns Hopkins University at the time, she was not awarded the degree at the completion of her studies. She was finally awarded her PhD in 1926, at the age of seventy-eight.

# 6   Soundness & Completeness

In the previous chapter, we introduced models and used them to define a notion of validity in terms of models in addition to our original notion of validity in terms of provability. Provability is motivated by trying to give an account of what makes good arguments good. An argument is good if it can be analyzed into a *proof*, where the premises lead to the conclusion using the basic rules governing the concepts involved. Validity is motivated by trying to give an account of what makes arguments *fail* to be good. An argument fails when it has some counterexample. If it does not fail, it is good in the sense of failing to have a counterexample.

This raises the question of how validity according to models, $X \vDash_{CL} A$, relates to provability in our natural deduction system, $X \vdash_I A$. Do these notions coincide? This is a general question for any notions of validity according to *provability* ($\vdash$) and validity according to *models* ($\vDash$). In fact, these are two separate questions, relating proofs and models.[54]

[SOUNDNESS] If the argument $X \succ A$ is provable, that is, $X \vdash A$ (if it has some proof), is it also valid, that is, $X \vDash A$ (does it have no counterexample)?

[COMPLETENESS] If the argument $X \succ A$ is valid, that is, $X \vDash A$ (if it has no counterexample), is it also provable? $X \vdash A$ (is there some proof)?

"Soundness" is a kind of consistency criterion. We don't have both a proof and a counterexample for a single argument. "Completeness" is the opposite. For every argument, we have *either* a proof *or* a counterexample. Soundness and completeness (for the notions $\vdash$ and $\vDash$) are the claims that our notions of proof and counterexample are mutually exclusive and exhaustive. We will now consider the soundness and completeness questions for our notions $\vdash_I$ and $\vDash_{CL}$.

We need two different methods to answer these two questions. The technique for settling the soundness question is a *structural induction*. We have seen that kind of argument in previous chapters, and in this chapter, we will go into structural induction in some detail. After we address soundness, we will turn to the question of completeness.

---

54. One can raise questions of soundness and completeness for any two consequence relations, although it is most often done where one of the consequence relations is defined in terms of proofs and the other in terms of models.

## 6.1 Soundness

In this section, we will sketch the proof of soundness for our intuitionistic natural deduction system with respect to Boolean valuations.

We will rely on the fact that the formal proofs are inductively defined. We will show that if $\Pi$ is a proof for $X \succ A$, then this argument is *truth preserving* in all models. That is, for all Boolean models $v$, if $v(X) = 1$, then $v(A) = 1$, that is, $X \models_{CL} A$. We will show this by demonstrating the base case (that a proof consisting of a single assumption is truth preserving), and for the induction step, we show that if we have proofs that are truth preserving, and we make a new proof out of them using one of our rules, then the result is truth preserving too. We will prove this theorem:

**Theorem 15 (Soundness)** *If an argument is valid in intuitionistic natural deduction, that is, $X \vdash_I A$, then it is valid according to Boolean models, $X \models_{CL} A$.*

*Proof.* The proof breaks into several cases, one for each rule.

The assumption rule provides the base case. The atomic proof, $A$, is a proof for $A \succ A$. Let $v$ be a Boolean model such that $v(A) = 1$. Right away we have $v(A) = 1$, so we are done. It is truth preserving.

For the inductive hypothesis, we assume that we are given proofs that are truth preserving, and we wish to show that whatever proofs we make using an inference step are also truth preserving. There are quite a few cases. We will go through many of them here, illustrating all of the techniques we need to use, and leave the rest to the questions at the end of the chapter.

First, the $\wedge I$ case. We assume we have proofs $\Pi_1$ for $X \succ A$ and $\Pi_2$ for $Y \succ B$ and that both are truth preserving. Suppose we form a new proof, using the rule $\wedge I$.

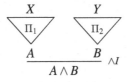

We want to show that $X, Y \models_{CL} A \wedge B$. Suppose that $v(X, Y) = 1$. From the inductive hypothesis, if $v(X) = 1$, then $v(A) = 1$. Since $v(X, Y) = 1$ implies $v(X) = 1$, $v(A) = 1$. Similarly, by the inductive hypothesis, if $v(Y) = 1$, then $v(B) = 1$. As $v(X, Y) = 1$ implies $v(Y) = 1$, $v(B) = 1$. Since have $v(A) = v(B) = 1$, $v(A \wedge B) = 1$, as desired.

Next, consider the $\vee I$ case. We assume we have a proof $\Pi$ for $X \succ A$. We then form a new proof using the $\vee I$ rule.

$$\frac{\begin{array}{c} X \\ \diagdown\Pi\diagup \\ A \end{array}}{A \vee B} \vee I$$

We need to show that $X \models_{CL} A \vee B$. Assume there is a counterexample, so there is a Boolean model $v$ such that $v(X) = 1$ and $v(A \vee B) = 0$. By definition, $v(A \vee B) = 0$ iff $v(A) = 0$ and

$v(B) = 0$. By the inductive hypothesis, $X \models_{CL} A$. Since $v(X) = 1$, $v(A) = 1$, which is a contradiction. Therefore, there is no counterexample. The other $\lor I$ case, where the conclusion is $B \lor A$, is similar.

Next, consider the $\to I$ case. We assume we have a proof $\Pi$ for $X, A \succ B$. We then form a new proof using the $\to I$ rule.

$$[A]^1 \quad X$$
$$\Pi$$
$$\frac{B}{A \to B} \to I^1$$

We need to show that $X \models_{CL} A \to B$. Assume there is a counterexample, so there is a Boolean model $v$ such that $v(X) = 1$ and $v(A \to B) = 0$. By definition, $v(A \to B) = 0$ iff $v(A) = 1$ and $v(B) = 0$. The assumption then implies $v(A) = 1$ and $v(B) = 0$, so $v(X, A) = 1$. By the inductive hypothesis, $X, A \models_{CL} B$, so $v(B) = 1$, which is a contradiction. Therefore, there is no counterexample.

Next, we will do the $\to E$ case. We assume we have proofs $\Pi_1$ for $X \succ A \to B$ and $\Pi_2$ for $Y \succ A$, and we form a new proof using the rule $\to E$.

$$X \qquad Y$$
$$\Pi_1 \qquad \Pi_2$$
$$\frac{A \to B \qquad A}{B} \to E$$

We want to show that $X, Y \models_{CL} B$. By the inductive hypothesis, $X \models_{CL} A \to B$ and $Y \models_{CL} A$. Suppose that there is a counterexample, so $v(X, Y) = 1$ and $v(B) = 0$. Since $v(X, Y) = 1$, then $v(X) = 1$. It follows from the inductive hypothesis that $v(A \to B) = 1$. Similarly, $v(Y) = 1$, so it follows from the inductive hypothesis that $v(A) = 1$. But, $v(A \to B) = 1$ iff either $v(A) = 0$ or $v(B) = 1$. Since $v(A) \neq 0$, $v(B) = 1$, which contradicts the assumption. Therefore, there is no counterexample.

As our last example, we'll do the $\neg E$ case. We assume we have proofs $\Pi_1$ for $X \succ \neg A$ and $\Pi_2$ for $Y \succ A$, and we form a new proof using the rule $\neg E$.

$$X \qquad Y$$
$$\Pi_1 \qquad \Pi_2$$
$$\frac{\neg A \qquad A}{\bot} \neg E$$

We want to show that $X, Y \models_{CL} \bot$. By the inductive hypothesis, $X \models_{CL} \neg A$ and $Y \models_{CL} A$. Suppose that there is a counterexample, so $v(X, Y) = 1$ and $v(\bot) = 0$. Since $v(X, Y) = 1$, then $v(X) = 1$. It follows from the inductive hypothesis that $v(\neg A) = 1$. Similarly, $v(Y) = 1$ and so it follows from the inductive hypothesis that $v(A) = 1$. But $v(\neg A) = 0$ iff $v(A) = 1$, so we have $v(\neg A) = 1$ and $v(\neg A) = 0$, which yields a contradiction. Therefore, there is no counterexample.

There are several cases left to do before the *soundness theorem* is exhaustively demonstrated. We are going to leave them for the reader to do to test their understanding, with the easiest cases being $\land E$ and $\bot E$ and the hardest being $\lor E$. Despite there being some cases left to work through, we will consider soundness proved. $\qquad \square$

## 6.2  Completeness

Let us now turn to the question of completeness. To answer this in the affirmative, we need to show that whenever an argument is valid according to Boolean models, $X \models_{CL} A$, there is a proof of $A$ from $X$, $X \vdash_I A$. To settle the question in the negative, we need to produce an argument that is valid according to Boolean models for which there is no intuitionistic proof.

The law of excluded middle is valid according to Boolean models.

**Lemma 7**   $\models_{CL} p \vee \neg p$.

*Proof.*   The following truth table shows that $p \vee \neg p$ is a tautology.

| $p$ | $p$ | $\vee$ | $\neg$ | $p$ |
|---|---|---|---|---|
| 0 | 0 | 1 | 1 | 0 |
| 1 | 1 | 1 | 0 | 1 |

$\square$

Recall that in chapter 4, we showed that the law of excluded middle is not derivable in intuitionistic logic, $\nvdash_I p \vee \neg p$. So, while $\vdash_I$ is sound for $\models_{CL}$, $\vdash_I$ is *not* complete for $\models_{CL}$.

**Theorem 16**   *Intuitionistic logic is not complete with respect to validity according to Boolean models.*

There is a gap between our accounts of proofs and of counterexamples. Some arguments (like $\succ p \vee \neg p$) have neither proofs nor counterexamples.

* * *

The fact that intuitionistic logic is not complete for validity according to Boolean models raises two questions.

- Is any system of proofs complete for validity according to Boolean models?
- Are there other kinds of models that can be used to define a notion of validity for which intuitionistic provability is complete?

To answer the first question in the affirmative, we would need to expand our system of proofs to provide proofs for every argument that has no Boolean counterexample. To answer the second question in the affirmative, we would need to expand our collection of models to provide counterexamples to any arguments that don't have proofs. We can do *both* of these things.

* * *

In chapter 4, we introduced *classical logic* by adding the rule $\dfrac{\neg\neg A}{A}$ *DNE* to the intuitionistic natural deduction system. Adding *DNE* is one way to answer our first question in the affirmative. Classical proofs are sound and complete with respect to validity according to Boolean models. For soundness, we need to show that the new rule, *DNE*, preserves truth.

The following line of argument establishes that.

$$v(\neg\neg A) = 1 \text{ iff } v(\neg A) = 0 \text{ iff } v(A) = 1$$

So, $\neg\neg A$ and $A$ must have the same truth value in any Boolean model. Thus, the rule *DNE* preserves truth.

Proving completeness for classical logic involves introducing a number of more technical ideas, so we will defer the proof to the last part of this section. Until you are confident with the material in the rest of this chapter, leave that section until later, and read on.

### Theorem 17 (Soundness and completeness for classical logic)

$$X \vdash_C A \text{ iff } X \models_{CL} A.$$

The answer to the second question, whether we can expand our collection of models to provide counterexamples for any arguments that have no intuitionistic proofs, is also affirmative. The two most common kinds of models that can do this job are *Kripke models* and *Heyting algebras*. Kripke models for intuitionistic logic will be easier to introduce after we learn some basic modal logic next chapter, so we will defer that task for a moment. Instead, we will briefly look at Heyting algebras.

The intuitive reason that the law of the excluded middle is valid in Boolean models is that there are not enough values for formulas to take to obtain a counterexample. As an example, consider the instance of excluded middle $p \vee \neg p$. If $v(p) = 1$, then the disjunction gets value 1, and if $v(p) = 0$, then the right disjunct gets value 1, as does the disjunction. There are many ways to generalize the truth tables by adding additional values. Heyting algebras are one way of doing so that line up neatly with intuitionistic logic.

Heyting algebras generalize Boolean models by permitting more semantic values for our interpretations. We will define Heyting algebras in section 6.4. For the moment, we will note that they are sets of values, partially ordered by a relation $\leq$ (intuitively leading from *less* true to *more* true), and always have a greatest element, 1 (the *most* true something can be), and a least element, 0 (the *least* true).

A VALUATION in a Heyting algebra is a function $v$, from the set of atoms to the set of values, that obeys the truth tables for the particular Heyting algebra. We can use these valuations to define validity, as before:

### Definition 32 (Counterexample, validity)   *In a Heyting algebra, if X is the set of formulas $B_1, \ldots, B_n$, then $v(X) = v(B_1 \wedge \ldots \wedge B_n)$, and if $X = \emptyset$, $v(X) = 1$.[55]*
*A* HEYTING COUNTEREXAMPLE *v to an argument $X \succ A$ is a valuation v on a Heyting algebra such that it is not the case that $v(X) \leq v(A)$. An argument $X \succ A$ is Heyting valid iff it has no Heyting counterexamples. When the argument $X \succ A$ is Heyting valid, we write $X \models_H A$.*

The two values we have already seen constitute a Heyting algebra, when we think of the values $\{0, 1\}$ ordered by setting $0 \leq 1$ (and $1 \not\leq 0$), and where conjunction, disjunction, the conditional, and negation are defined in the way given by standard truth tables. A Boolean counterexample is one where the conclusion is assigned a value that is not at least as great

---

55. This definition agrees with the use of the notation "$v(X)$" in Boolean valuations but it is more general. Using some concepts introduced in appendix 6.4, we can state it without inserting conjunctions.

as the value assigned to the premises. The only way for that to happen is for the premises to take value 1 and for the conclusion to take value 0, since there are only two values.

The simplest (nontrivial) Heyting algebra that is *not* that two-valued Boolean model has *three* values, $\{1, \frac{1}{2}, 0\}$, with $0 < \frac{1}{2} < 1$. We will call this Heyting algebra $H_3$. The truth tables for $H_3$ are the following, with $v(\bot) = 0$.

| $\wedge$ | 0 | $\frac{1}{2}$ | 1 | | $\vee$ | 0 | $\frac{1}{2}$ | 1 | | $\rightarrow$ | 0 | $\frac{1}{2}$ | 1 | | $\neg$ | |
|---|---|---|---|---|---|---|---|---|---|---|---|---|---|---|---|---|
| 0 | 0 | 0 | 0 | | 0 | 0 | $\frac{1}{2}$ | 1 | | 0 | 1 | 1 | 1 | | 0 | 1 |
| $\frac{1}{2}$ | 0 | $\frac{1}{2}$ | $\frac{1}{2}$ | | $\frac{1}{2}$ | $\frac{1}{2}$ | $\frac{1}{2}$ | 1 | | $\frac{1}{2}$ | 0 | 1 | 1 | | $\frac{1}{2}$ | 0 |
| 1 | 0 | $\frac{1}{2}$ | 1 | | 1 | 1 | 1 | 1 | | 1 | 0 | $\frac{1}{2}$ | 1 | | 1 | 0 |

A Heyting valuation assigns values to all atomic formulas, and the value of complex formulas is determined using the tables above, just as with classical truth tables and Boolean valuations.[56] We can show that the law of the excluded middle is not valid according to Heyting algebras by giving it a value besides 1 in $H_3$. For that, take a Heyting valuation $v$ with $v(p) = \frac{1}{2}$. Then $v(\neg p) = 0$, so the disjunction $\frac{1}{2} \vee 0$ is $\frac{1}{2}$, as you see by following these equations:

$$v(p \vee \neg p) = \frac{1}{2} \vee v(\neg p) = \frac{1}{2} \vee 0 = \frac{1}{2}$$

This establishes $\not\models_H p \vee \neg p$, since being Heyting valid requires getting value 1 in all Heyting valuations. (For the full definition of Heyting validity, $\models_H$, see section 6.4.)

This single, three-valued Heyting algebra won't pin down intuitionistic logic exactly. To do that, we need to consider Heyting algebras with more values than just *three*. Studying these algebras, however, is a field all its own, with extended treatments by Helena Rasiowa[57] (1974) and Leo Esakia (2019), among others.[58]

We will leave Heyting algebras here and return to proving the completeness theorem for classical logic. If this is your first time through, skip the rest of this section, and go ahead to section 6.3 on page 101, where we turn to the relationship between proofs and models. If you have already grasped the rest of this chapter, and you would like to see how theorem 17 is proved, read on, but beware. The rest of this section comes with a   WARNING LABEL  .

* * *

To complete the proof of theorem 17, we want to show that if $X \models_{CL} A$, then $X \vdash_C A$. The way we will tackle this is by showing the *contrapositive*, that if $X \not\vdash_C A$ (that is, if there is no proof from $X$ to $A$), then $X \not\models_{CL} A$ (that is, there is some counterexample to the argument from $X$ to $A$). To do this, we need to find some model $v$ that assigns 1 to each

---

56. For more on Heyting algebras, see section 6.4 later in this chapter. When we get to the general definition, we will use slightly different notation for the first three truth tables, in particular, the symbols "⊓," "⊔," and "⇒."

57. Helena Rasiowa (1917–1994) was a Polish logician who made many contributions to algebraic logic and the foundations of mathematics. Her PhD thesis, supervised by Andrzej Mostowski, was titled *Algebraic Treatment of the Functional Calculi of Lewis and Heyting*. In addition to her ground-breaking monograph *An Algebraic Approach to Non-Classical Logics* (1974), she also wrote *The Mathematics of Metamathematics* (1963) with Roman Sikorski. https://en.m.wikipedia.org/wiki/Helena_Rasiowa

58. Heyting algebras are just one sort of algebraic structure that is useful in the study of nonclassical logics. See Dunn and Hardegree (2001) for a useful overview of algebraic techniques for philosophical logic.

member of $X$ and assigns 0 to $A$. A model sorts the formulas in our language into the *true* and the *false*. $X$ and $A$ give us a start on specifying our model: we want each member of $X$ to be on the *true* side and $A$ to be on the false side. But our model will decide for *every* formula in our language whether it's on one side or another. Usually, there will be many more formulas that are counted as *true* other than those in $X$. For example, $X$ might contain $B$ and $C$ without containing $B \wedge C$. But if $X \not\vdash_C A$, then $X, B \wedge C \not\vdash_C A$ too, since any proof of $A$ using $B \wedge C$ as a premise could be extended into one that instead uses $B$ and $C$. So, if our set $X$ contains $B$ and $C$, and isn't enough to prove $A$, then adding in $B \wedge C$ as an extra member wouldn't change that fact. The larger set still isn't enough to prove $A$. The larger set looks a little bit more like the set of formulas made true by some model.

Our proof is going to go like this: if we have some set $X$ where $X \not\vdash_C A$, then we will expand $X$ out into a much bigger set $X'$, which still doesn't prove $A$, but is as big as can be. It will be a *maximal $A$-avoiding* set of formulas. We will then show that for any maximal $A$-avoiding set of formulas, the formulas *in* that set act exactly like the formulas that are *true* according to some model, so from our maximal $A$-avoiding set, we can find a model that makes every member of $X$ true and $A$ false.

The techniques we will use here are quite general. We will use them again in later chapters, where we construct other kinds of models for other kinds of logics. In fact, we will be using these results to apply to intuitionistic logic, too, so instead of proving everything again and again for each different system, we will state things in terms of a *logic* L, where L could be classical logic, intuitionistic logic, or something in between.

**Definition 33 (Maximal $A$-avoiding sets)**  *For any logic L, a set $X'$ of formulas is a* MAXIMAL $A$-AVOIDING SET *according to the logic L (sometimes called an L-maximal $A$-avoiding set) iff $X'$ is a set of formulas such that $X' \not\vdash_L A$, but for any formula B not in $X'$, $X', B \vdash_L A$.*

The definition of "maximal $A$-avoiding set" involves a logic L. One may have a set of formulas $X$ that is maximal $A$-avoiding according to $L_1$ but not maximal $A$-avoiding according to $L_2$, which is to say $L_1$-maximal but not $L_2$-maximal. Indeed, we will see this in the completeness proof for intuitionistic logic in the next chapter. When discussing multiple logics, the logics under consideration need to be specified explicitly. In this chapter, we will focus on C, so we will leave the dependence implicit.

Before showing you how to construct maximal $A$-avoiding sets, let's explore some of their properties. This way, we will see why they are worth looking for. The first result is that maximal $A$-avoiding sets contain all of their consequences.

**Lemma 8**  *If $X'$ is a maximal $A$-avoiding set according to the logic L, then whenever $X' \vdash_L B$, $B \in X'$.*

*Proof.*  Let $X'$ be a maximal $A$-avoiding set. Suppose that for some $B$, $X' \vdash_L B$, but $B \notin X'$. Since $X'$ is a *maximal $A$-avoiding* set, if $B \notin X'$, then $X', B \vdash_L A$. Since we supposed that $X' \vdash_L B$, we can chain these two proofs together to produce a proof that $X' \vdash_L A$, which contradicts the hypothesis that $X'$ was $A$-avoiding. Therefore, $B \in X'$.  $\square$

Second: The following is what we can say about maximal $A$-avoiding sets where the logic is intuitionistic logic, or stronger.

**Lemma 9** *If $X'$ is a maximal A-avoiding set in a logic L that contains all of the rules of intuitionistic logic, then $X'$ satisfies the following conditions:*

1. $\bot \notin X'$.
2. $B \wedge C \in X'$ *if and only if* $B \in X'$ *and* $C \in X'$.
3. $B \vee C \in X'$ *if and only if* $B \in X'$ *or* $C \in X'$.
4. *If* $\neg B \in X'$, *then* $B \notin X'$.
5. *If* $B \to C \in X'$, *then either* $B \notin X'$ *or* $C \in X'$.

*Proof.* For (1), since $X'$ is $A$-avoiding in L, we have $X' \nvdash_L A$. If $\bot$ were in $X'$, we would have $X' \vdash_L A$, since $\bot \vdash_L A$. So, we have $\bot \notin X'$.

For (2), if $B \wedge C \in X'$, we wish to show that $B \in X'$ (and similarly for $C$). This follows immediately from lemma 8, since $B \wedge C \vdash_L B$ and $B \wedge C \vdash_L C$.

Similarly, if $B \in X'$ and $C \in X'$, we must have $B \wedge C \in X'$, by lemma 8, since $B, C \vdash_L B \wedge C$.

For (3), if $B \in X'$ (or if $C \in X'$), then we must have $B \vee C \in X'$ for similar reasons: $B \vdash_L B \vee C$ and $C \vdash_L B \vee C$.

For the other direction, if $B \vee C \in X'$, then we cannot have $B \notin X'$ and $C \notin X'$, since if neither $B$ nor $C$ were in $X'$, we would have $X', B \vdash_L A$ and $X', C \vdash_L A$. Call these proofs $\Pi_1$ and $\Pi_2$. Combine these two proofs like this:

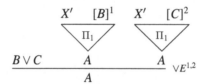

and the result is a proof from $X'$ to $A$, since $B \vee C$ is in $X'$. But there is no proof from $X'$ to $A$, as $X'$ is a maximal $A$-avoiding set. It follows that there is no such $\Pi_1$ or no such $\Pi_2$, and hence either $B \in X'$ or $C \in X'$.

The remaining facts fall short of what we would like for two-valued classical valuations. For (4), it is straightforward to show that if $\neg B \in X'$, then $B \notin X'$, since if $B, \neg B \in X'$, then $X'$ would not be $A$-avoiding since $B, \neg B \vdash_L A$. However, we cannot prove the converse. There is no way to show that—if we are using intuitionistic logic, at least—if $B \notin X'$, then $\neg B \in X'$. Here is why: $\neg\neg p \nvdash_I p$. We will be able to find a maximal $p$-avoiding set containing $\neg\neg p$. That set won't contain $p$, since it's $p$-avoiding. But since it contains $\neg\neg p$, it can't contain $\neg p$ too, since $\neg\neg p, \neg p \vdash_I \bot$, and, hence, $\neg\neg p, \neg p \vdash_I p$. So, (4) is as strong as we can do if we are not going to add *DNE*.

Similarly, we can prove (5), since if $B \to C \in X'$ and $B \in X'$ then we must have $C \in X'$ since if $X', C \vdash_L A$, then we would have $X' \vdash_L A$ too, since $B \to C, B \vdash_L C$, and $B \to C$ and $B$ are both in $X'$. So, it follows that if $B \to C \in X'$, then either $B \notin X'$, or if $B \in X'$, we must have $C \in X'$ too. $\qquad\qquad\square$

A set of formulas $X$ having the feature that whenever $X \vdash_L B$, $B \in X$ is sometimes called an L-THEORY, or theory in L. Theories are special sets of formulas that have a wide range of uses in logic. Some theories, such as the theory of Peano arithmetic, have been studied extensively. The set of theorems of a logic L forms an L-theory. Maximal $A$-avoiding sets according to L are rich L-theories. The properties in the preceding lemma are important properties that theories may have. In particular, when a theory obeys the condition in (2), it is called a *prime theory*. We have seen one prime theory, namely, the set of theorems of intuitionistic logic. Prime theories are useful in proving completeness results. Taking maximal $A$-avoiding sets is one way of ensuring that the theories under consideration are prime. For some logics, we do not need to use *maximal $A$-avoiding* sets to prove completeness. Instead, merely using *prime $A$-avoiding* sets will work.[59]

We can say more about our sets if the logic contains the rule *DNE* and strengthen (4) and (5) to the full-strength classical conditions.

**Lemma 10** *If $X'$ is a maximal $A$-avoiding set in a logic $L$ that at least contains all of the rules of classical logic, then $X'$ also satisfies the following conditions:*

6. $\neg B \in X'$ *if and only if $B \notin X'$.*
7. $B \to C \in X'$ *if and only if $B \notin X'$ or $C \in X'$.*

*Proof.* For (6), we already have that if $\neg B \in X'$, then $B \notin X'$ by (4) from the previous lemma. So, we aim to show that if $B \notin X'$, then $\neg B \in X'$. If $B \notin X'$, then we have $X', B \vdash_L A$. If we had $\neg B \notin X'$, then we would also have $X', \neg B \vdash_L A$. But this cannot happen, because if it does, we have $X' \vdash_L A$. Here is why. Suppose $\Pi_1$ is a proof for $X', B \succ A$ and $\Pi_2$ is a proof for $X', \neg B \succ A$. We can make a proof for $X' \succ A$ like this:

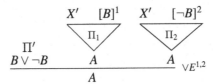

where $\Pi'$ is a proof of the law of the excluded middle. But our assumption was that $X'$ is $A$-avoiding, so if $B \notin X'$ (if adding $B$ to $X'$ would be enough to prove $A$), we must have $\neg B \in X'$. The reason is that if adding $\neg B$ would also be enough to prove $A$, by using the displayed proof, we could prove $A$ regardless.

The reasoning for (7) can piggyback on the reasoning for (6). First, we show that if $C \in X'$, then $B \to C \in X'$. That is not difficult, because if $C \in X'$ since $C \vdash_L B \to C$, we must have $B \to C \in X'$ too. Second, we show that if $B \notin X'$, then $B \to C \in X'$. Now we can appeal to (6) and say that since $B \notin X'$, we have $\neg B \in X'$, and since $\neg B \vdash_L B \to C$, we must have $B \to C \in X'$ too, as desired. □

The left-to-right half of (6) is a consistency condition. The converse, the right-to-left half, is a kind of completeness condition, which we will call NEGATION-COMPLETENESS, to avoid confusion with the completeness of the completeness theorem.

So, given the power of classical logic, any maximal $A$-avoiding set acts just like the set of formulas true in a valuation. We have this result:

**Lemma 11** *If $X'$ is a maximal $A$-avoiding set in a logic $L$ that contains all of the rules of classical logic, then there is some model $v_{X'}$ where for any formula $B$, $v_{X'}(B) = 1$ if and only if $B \in X'$.*

*Proof.* Define $v_{X'}$ by setting $v_{X'}(p) = 1$ if and only if $p \in X'$, for each atom $p$. Then, it is a simple proof by induction on the structure of the formula $B$, appealing to (1)–(3), (6), and (7), that $v_{X'}(B) = 1$ if and only if $B \in X'$. □

---

59. The completeness proof presented by Restall (2000, chap. 11) uses prime theories in the way described.

So, we need one final piece of the puzzle to finish our proof of the completeness theorem. We need to show that for any set $X$ of formulas such that $X \nvdash_L A$, there is some maximal $A$-avoiding set $X'$ extending the set $X$.

**Lemma 12** *For any logic L, if $X \nvdash_L A$, then there is some maximal A-avoiding set $X'$ where $X \subseteq X'$.*

*Proof.* The crucial idea behind the proof is this. We start with our formulas $X$ in a pile and consider each formula in the language one-by-one, and if we could add it to our pile, without making $A$ provable from that pile, we add it to the pile. If we can't, we throw it away. Keep going, and at the "end" of this ongoing process, the result is our set $X'$. That will be a maximal $A$-avoiding set.

Here are the details. First, we select an *enumeration* of the language. That is, we number off the formulas in a list like this:

$$B_1, B_2, B_3, B_4 \ldots$$

The requirement for the list is that every formula in the language appears on the list somewhere. Each formula in our language appears as $B_n$ for some number $n$. There are plenty of different ways to enumerate our language. One is to use the UNICODE standard[60] for representing text in code. Type your formula into some app, and the underlying representation is a number. Take all the formulas and list them in the order of the numbers that represent them in UNICODE. There are plenty of numbers that aren't codes for *formulas*, of course, but that big list of formulas in order will be a list of all of the formulas in the language, as every formula in our language has some UNICODE code number. That's just one enumeration. You could use another. If you have your enumeration, now we define the process for building $X'$.

We define $X'$ step-by-step. We start with $X$, which we will call $X_0$. To define $X_n$ for each number $n > 0$, we make the move we saw at the beginning. Consider the formula $B_n$. If $X_{n-1}, B_n \nvdash_L A$, then $X_n = X_{n-1} \cup \{B_n\}$. Otherwise, $X_n = X_{n-1}$. So, we have this increasing sequence

$$X = X_0 \subseteq X_1 \subseteq X_2 \subseteq X_3 \subseteq \cdots$$

of sets of formulas. At each step of the way, the set $X_n$ is $A$-avoiding, since $X_0$ starts off avoiding $A$, and each $X_n$ is defined to be the same as $X_{n-1}$ unless the new formula under consideration (that is, $B_n$) can be added without also requiring that $A$ be included. So, each $X_n$ is $A$-avoiding. The set $X'$ is defined as the union of $\bigcup_{n=0}^{\infty} X_n$[61] of all of the sets in the series. It contains the starting set $X$ and any formula that is added to any of the sets along the way. In other words, to decide whether a formula $B_n$ is in the set $X'$, simply go through the process of constructing the sets $X_m$ up until stage $n$. $B_n$ is in $X'$ if and only if $B_n \in X_n$.

It remains to show that $X'$, defined like this, is a maximal $A$-avoiding set. First, $X'$ is $A$-avoiding, since if we had $X' \vdash_L A$, then there is some proof from $X'$ to $A$. But a proof is finite, so the assumptions used in that proof, since they are all in $X'$, are all in some set $X_n$ constructed along the way. But each $X_n$ is $A$-avoiding, so there is no proof from $X'$ to $A$ either.

---

60. See https://www.unicode.org/ for details.
61. The notation here means that we are taking the union of all the sets $X_0, X_1, X_2, \ldots$ that we constructed.

Now, it remains to show that $X'$ is a *maximal $A$-avoiding set*. We want to show that if $B \notin X'$, then $X', B \vdash_L A$. This is easy to show since $B$ appears on our list somewhere; let's say it is $B_n$. Since $B_n \notin X'$, it follows that $B_n \notin X_n$, which means that when $B_n$ came up for consideration, it was not added to $X_n$. That is, we saw that $X_{n-1}, B_n \vdash_L A$. So, since $X_{n-1} \subseteq X'$, then $X', B_n \vdash_L A$ too. So, $X'$ is a *maximal $A$-avoiding set*, as we wanted to show. $\square$

Now we have all of the pieces of our puzzle. We can put them together to prove the *completeness* half of theorem 17. We can show that if $X \models_{CL} A$, then $X \vdash_C A$.[62]

*Proof.* Let's suppose $X \nvdash_C A$. By lemma 12, there is some $X'$ such that $X \subseteq X'$ and $X'$ is a maximal $A$-avoiding set according to $C$. By lemma 11, there is a model $v_{X'}$ that assigns 1 to each member of $X$ and 0 to $A$. So, it follows that $X \nmodels_{CL} A$. $\square$

That is the proof of completeness. The techniques we used here will be used again in future chapters, including when we prove completeness for *intuitionistic* logic using models that we will define in chapter 7.

## 6.3 Proofs First or Models First?

This mismatch between intuitionist proofs and Boolean models can be viewed in different ways—and it has been, throughout the history of twentieth-century logic and into the twenty-first century. We can think of any mismatch between a system of proofs and a system of models as a problem with the proofs, or a problem with the models. These differences are different views on which, if any, of proofs and models have the explanatory priority. If the job of models is to provide counterexamples of arguments for which there are no proofs—if proofs have the priority—then the failure of completeness is the failure to have enough models to serve as counterexamples.

If, on the other hand, it is the job of proofs to underwrite all of those arguments for which there are no counterexamples—if models have the priority—then the failure of completeness is the failure of all of our proofs to underwrite all the valid arguments.

This is a general methodological distinction in different views of the explanatory significance of logical tools and the way formal logical tools are related to more general discussions of *semantics*, the meanings of expressions in our thought and talk. One way to understand this distinction is the distinction between *representationalist* and *inferentialist* approaches to meaning.[63]

A representationalist gives an account of the meanings of expressions in terms of a *representation* relation. We interpret items from a language by way of a function from the language to the "world." A valuation could be understood as a possible representation relation respecting the meanings of the vocabulary involved. On this understanding, an argument is invalid if there is some interpretation of the language that renders the premises true and the conclusion false. (Understanding logic and semantics in terms of models and representation relations is the dominant approach, so examples of this perspective are easy

---

62. This method of proving completeness is known as the Henkin method, and it was pioneered by Leon Henkin (1949).

63. This distinction is due to Robert Brandom. See his *Articulating Reasons* (2000), especially chapter 1, for his account of the difference between these approaches to semantics.

to find. Some different examples of this approach throughout the past eighty years are due to Tarski [1944], Montague [1970], and Williamson [2013].)

An inferentialist, on the other hand, gives priority in semantics to the norms of inference governing our concepts. Following the slogan from Wittgenstein, meaning conditions are given in terms of norms governing language *use*, and an important aspect of the use of our logical concepts (like any concepts) is their use in our inferential practice. For an inferentialist, rules like the rules of inference in our natural deduction systems are a ready example for the norms governing all of our concepts. On this view, an argument is valid if the inference from the premises to the conclusions is mandated by our correct practice, if it arises out of the fundamental rules governing those concepts. (For examples of inferentialist approaches to meaning, see Brandom's *Articulating Reasons* [2000], and Dummett's *The Logical Basis of Metaphysics* [1991].)

* * *

This methodological difference—between proof-first and model-first understandings of logic, as well as their connections to approaches to semantics and the philosophy of language—is connected to, but is not the same as, the debate between partisans of intuitionistic logic and classical logic. It is true that most of those who have defended classical logic take Boolean valuations (and richer model structures, of the kind we will see in the second half of the book) to have explanatory priority, and they take systems of proofs to be secondary and to have derivative significance as a means of studying validity. Those who favor intuitionistic logic have tended to take proofs to be primary and models secondary. However, there is no need that these be the only options. You could take proofs as primary and prefer classical logic (this would require defending *DNE* or another system of proofs that would somehow be connected most deeply to practice and the meanings of our logical concepts). Or you could take models as primary and find room for Heyting algebras (or Kripke models, or some other model for intuitionistic logic) to do the job of interpreting our language. All options remain open to us.

The tools of logic are useful in providing frameworks and structures that can be of use to semantics and the philosophy of language. The interaction between philosophy and logic has proved mutually fruitful. Different ways of understanding meaning have motivated logicians' and philosophers' constructions of logical tools. These tools have then been of use, not only in helping us give an account of good and bad arguments but also in exploring theories of meaning and connections between meaning and other concerns, such as reference and inference, metaphysics, and epistemology.[64] Soundness and completeness results are useful in that they show how representational concerns and inferential considerations

---

64. For a discussion of the significance of these logical techniques in the midst of a wide-ranging *plurality* of different formal systems, see Susan Haack's *Deviant Logic* (1974). Susan Haack (1945– ) is a British philosopher of science and logic, formerly Professor of Philosophy at the University of Warwick, and now at the University of Miami. Her work in philosophical logic set the scene for much of the twentieth-century understanding of the range of logical systems. Her account of the territory places classical logic at the center. Logics are *deviant* in different ways, either by going beyond classical logic, perhaps with new connectives or quantifiers, as we will discuss in coming chapters, or by restricting various classical principles, as we have seen with intuitionistic logic, not validating double negation elimination.

can be connected, and these connections are of value, whether you think representation is primary, or inference is, or something else, or some mixture of these.

## 6.4  Appendix: Heyting Algebras

The definition of a Heyting algebra requires us to define a few concepts. First, the concept of a *lattice*.

**Definition 34 (Bounds, lattice)**  *Let $\langle \mathcal{V}, \leq \rangle$ be a partially ordered set and $\sqcap$ and $\sqcup$ binary operations on the set $\mathcal{V}$.*

- *An element y is an* UPPER BOUND *of a subset $X \subseteq \mathcal{V}$ iff for all $x \in X, x \leq y$.*
- *An element y is a* LOWER BOUND *of $X \subseteq \mathcal{V}$ iff for all $x \in X, y \leq x$.*
- *An element $z \in \mathcal{V}$ is the* LEAST UPPER BOUND *of $X \subseteq \mathcal{V}$ iff z is an upper bound of X and $z \leq y$, for any upper bound y of X.*
- *An element $z \in \mathcal{V}$ is the* GREATEST LOWER BOUND *of $X \subseteq \mathcal{V}$ iff z is a lower bound of X and $y \leq z$, for any lower bound y of X.*
- *A partially ordered set $\langle \mathcal{V}, \leq \rangle$ is a* LATTICE *iff every pair of elements $x, y \in V$ has a least upper bound, the join $x \sqcup y$, and a greatest lower bound, the meet $x \sqcap y$.*
- *A lattice $\langle \mathcal{V}, \leq \rangle$ is* DISTRIBUTIVE *iff for all elements $x, y, z \in \mathcal{V}$,*

$$x \sqcap (y \sqcup z) = (x \sqcap y) \sqcup (x \sqcap z).$$

It follows from the definitions that in a lattice, the least upper bound of a pair of elements is unique, as is the greatest lower bound. Note that not every partially ordered set is a lattice. Some partially ordered sets may have pairs of elements that have *no* upper bounds, or they may not have *least* upper bounds, for example.

Let us illustrate the lattice concepts with two examples.

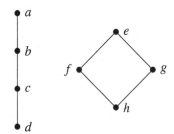

These two diagrams represent distributive lattices.[65] The left-hand one, $L$, has $\mathcal{V} = \{a, b, c, d\}$ with the ordering $d < c < b < a$, and the right-hand one, $R$, has $\mathcal{V} = \{e, f, g, h\}$ with the ordering $h < g < e$ and $h < f < e$.

The diagram represents the *covering relation*, $\leq_1$, of the order $\leq$, where $x \leq_1 y$ iff $x \leq y$ and there is no $z \in V$ such that $z \neq x$, $z \neq y$, $x \leq z$, and $z \leq y$. If $x \leq_1 y$, then $x$ appears below $y$ in the diagram with a line connecting them. We can recover the ordering $\leq$ from the diagram. If $x \leq_1 y$, then $x \leq y$. We add the reflexive relations, $x \leq x$, for each $x$, and also use *transitivity*, if $x \leq y$ and $y \leq z$, then $x \leq z$. Transitivity says that if a point $x$ is connected by a multistep path upward through the diagram to another point $y$, then $x \leq y$.

---

65. Diagrams for lattices and partial orderings are known as Hasse diagrams.

In $L$, the set $\{b,c\}$ has $a$ and $b$ as upper bounds, with $b$ as the least upper bound. The set $\{a,c\}$ has $c$ and $d$ as lower bounds, with $c$ as the greatest lower bound. In $R$, the set $\{f,h\}$ has $f$ and $e$ as upper bounds, with $f$ as the least upper bound. The set $\{f,g\}$ has $h$ as its only lower bound, and it has $e$ as its only upper bound. In $R$, every point is an upper bound of the empty set, $\emptyset$, so the least upper bound of $\emptyset$ is $h$. In $R$, every point is a lower bound of $\emptyset$, so the greatest lower bound of $\emptyset$ is $e$.

With the definition of a distributive lattice in hand, we can define Heyting algebras.

**Definition 35 (Heyting algebra)**  *A* HEYTING ALGEBRA *is a quintuple* $\langle \mathcal{V}, \leq, \Rightarrow, 0, 1 \rangle$ *such that* $\langle \mathcal{V}, \leq \rangle$ *is a distributive lattice,* $\Rightarrow$ *is a binary operation on* $\mathcal{V}$*, with* $0, 1 \in \mathcal{V}$*, and the following conditions are satisfied:*

- *for all* $x \in \mathcal{V}$, $0 \leq x$,
- *for all* $x \in \mathcal{V}$, $x \leq 1$*, and*
- *for all* $x, y, z \in \mathcal{V}$, $x \sqcap y \leq z$ *iff* $x \leq y \Rightarrow z$.

In a Heyting algebra, we can define a negation operation in terms of $\Rightarrow$ and $0$, by setting $\neg x$ to be $x \Rightarrow 0$. (This should remind you of the connection between $\neg A$ and $A \to \bot$ in proofs in intuitionistic logic.)

A VALUATION on a Heyting algebra is a function $v$ from Atom to $\mathcal{V}$ such that $v(\bot) = 0$, and it is extended to the whole language using the following clauses:

- $v(A \wedge B) = v(A) \sqcap v(B)$,
- $v(A \vee B) = v(A) \sqcup v(B)$,
- $v(A \to B) = v(A) \Rightarrow v(B)$, and
- $v(\neg A) = v(A) \Rightarrow 0$.

We will repeat the definitions of counterexample and validity from the chapter.

**Definition 36 (Counterexample, validity)**  *Let $X$ be a finite set of formulas and $v$ a valuation on a Heyting algebra. Then, $v(X)$ is the greatest lower bound of the set $\{v(A) : A \in X\}$.*

*A* HEYTING COUNTEREXAMPLE $v$ *to an argument* $X \succ A$*, with $X$ a finite set of formulas, is a valuation $v$ on a Heyting algebra such that $v(X) \not\leq v(A)$.*[66]

*An argument $X \succ A$ is Heyting valid iff it has no Heyting counterexamples. When the argument $X \succ A$ is Heyting valid, we write $X \vDash_H A$.*

Let us turn to some examples of Heyting algebras. Every Boolean valuation is a Heyting algebra with $\mathcal{V} = \{0, 1\}$ and $0 < 1$. The three-valued example $H_3$ from the chapter, repeated here with the notation adjusted, is a Heyting algebra. In the example, $\mathcal{V} = \{0, \frac{1}{2}, 1\}$ and the ordering $0 < \frac{1}{2} < 1$, which we represent diagrammatically as follows.

---

66. We can relax the restriction that $X$ is finite by adding a condition that the Heyting algebras under consideration are *complete*, which is to say that all subsets of $\mathcal{V}$ have least upper bounds and greatest lower bounds. Alternatively, we could restrict to finite Heyting algebras, which are all complete and still adequate for l, or we could adjust the definition of $v(X)$ and the condition for being a counterexample when $X$ is infinite. These options all add some complications. Since l is compact, we can focus our attention on finite sets of premises without loss.

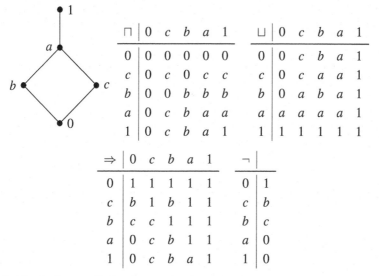

| ⊓ | 0 | ½ | 1 |
|---|---|---|---|
| 0 | 0 | 0 | 0 |
| ½ | 0 | ½ | ½ |
| 1 | 0 | ½ | 1 |

| ⊔ | 0 | ½ | 1 |
|---|---|---|---|
| 0 | 0 | ½ | 1 |
| ½ | ½ | ½ | 1 |
| 1 | 1 | 1 | 1 |

Notice also that the meet and join tables can be read off the diagram. In this case, the meet of two elements is the lower of the two in the diagram, and the join of two elements is the higher of the two. We add to this lattice information the implication and negation tables.

| ⇒ | 0 | ½ | 1 |
|---|---|---|---|
| 0 | 1 | 1 | 1 |
| ½ | 0 | 1 | 1 |
| 1 | 0 | ½ | 1 |

| ¬ | |
|---|---|
| 0 | 1 |
| ½ | 0 |
| 1 | 0 |

We are not restricted to three values. As an example of a larger Heyting algebra, we can take $\mathcal{V} = \{0, 1, a, b, c\}$, with the ordering $0 < b < a < 1$ and $0 < c < a < 1$. We can represent the ordering diagrammatically as follows:

| ⊓ | 0 | c | b | a | 1 |
|---|---|---|---|---|---|
| 0 | 0 | 0 | 0 | 0 | 0 |
| c | 0 | c | 0 | c | c |
| b | 0 | 0 | b | b | b |
| a | 0 | c | b | a | a |
| 1 | 0 | c | b | a | 1 |

| ⊔ | 0 | c | b | a | 1 |
|---|---|---|---|---|---|
| 0 | 0 | c | b | a | 1 |
| c | 0 | c | a | a | 1 |
| b | 0 | a | b | a | 1 |
| a | a | a | a | a | 1 |
| 1 | 1 | 1 | 1 | 1 | 1 |

| ⇒ | 0 | c | b | a | 1 |
|---|---|---|---|---|---|
| 0 | 1 | 1 | 1 | 1 | 1 |
| c | b | 1 | b | 1 | 1 |
| b | c | c | 1 | 1 | 1 |
| a | 0 | c | b | 1 | 1 |
| 1 | 0 | c | b | a | 1 |

| ¬ | |
|---|---|
| 0 | 1 |
| c | b |
| b | c |
| a | 0 |
| 1 | 0 |

Let us call this algebra $H_5$.[67] Consider a Heyting valuation $v$ on $H_5$ with $v(p) = a$, $v(q) = b$, $v(r) = c$. This provides counterexamples to some classically valid arguments that are not intuitionstically valid, such as $\neg\neg p \succ p$, $\succ q \vee \neg q$, and $\succ (q \to r) \vee (r \to q)$. For the first, note that $v(\neg\neg p) = \neg\neg a = \neg 0 = 1$, but it is not the case that $1 \leq v(p)$. For the second, $v(\neg q) = c$, and $b \sqcup c = a$. For an argument with no premises to be valid, there cannot be a valuation that assigns the conclusion a value apart from 1. So, we have a counterexample. For the third, $v(q \to r) = b \Rightarrow c = c$ and $v(r \to q) = c \Rightarrow b = b$. As $b \sqcup c = a$, we have a counterexample.

---

67. Despite what this name may suggest, there are other five-valued Heyting algebras. Can you construct any others?

Heyting algebras provide counterexamples to classically valid arguments that do not have intuitionistic proofs. The three-valued Heyting algebra does not provide counterexamples to all the classically valid arguments that lack intuitionistic proofs, and the five-valued Heyting algebra above does not provide enough counterexamples either. There is a classically valid tautology, $(p \rightarrow q) \vee ((q \rightarrow r) \vee (r \rightarrow p))$, that has no counterexamples on either Heyting algebra. A natural question is whether any Heyting algebra with finitely many values provides enough counterexamples to pin down intuitionistic logic exactly. The answer is negative. For any Heyting algebra with finitely many values, one can find a classical tautology that lacks both an intuitionistic proof and a counterexample on that algebra.

Just like classical logic is sound and complete with respect to Boolean validity, intuitionistic logic is sound and complete with respect to Heyting validity. The methods to prove these facts are similar to those used in the classical case. We will provide some detail on the proof of soundness here. It is worth noting that the proof of soundness given earlier in this chapter does not settle the issue here, since there are more possible counterexamples when one moves from Boolean models to Heyting algebras. Before going through the details of soundness, it will be useful to prove some facts about lattices.

**Lemma 13**   *Let $\langle \mathcal{V}, \leq \rangle$ be a lattice. The following are true, where $x, y, z \in \mathcal{V}$.*

1. $x \sqcap y \leq x$ *and* $x \sqcap y \leq y$.
2. *If* $x \leq y$, *then* $x \sqcap z \leq y$.
3. *If* $x \leq y$ *and* $x \leq z$, *then* $x \leq y \sqcap z$.
4. $x \leq x \sqcup y$ *and* $x \leq y \sqcup x$.
5. *If* $x \leq y$, *then* $x \leq y \sqcup z$.
6. *If* $x \leq z$ *and* $y \leq z$, *then* $x \sqcup y \leq z$.

*Proof.*   For (i), $x \sqcap y$ is the greatest lower bound of $\{x, y\}$. Therefore, $x \sqcap y \leq x$ and $x \sqcap y \leq y$, from the definition of lower bound.

For (ii), suppose $x \leq y$. Since $x \sqcap z \leq x$, by the transitivity of $\leq$, $x \sqcap z \leq y$.

For (iii), suppose $x \leq y$ and $x \leq z$. Then $x$ is a lower bound of $\{y, z\}$. Since $y \sqcap z$ is the greatest lower bound of $\{y, z\}$, $x \leq y \sqcap z$.

The arguments for (iv)–(vi) are similar, using corresponding definitions for upper bounds.                                                                                                           □

**Theorem 18 (Soundness with respect to Heyting validity)**   *If an argument is valid in intuitionistic natural deduction, $X \vdash_I A$, then it is Heyting valid, $X \models_H A$.*

*Proof.*   The argument proceeds by induction on the structure of the proof, as in the proof of soundness with respect to Boolean validity. The base case is the assumption rule, $A \succ A$. From the fact that $\leq$ is reflexive, we have $v(A) \leq v(A)$, for any valuation in any Heyting algebra. There are, then, several inductive steps. We will do some of them and leave the rest for exercises.

For the inductive hypothesis, we assume that we are given proofs that are *value preserving* in the sense that for any valuation $v$ in any Heyting algebra, the value $v$ assigns to the premises is less than or equal to the value $v$ assigns to the conclusion, and we wish to show that the proofs we make using an inference step are also value preserving in that sense.

Let us begin with the $\wedge I$ case. We assume we have proofs $\Pi_1$ for $X \succ A$ and $\Pi_2$ for $Y \succ B$ and that both are value preserving. Suppose we form a new proof, using the rule $\wedge I$.

$$\frac{\overset{X}{\underset{A}{\diagdown \Pi_1 \diagup}} \qquad \overset{Y}{\underset{B}{\diagdown \Pi_2 \diagup}}}{A \wedge B} {\scriptstyle \wedge I}$$

We want to show that $X, Y \models_H A \wedge B$. Suppose that there is a counterexample, so there is a Heyting algebra and valuation $v$ such that $v(X, Y) \not\leq v(A \wedge B)$. From the inductive hypothesis, $v(X) \leq v(A)$ and $v(Y) \leq v(B)$. By lemma 13, $v(X, Y) \leq v(X)$ and $v(X, Y) \leq v(Y)$, so from the transitivity of $\leq$, it follows that $v(X, Y) \leq v(A)$ and $v(X, Y) \leq v(B)$. By lemma 13, $v(X, Y) \leq v(A) \sqcap v(B)$, but $v(A) \sqcap v(B) = v(A \wedge B)$, so $v(X, Y) \leq v(A \wedge B)$, which contradicts an assumption. Therefore, there is no counterexample.

Next is the $\rightarrow I$ case. We assume we have a proof $\Pi$ for $X, A \succ B$. We then form a new proof using the $\rightarrow I$ rule.

$$\frac{\overset{[A]^1 \quad X}{\underset{B}{\diagdown \Pi \diagup}}}{A \rightarrow B} {\scriptstyle \rightarrow I^1}$$

From the inductive hypothesis, we have $X, A \models_H B$, and we need to show that $X \models_H A \rightarrow B$. Suppose that $v$ is an arbitrary valuation on an arbitrary Heyting algebra $H$. By the inductive hypothesis, $v(X, A) \leq v(B)$. From the definition of Heyting algebra, $v(X) \leq v(A \rightarrow B)$. Since $v$ and $H$ were arbitrary, we can conclude $X \models_H A \rightarrow B$.

Next, we will do the $\rightarrow E$ case. We assume we have proofs $\Pi_1$ for $X \succ A \rightarrow B$ and $\Pi_2$ for $Y \succ A$, and we form a new proof using the rule $\rightarrow E$.

$$\frac{\overset{X}{\underset{A \rightarrow B}{\diagdown \Pi_1 \diagup}} \qquad \overset{Y}{\underset{A}{\diagdown \Pi_2 \diagup}}}{B} {\scriptstyle \rightarrow E}$$

By the inductive hypothesis, $X \models_H A \rightarrow B$ and $Y \models_H A$. Suppose $H$ is an arbitrary Heyting algebra and $v$ is an arbitrary valuation on $H$. From the inductive hypothesis, $v(X) \leq v(A \rightarrow B)$, so by the definition of Heyting algebra, $v(X, A) \leq v(B)$. From the inductive hypothesis, $v(Y) \leq v(A)$, so by lemma 13, $v(X, Y) \leq B$, which suffices to conclude that $X, Y \models_H B$, as desired.

Finally, we will do the $\vee E$ case. Assume we have proofs $\Pi_1$ for $X \succ A \vee B$, $\Pi_2$ for $Y, A \succ C$, and $\Pi_3$ for $Z, B \succ C$. We form a new proof as follows using $\vee E$.

$$\frac{\overset{X}{\underset{A \vee B}{\diagdown \Pi_1 \diagup}} \qquad \overset{Y \quad [A]^1}{\underset{C}{\diagdown \Pi_2 \diagup}} \qquad \overset{Z \quad [B]^2}{\underset{C}{\diagdown \Pi_3 \diagup}}}{C} {\scriptstyle \vee E^{1,2}}$$

From the inductive hypothesis, we have (i) $X \vDash_H A \vee B$, (ii) $Y, A \vDash_H C$, and (iii) $Z, B \vDash_H C$. Suppose that there is a Heyting algebra and a valuation $v$ such that $v(X, Y, Z) \not\le v(C)$. From (ii) and (iii), $v(Y, A) \le v(C)$ and $v(Z, B) \le v(C)$. From lemma 13, $v(Y, A) \sqcup v(Z, B) \le v(C)$. Since $v(Y, A) = v(Y) \sqcap v(A)$ and $v(Z, B) = v(Z) \sqcap v(B)$, using lemma 13, we have $(v(Y, Z) \sqcap v(A)) \sqcup (v(Y, Z) \sqcap v(B)) \le v(C)$. Since every Heyting algebra is a distributive lattice,

$$(v(Y, Z) \sqcap v(A)) \sqcup (v(Y, Z) \sqcap v(B)) = v(Y, Z) \sqcap (v(A) \sqcup v(B)).$$

As $v(A) \sqcup v(B) = v(A \vee B)$, using lemma 13 again, we obtain $v(Y, Z) \sqcap (v(A \vee B)) \le v(C)$. From (i), $v(X) \le v(A \vee B)$, so by lemma 13, we have $v(X, Y, Z) \le v(C)$, which contradicts an assumption. Thus, there is no counterexample.

The remaining cases are left as exercises.                                            □

The soundness theorem ensures that if we have a counterexample to an argument, there is no proof in our intuitionistic natural deduction system for that argument. This justifies our use of counterexamples in Heyting algebras to show that certain arguments have no proofs. Unlike with validity defined in terms of Boolean valuations, intuitionistic logic is complete with respect to validity defined in terms of Heyting algebras.

**Theorem 19 (Completeness with respect to Heyting validity)** *If an argument is Heyting valid, $X \vDash_H A$, then the argument is valid in intuitionistic natural deduction, $X \vdash_I A$.*

We will not prove completeness here. For details, see Dummett (2000, 119–129). In the appendix to chapter 7, we will prove completeness for intuitionistic logic with respect to another definition of validity, and that proof will provide the tools for proving completeness here.

One might wonder if there are any philosophical interpretations of the values in the Heyting algebras. It is common to think of 0 and 1 as representing false and true, respectively. What could $a$, $b$, and $c$ represent in $H_5$? It is not clear what one might say about that. This can lead one to wonder whether there is any philosophical interpretation one can give to models with more than two semantic values. In the next appendix, we will briefly look at two three-valued logics that have attracted a lot of philosophical attention, K3 and LP.

## 6.5  Appendix: Some Three-Valued Logics

The Heyting algebra $H_3$ provides one example of how to provide truth tables for three semantic values. There are other options for defining truth tables on three values. In fact, once one has more than two semantic values, there are more options for defining counterexamples. In this appendix, we will present two prominent three-valued logics.

These logics have the same set of values, $\{1, 0, n\}$, and the same truth tables.

| $\wedge$ | 0 | $n$ | 1 | | $\vee$ | 0 | $n$ | 1 | | $\rightarrow$ | 0 | $n$ | 1 | | $\neg$ | | | $\perp$ |
|---|---|---|---|---|---|---|---|---|---|---|---|---|---|---|---|---|---|---|
| 0 | 0 | 0 | 0 | | 0 | 0 | $n$ | 1 | | 0 | 1 | 1 | 1 | | 0 | 1 | | 0 |
| $n$ | 0 | $n$ | $n$ | | $n$ | $n$ | $n$ | 1 | | $n$ | $n$ | $n$ | 1 | | $n$ | $n$ | | |
| 1 | 0 | $n$ | 1 | | 1 | 1 | 1 | 1 | | 1 | 0 | $n$ | 1 | | 1 | 0 | | |

**Definition 37 (Trivaluation)** *A* TRIVALUATION *v is a function from* Atom *to* $\{0, 1, n\}$ *such that* $v(\bot) = 0$. *The value of a complex formula in a given trivaluation v is determined using the truth tables above.*

The logics differ on how they define counterexamples.

**Definition 38 (Counterexample)** *A trivaluation v is a* K3-COUNTEREXAMPLE *to an argument* $X \succ A$ *iff* $v(X) = 1$ *and* $v(A) \neq 1$.
  *A trivaluation v is an* LP-COUNTEREXAMPLE *to an argument* $X \succ A$ *iff for all* $B \in X$, $v(B) \neq 0$, *and* $v(A) = 0$.

A trivaluation is a K3-counterexample to an argument just in case it assigns all the premises 1 and assigns the conclusion either $n$ or 0. A trivaluation is an LP-counterexample to an argument just in case it assigns each of the premises either 1 or $n$ and assigns the conclusion 0. We can rephrase these definitions using a concept common in the study of many-valued logics, the concept of *designated values*. A set of designated values $D$ of a set of values $V$ is a nonempty subset of $V$. A $D$-counterexample to an argument is a valuation $v$ such that all the premises take values in $D$ and the conclusion takes a value outside of $D$.[68] For LP, the set of designated values $D_{LP} = \{1, n\}$, and for K3, $D_{K3} = \{1\}$.

The different definitions of counterexample yield different definitions of validity.

**Definition 39 (Validity)** *An argument* $X \succ A$ *is* K3-VALID, $X \models_{K3} A$, *iff there is no* K3-*counterexample to it.*
  *An argument* $X \succ A$ *is* LP-VALID, $X \models_{LP} A$, *iff there is no* LP-*counterexample to it.*

Proponents of these logics interpret the trivaluations differently. Proponents of K3 often take the value $n$ to represent being *neither* true nor false, indeterminacy, or a truth-value gap. If there is no fact of the matter whether $p$ holds, it is natural to assign $p$ value $n$. Why might one want to have such a value? We will mention one common motivation: vague expressions, such as "forms a heap of sand." It is plausible that 100,000 grains of sand form a heap, and that three grains of sand do not form a heap. Imagine that we remove grains one by one from our heap. After removing two grains, we still have a heap. At some point, we no longer have a heap. In between, it is natural to think that it is indeterminate whether some of the intermediate amounts of sand form a heap.[69]

Proponents of LP often take the value $n$ to represent being *both* true and false, overdeterminacy, or a truth-value glut. A common motivation for this view is the semantic paradoxes, such as the liar paradox. For the liar paradox, we will suppose that there is a formula, $L$, that is equivalent to its own negation, so that $(L \to \neg L) \land (\neg L \to L)$ is a logical truth. We can then argue as follows, where $\Pi_1$ is a proof of $L \lor \neg L$.

$$\cfrac{\Pi_1 \quad L \lor \neg L \qquad \cfrac{\cfrac{L \to \neg L \quad [L]^1}{\neg L} \to E \qquad [L]^1}{\bot} \neg E \qquad \cfrac{[\neg L]^2 \quad \cfrac{\neg L \to L \quad [\neg L]^2}{L} \to E}{\bot} \neg E}{\cfrac{\bot}{p} \bot E} \lor E^{1,2}$$

---

68. Recently, there has been investigation into logics that permit one set of designated values for the premises and another for the conclusion. See Ripley (2013a) and French (2016) for more details.
69. See N. J. J. Smith (2008), for example.

We see that if $L$ is equivalent to its own negation, then we end up in a contradiction. If we add the resources to talk about *truth* in the object language, it turns out there will be formulas like $L$ around.

Since we might want to talk about truth, and the logic of truth, we need to offer a response to the semantic paradoxes. There are many different theories out there. One response, offered by Graham Priest (1979), is to use LP to deny that the move from $L$ and $\neg L$ to $\bot$ is valid. Another response, offered by Kripke (1975), is to use K3 and deny that excluded middle holds for paradoxical formulas like $L$. This means that some step of the derivation of excluded middle for $L$ must be rejected.

Both of the logics K3 and LP differ from classical logic. In K3, the argument $p \succ q \vee \neg q$ is invalid. Take a trivaluation $v$ such that $v(p) = 1$ and $v(q) = n$. On this valuation, $v(\neg q) = n$, so $v(q \vee \neg q) = n$. Therefore, the premise has value 1 and the conclusion does not, so the argument is not K3-valid. In LP, the argument $p \wedge \neg p \succ q$ is invalid. Take a trivaluation $v$ with $v(p) = n$ and $v(q) = 0$. Since $v(\neg p) = n$, the premise takes value $n$ and the conclusion takes value 0. So, the argument is not LP-valid.

Both K3 and LP are nonclassical logics based on the use of trivaluations. There are other logics one can obtain using trivaluations, either by altering the truth tables, as with WK,[70] or adding additional connectives, such as Ł3 or RM3.[71] One can also permit additional semantic values, as with the four-valued logic FDE.[72] In chapter 12, we provide some references to related work on many-valued logics.

## 6.6  Key Concepts and Skills

☐ You need to understand—and to clearly state for yourself—the definitions of soundness and completeness and the difference between them.

---

70. See, for example, Bochvar and Bergmann (1981).

71. See Priest (2008, chap. 7), especially §3 and §4.

72. See Priest (2008, chap. 8) for an overview, and see Omori and Wansing (2019) for several early and contemporary papers on FDE.

FDE is sometimes called Belnap–Dunn logic, in honor of Nuel Belnap and J. Michael Dunn for their early work on the logic. The name FDE stands for "first-degree entailment," because it is the first-degree fragment of the logic E of entailment, defended by Anderson and Belnap (1975). In addition to the four-valued models, FDE also has worlds models that use an operation known as the Routley star, first presented by Richard Sylvan and Val Plumwood, then writing as Richard and Valerie Routley (1972).

J. Michael Dunn (1941–2021) was an American logician known for his work on relevant logic and FDE. His PhD thesis, *The Algebra of Intensional Logics* (1966), supervised by Nuel Belnap, used algebraic techniques in the study of nonclassical logics, including four-valued models for FDE. Some of his recent work includes *Generalized Galois Logics* (2008) and "On the Decidability of Implicational Ticket Entailment" (2013), both coauthored with Katalin Bimbó.

Richard Sylvan (1935–1996), born Routley, was a New Zealand logician and philosopher who worked in Australia. Sylvan made many contributions to nonclassical logics, particularly relevant logics. He developed the Routley star operation to model FDE with his then-wife Val Routley (later Plumwood), and he developed ternary relational models for relevant logics with Robert K. Meyer. He was prolific, authoring many books, among which are the coauthored *Relevant Logics and Their Rivals* (Routley et al. 1982, vol. 1), with Plumwood, Meyer, and Ross Brady, and his *Exploring Meinong's Jungle* (Routley 2018).

Val Plumwood (1939–2008), born Morell, was an Australian logician and philosopher. She worked on nonclassical logics, environmental philosophy, and feminist philosophy. In the area of logic, she is known for her still unpublished "Some False Laws of Logic," which is to be published in a special issue of the *Australasian Journal of Logic*, and the coauthored works mentioned above, and she published many books in environmental philosophy and feminist philosophy, including her *Feminism and the Mastery of Nature* (1994).

☐ You should understand the proof of soundness of intuitionistic and classical proofs for Boolean validity. In particular, you should understand the shape of the argument (an induction on the construction of the proof in question), and you should be able to prove particular instances for yourself.

☐ You understand what arguments count as counterexamples to the completeness of intuitionistic provability for Boolean validity and how adding *DNE* strengthens the system to give completeness.

☐ You can evaluate formulas and arguments in the example Heyting algebras, when given the truth tables to work from.

☐ You understand the connections between proofs and models, and between inferentialism and representationalism.

### 6.7 Questions for You

### Basic Questions

1. Complete the proof of the soundness theorem by completing the cases for $\wedge E$, $\perp E$, and $\vee E$. (For example, for $\wedge E$, we want to show that if $\Pi$, a proof for $X \succ A_1 \wedge A_2$ is truth preserving, then the proof of $A_i$ (whether $i = 1$ or 2), given by extending $\Pi$ with an $\wedge E$ step, is also truth preserving.)

2. Consider the rules for the biconditional ($\leftrightarrow$) given in section 2.3. Give Boolean valuation rules for biconditional formulas (explain when $v(A \leftrightarrow B) = 1$, and when it takes the value 0), such that the soundness theorem still holds for the proofs and the models.

3. We have seen that adding *DNE* to our proof system is enough to make it complete for Boolean validity. Would adding the following negation rule suffice instead? Why or why not?

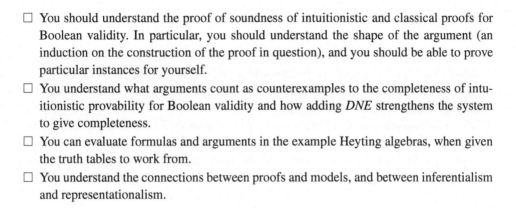

$$\begin{array}{c} [\neg A]^i \\ \Pi \\ \dfrac{\perp}{A} \ \textit{reductio}^i \end{array}$$

4. Consider the following arguments. Which have counterexamples in the Heyting algebra $H_3$, given on page 96? Which have counterexamples in $H_5$, given on page 105?

   i. $\neg\neg p \succ p$

   ii. $\neg(p \wedge q) \succ \neg p \vee \neg q$

   iii. $\neg p \vee q \succ p \rightarrow q$

   iv. $p \rightarrow q \succ \neg p \vee q$

   v. $\succ \neg p \vee \neg\neg p$

   vi. $(p \rightarrow q) \rightarrow p \succ p$

5. Consider the following arguments. Which have K3-counterexamples? Which have LP-counterexamples? Provide appropriate counterexamples, if there are any. If an argument is K3-valid or LP-valid, provide details showing that it is.

   i. $\neg\neg p \succ p$

   ii. $\neg(p \wedge q) \succ \neg p \vee \neg q$

   iii. $p \vee q, \neg p \succ q$

   iv. $p \succ (p \wedge q) \vee (p \wedge \neg q)$

v. $p \wedge \neg p \succ q \vee \neg q$

vi. $(p \rightarrow q) \rightarrow p \succ p$

6. Complete the proof of theorem 18 by doing the cases for $\wedge E$, $\vee I$, $\neg I$, $\neg E$, and $\perp E$.

7. Show that $\wedge I$, $\wedge E$, $\vee I$, and $\vee E$ are all sound for K3-validity and for LP-validity.

**Challenge Questions**

1. Design a two-place or three-place connective of your own.[73] Start by specifying an introduction rule and an elimination rule for your connective, $\sharp$. Like the other introduction and elimination rules, your rules should not use any of the *other* connectives. The rules for $A \sharp B$, for example, should tell us how to deduce something from $A \sharp B$ (in terms of $A$ and $B$ and any other premises or conclusions) and how to infer $A \sharp B$ (in terms of $A$ and $B$ and any other premises or conclusions). Show how to eliminate detours in a proof arising out of your introduction and elimination rules.

2. Continuing on from the previous question, find a way to interpret your connective using *Boolean valuations*. Show that your introduction and elimination rules are *sound* for the valuation rules you chose.

3. The Boolean compactness theorem of the previous chapter says: The validity relation $\models_{CL}$ is *compact* in the following sense. If $X \models_{CL} A$—even if the set $X$ is infinitely large— then there is some *finite* subset $X' \subseteq X$ where $X' \models_{CL} A$. Prove this claim. Hint: Use the soundness and completeness theorems.

4. The disjunction used in this book, $\vee$, is inclusive, since $v(A \vee B) = 1$ if $v(A) = 1$ and $v(B) = 1$. One can give truth tables for *exclusive disjunction*, $\underline{\vee}$, which differs from $\vee$ when both disjuncts take value 1. It has the following truth table:

| $\underline{\vee}$ | 0 | 1 |
|---|---|---|
| 0 | 0 | 1 |
| 1 | 1 | 0 |

Provide introduction and elimination rules for $\underline{\vee}$. Prove that your rules are sound for the truth table for $\underline{\vee}$. Can you eliminate detours with your rules for $\underline{\vee}$?

5. Show that every argument that has a counterexample in $H_3$ also has a counterexample in $H_5$.

6. Show that minimal logic is not complete with respect to Heyting validity by providing an argument that does not have a proof in minimal logic but also does not have a counterexample in any Heyting algebra.

7. Consider the rules $\neg E$, $\rightarrow I$, and $\rightarrow E$. For each of these rules, either show that it is sound for K3-validity or provide a K3-counterexample. Then, for each rule, either show that it is LP-valid or provide an LP-counterexample.

8. Arthur Prior (1961) introduced the binary connective tonk, for which we will use the symbol "$\pitchfork$." Prior specified the meaning of tonk by providing rules for it.

$$\frac{A}{A \pitchfork B} \; \pitchfork I \qquad \frac{A \pitchfork B}{B} \; \pitchfork E$$

---

73. Thanks to Dave Ripley for sharing this exercise with us.

First, show that the addition of these rules results in a nonconservative extension of intuitionistic logic, which is to say there is an argument $X \succ A$, where tonk does not occur in $X$ or in $A$, such that there is a proof using the tonk rules but $X \nvdash_i A$ (there is no proof without the tonk rules). Next, show that you cannot provide a two-valued truth table for tonk.

9. Say that an argument $X \succ A$ is valid in a class of Heyting algebras, if for any valuation $v$ on an algebra $H$ from that class, $v(X) \leq v(A)$. Show that intuitionistic logic is not complete for the class of $n$-valued Heyting algebras, for any $n \geq 1$. Hint: Find an argument that is valid in that class but lacks a proof in intuitionistic logic. You can appeal to the fact that intuitionistic logic is valid in the class of all Heyting algebras, if needed.

**II** MODAL LOGIC

# 7 Necessity & Possibility

When we introduced Boolean valuations, we introduced them by explaining how they model situations. These situations serve as counterexamples to invalid arguments. If we have a valuation that assigns "true" to each member of the set $X$ of formulas and "false" to $A$, then we can see that the argument from $X$ to $A$ is invalid because the valuation specifies a way to make every member of $X$ true while not making $A$ true too. If the world happened to be like the situation described by that valuation, the premises would be true and the conclusion would be false. So, we need more than the truth of the premises to ensure the truth of the conclusion.

In this explanation, we used words like *"ensure"* and *"happened to be like."* We are not talking merely about what *is* true and *is* false, but different ways things *could* be and ways things *must* be. These are *modal* notions: notions of possibility and necessity.

A natural way to understand the relationship between valuations is that each valuation represents a *possibility*. The concept of possibility and its partner concept, *necessity*, are important for philosophy beyond the study of logic, so in this next section of our course, we will spend some time looking at these concepts and the logical methods that have been introduced to clarify and study them.

To start thinking about necessity and possibility, we will use an idea due to the Enlightenment philosopher, Gottfried Leibniz.[74] In Leibniz's view, it is necessary that $p$ if and only if $p$ is true in all possible worlds. It is possible that $p$ iff $p$ is true in some possible world. We move between the modal notions of necessity and possibility and the notion of truth *in a world*. Necessity is truth in every world (including this one), and possibility is truth in some world or other (maybe this one, or maybe some other possible world). In the rest of this chapter, we will formalize this analysis of necessity and possibility and explore some of its consequences.

## 7.1 Possible Worlds Models

We begin by adding two new operators to our language: □ (for necessity) and ◊ (for possibility).

---

74. Gottfried Wilhelm Leibniz (1646–1716). He was influential in the development of the infinitesimal calculus, in other areas of science, and in philosophy.

**Definition 40 (Modal language)** *The formulas of our modal language* MForm *are given in the usual way. We start with our set* Atom *of* ATOMIC FORMULAS *(so each atomic formula, including* ⊥, *is a formula). To make new formulas out of old formulas, if A and B are formulas of* MForm, *then we can build new formulas of* MForm *like this:*

$$(A \wedge B) \qquad (A \vee B) \qquad (A \rightarrow B) \qquad \neg A \qquad \Box A \qquad \Diamond A$$

*These are also formulas of* MForm, *no matter what formulas A and B are. Nothing else is in* MForm.

The box ($\Box$) and diamond ($\Diamond$) are *operators* in the same way that negation is. They bind tightly, so to say that a conditional $A \rightarrow B$ is necessary, we don't write

$$\Box A \rightarrow B$$

because this formula is the conditional with the antecedent $\Box A$ and consequent $B$. The formula $\Box A \rightarrow B$ says that if $\Box A$ is true, then so is $B$. Here, the conditional is the main connective in the same way that the conditional is the main connective of $\neg A \rightarrow B$. If we wish to say that $A \rightarrow B$ is necessary, we write

$$\Box(A \rightarrow B)$$

which says that the conditional $A \rightarrow B$ holds necessarily. Similarly, if we wish to say that the conjunction $A \wedge B$ is possible—to say that $A$ and $B$ are *jointly possible*, or can be true together—we write

$$\Diamond(A \wedge B)$$

and not $\Diamond A \wedge B$, because this second formula is a conjunction of $\Diamond A$ with $B$. It says that $A$ is possible and that $B$ is true.

Let's start thinking about how Boolean valuations can be used to model possibility and necessity, using Leibniz's idea that something is necessary iff it is true in every possible world, and something is possible iff it is true in some possible world. Consider four different valuations, for the four different combinations of values for the statements $p$ and $q$. If we look at the truth table for the three statements $q \wedge \neg q$, $p \vee q$, and $p \vee \neg p$, we see

| $p$ | $q$ | $q$ | $\wedge$ | $\neg$ | $q$ | $p$ | $\vee$ | $q$ | $p$ | $\vee$ | $\neg$ | $p$ |
|---|---|---|---|---|---|---|---|---|---|---|---|---|
| 0 | 0 | 0 | 0 | 1 | 0 | 0 | 0 | 0 | 0 | 1 | 1 | 0 |
| 0 | 1 | 1 | 0 | 0 | 1 | 0 | 1 | 1 | 0 | 1 | 1 | 0 |
| 1 | 0 | 0 | 0 | 1 | 0 | 1 | 1 | 0 | 1 | 1 | 0 | 1 |
| 1 | 1 | 1 | 0 | 0 | 1 | 1 | 1 | 1 | 1 | 1 | 0 | 1 |

that $q \wedge \neg q$ is true in no valuations, that $p \vee q$ is true in some but not all, and that $p \vee \neg p$ is a true in all valuations. If we take Leibniz's idea seriously and think of each of these valuations as a possibility or a possible world (a way that the world *might be*), we can represent the possibilities in a diagram.

In this diagram, we have depicted each world as an ellipse, with a label (its name, $w_1$, $w_2$, $w_3$, $w_4$) containing the atoms or negations of atoms that are true in that world. These formulas tell us how the atoms in our language are evaluated at the world and, from these formulas, figure out the truth or falsity of any formula involving just $\wedge$, $\vee$, $\neg$, and $\rightarrow$.

We will use atoms and negations of atoms so frequently in what follows that we'll use a new name for the concept.

**Definition 41 (Literals)**  *A* LITERAL *is an atom (apart from $\bot$), such as p, q, r, and so on, or its negation, $\neg p$, $\neg q$, $\neg r$, and so on. Nothing else is a literal.*

So, in a world diagram like this one, we list each world together with the literals true at that world. Once we have done that, if we're interested in evaluating other formulas at worlds, we can list formulas or their negations beneath each world, according to the truth value of the formula at that world. So, to check the status of $q \wedge \neg q$, $p \vee q$, and $p \vee \neg p$, we have:

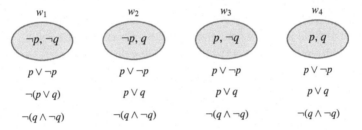

This much is just rewriting the truth table in a different way. But we can do more. Using the idea that a statement $\Box A$ is true at a world if $A$ is true at *all* of the worlds, we can evaluate $\Box(p \vee \neg p)$ and $\Box(p \vee q)$:

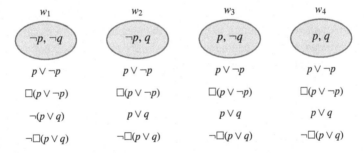

$\Box(p \vee \neg p)$ is true at *all* worlds, since $p \vee \neg p$ is true at all worlds, and $\Box(p \vee q)$ is true at *no* worlds since $p \vee q$ is not true at all worlds.

To evaluate claims of the form $\Diamond A$, we check whether $A$ is true at *some* world. We can evaluate $\Diamond(p \vee q)$ and $\Diamond(q \wedge \neg q)$, as in the following diagram.

$$w_1 \qquad w_2 \qquad w_3 \qquad w_4$$

$$\boxed{\neg p, \neg q} \qquad \boxed{\neg p, q} \qquad \boxed{p, \neg q} \qquad \boxed{p, q}$$

$$\neg(p \lor q) \qquad p \lor q \qquad p \lor q \qquad p \lor q$$

$$\Diamond(p \lor q) \qquad \Diamond(p \lor q) \qquad \Diamond(p \lor q) \qquad \Diamond(p \lor q)$$

$$\neg(q \land \neg q) \qquad \neg(q \land \neg q) \qquad \neg(q \land \neg q) \qquad \neg(q \land \neg q)$$

$$\neg\Diamond(q \land \neg q) \qquad \neg\Diamond(q \land \neg q) \qquad \neg\Diamond(q \land \neg q) \qquad \neg\Diamond(q \land \neg q)$$

So here, $\Diamond(p \lor q)$ is true everywhere because $p \lor q$ is true somewhere, $w_2$, and since $q \land \neg q$ is true *nowhere*, $\Diamond(q \land \neg q)$ is also true nowhere.

So far, we have been dealing with a model where we have one world for each different valuation on the atoms $p$ and $q$. This is simply *one* way to evaluate formulas. Suppose $p$ and $q$ were statements that—because of what they mean, or for some other reason—couldn't be true together. (Maybe $p$ is "it's Tuesday" and $q$ is "it's Thursday.") In that case, we don't have a world like $w_4$ where $p$ and $q$ are both true. A better way to interpret *this* language would be to use three worlds, like this, rather than four:

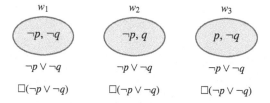

In *this* interpretation—with three possible worlds instead of the four of the previous example—every world makes $\neg p \lor \neg q$ true, and so, $\Box(\neg p \lor \neg q)$ holds at every world, too.

These examples should be enough to give you the idea that this is a much more general phenomenon. We can evaluate statements as not simply true or false, but as true or false over a range of different locations, which we here call *possible worlds*. Here is the general definition of a possible worlds model.

**Definition 42 (Possible worlds model)** *A* POSSIBLE WORLDS MODEL *is a pair* $\langle W, V \rangle$, *where $W$ is a nonempty set of* POSSIBLE WORLDS *and $V$ is a function that gives a truth value (from* $\{0, 1\}$*) to each choice of an atomic formula and a world, with $V(\bot, w) = 0$ for all worlds $w$. That is, if $p \in$ Atom and $w \in W$, $V(p, w)$ is 0 or 1. If $V(p, w) = 1$, we say that $p$ is true at $w$, and if $V(p, w) = 0$, we say that $p$ is false at $w$.*

These models are both like and unlike Boolean valuations. As with Boolean valuations, the function $V$ can be extended to give truth values to all formulas. What is different is that the assignment of a truth value is not absolute but is relative to a world. An atom might be true at one world and false at another. Here is how the values of complex formulas are determined. The clauses for the classical connectives remain the same, relative to the choice of a world.

$$V(A \land B, w) = 1 \quad \text{iff} \quad V(A, w) = 1 \text{ and } V(B, w) = 1$$
$$V(A \lor B, w) = 1 \quad \text{iff} \quad V(A, w) = 1 \text{ or } V(B, w) = 1$$
$$V(A \to B, w) = 1 \quad \text{iff} \quad V(A, w) = 0 \text{ or } V(B, w) = 1$$
$$V(\neg A, w) = 1 \quad \text{iff} \quad V(A, w) = 0$$
$$V(\bot, w) = 0 \qquad \text{always}$$

The novelty comes with the clauses for the modal operators.

$$V(\Box A, w) = 1 \quad \text{iff} \quad V(A, v) = 1 \text{ for all } v \in W$$
$$V(\Diamond A, w) = 1 \quad \text{iff} \quad V(A, v) = 1 \text{ for some } v \in W$$

In some books, you will see a different notation for possible worlds valuation functions. Instead of "$V(A, w) = 1$," you might see "$w \Vdash_V A$,"[75] and instead of "$V(A, w) = 0$," you would then see "$w \nVdash_V A$."[76]

Let's look at another example of a possible worlds model, and let's use it to evaluate some formulas.

**Example 6**  *Take this possible worlds model: it's the pair* $\langle W, V \rangle$ *where* $W = \{w, x, y\}$ *and where we define the function V like this:* $V(p, w) = 0$, $V(p, x) = V(p, y) = 1$, *and* $V(q, w) = V(q, x) = V(q, y) = 1$.

All of this information can be represented in a diagram like this:

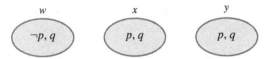

(Notice that $V$ assigns $p$ and $q$ the same values at the worlds $x$ and $y$. That is allowed in the definition of a possible worlds model. Nothing requires that different worlds evaluate formulas differently.)

Let's find the value of $\neg p \land \Diamond p$ at $w$. Since $V(p, w) = 0$, $V(\neg p, w) = 1$. To evaluate $\Diamond p$ at $w$ (or at any world), we need to consider the value of $p$ at all possible worlds. Since $V(p, x) = 1$, $V(\Diamond p, w) = 1$. So, $V(\neg p \land \Diamond p, w) = 1$. We could summarize this reasoning and fill in the values of $\neg p \land \Diamond p$ at the other worlds, like this:

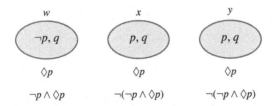

In the same model, let's consider $\Box p \lor \Box \neg p$ at $w$. Since $V(p, w) = 0$, it is not the case that, for all $u \in W$, $V(p, u) = 1$, so $V(\Box p, x) = 0$. Similarly, as $V(\neg p, x) = 0$, it is not the case that

---

75. Yes, logicians love their turnstiles. "⊢" and "⊨" relate sets of formulas on the left with individual formulas on the right, while "⊩" relates *worlds* on the left to formulas on the right. Humberstone (1988) investigates the logic of heterogeneous relations.

76. "$w \Vdash A$" is sometimes read as "$w$ verifies $A$" or "$A$ holds at $w$."

$V(\neg p, u) = 1$, for all $u \in W$, so $V(\Box \neg p, x) = 0$. Note that while $\Box p \vee \Box \neg p$ is not true at $x$, $\Box p \vee \neg \Box p$ is true at $x$. Filling in these values at each world, we have:

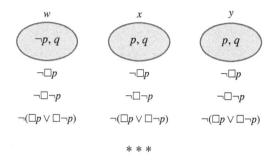

$$* * *$$

The *point* of models for our language is to find counterexamples to arguments involving modal concepts. We can use these models in exactly the same way we used Boolean models in the previous chapters, to define validity. So, let's do this.

## 7.2 Counterexamples and Validity

**Definition 43 (Counterexample, validity)**  *A possible worlds model* $\langle W, V \rangle$ *is a* COUNTEREX-AMPLE *to the argument* $X \succ A$ *iff there is a world* $w \in W$ *such that* $V(X, w) = 1$ *and* $V(A, w) = 0$, *where* $V(X, w) = 1$ *means that for all* $B \in X$, $V(B, w) = 1$.
    *The argument* $X \succ A$ *is* VALID *according to possible worlds models, iff there is no possible worlds model* $\langle W, V \rangle$ *that is a counterexample to it.*

In other words, a counterexample to an argument $X \succ A$ is a possible worlds model with a world where each member of $X$ is true and where $A$ is false. Let's see if the argument $\Box(p \vee q) \succ \Box p \vee \Box q$ has a counterexample. To find a counterexample for this argument, we want a model with a world where $\Box(p \vee q)$ is true and where $\Box p \vee \Box q$ is false. In other words, we want a model that looks like *this*.

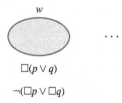

Is there any model with a world (which we're calling $w$ here) where $\Box(p \vee q)$ is true and $\Box p \vee \Box q$ is false? For $\Box(p \vee q)$ to be true, we need $p \vee q$ true at each world, and for the disjunction $\Box p \vee \Box q$ to be false, we need $\Box p$ to be false and $\Box q$ to be false. So let's write those in:

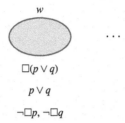

Now, since □$p$ is false at $w$, this means that $p$ isn't true at *every* world. We need to have some world where $p$ is false. Similarly, since □$q$ is false at $w$, this means that $q$ isn't true at every world. So we also need to have some world where $q$ is false. So, let's write these in. They don't have to be the same worlds, so let's add two new worlds: one where $p$ is false and one where $q$ is false. So, our picture expands into this:

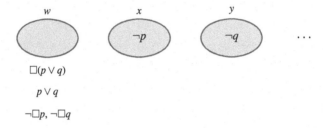

The first thing to notice is that since □($p \lor q$) is true at $w$, we need to add $p \lor q$ at the new worlds $x$ and $y$, so let's draw that in:

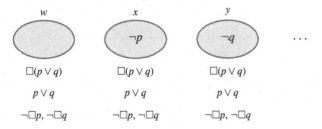

The second thing to notice is that since we want $p \lor q$ true in each world, we need to make one of $p$ and $q$ true at $w$, $x$, and $y$. The choice is forced on us for $x$ and $y$, but there are options for $w$.

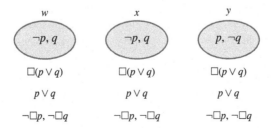

And now we are done, as we can pick any world for the premise to be true and the conclusion false. If $w$ and $x$ have $p$ false and $q$ true, and $y$ has $p$ true and $q$ false, we have made sure that $p \vee q$ is necessary, while $\Box p \vee \Box q$ is false. In fact, when we look at the picture we've drawn, we see that world $w$ adds nothing to the picture. (Everything we need true at $w$ is true at $x$ and at $y$.) So, we can get rid of $w$ and have a simple model with just two worlds:

Here the set of worlds is $W = \{x, y\}$; $V(p, x) = 0$, $V(p, y) = 1$, and $V(q, x) = 1$, $V(q, y) = 0$. The disjunction $p \vee q$ is true at both worlds, so $\Box(p \vee q)$ is true everywhere. Since $p$ is false at $x$, $\Box p$ is true nowhere. Since $q$ is false at $y$, $\Box q$ is true nowhere. So, $\Box p \vee \Box q$ is false everywhere too. So, we have a counterexample to our argument. World $x$ (as one example) has $\Box(p \vee q)$ true and $\Box p \vee \Box q$ false.

* * *

That was one example of how to construct a counterexample to a modal argument. Here is a general procedure you can follow to construct counterexamples for yourself.

- To find a counterexample for $X \succ A$, draw a world, and under it, write each formula in $X$ and $\neg A$. Your goal is to find a way to make each of the formulas you have written down true.
- Whenever you have written down a *literal* under a world, write it *inside* the world as well.
- If you write down a formula of the form $A \wedge B$ under a world, write down $A$ and $B$ under that world too.
- If you write $\neg(A \vee B)$, write down $\neg A$ and $\neg B$, too.
- If you write down $\neg(A \to B)$, write down $A$ and $\neg B$, too.
- If you write down $\neg\neg A$, write down $A$, too.
- If you write down $\Box A$, write down $A$ too, and add $A$ to every other world you have added.
- If you write down $\neg\Diamond A$, write down $\neg A$ too, and add $\neg A$ to every other world you have added.[77]
- If you write down $\Diamond A$, check if you already have a world where $A$ is true. If you don't yet have one, add a new world, and make $A$ true there.
- If you write down $\neg\Box A$, check if you already have a world where $\neg A$ is true. If you don't yet have one, add a new world, and make $\neg A$ true there.
- If you write down $A \vee B$ under a world (and you can't find anything else to do first), it looks like you should try two options: (1) try making $A$ true here or (2) try making $B$ true here.
- If you write down $\neg(A \wedge B)$ under a world (and you can't find anything else to do first), it looks like you should try two options: (1) try making $\neg A$ true here or (2) try making $\neg B$ true here.

---

77. Notice that $\neg\Diamond A$ functions just like $\Box\neg A$ and $\neg\Box A$ like $\Diamond\neg A$. We will return to these points below.

- If you write down $A \rightarrow B$ under a world (and you can't find anything else to do first), it looks like you should try two options: (1) try making $\neg A$ true here or (2) try making $B$ true here.
- If you ever have to write $A$ and $\neg A$ in the same world, stop writing formulas under that world. This option is impossible, so go on to the others. If no options are left, there is no counterexample to this argument.
- If you ever have to write $\bot$ in a world, stop writing formulas under that world. This option is impossible, so go on to the others. If no options are left, there is no counterexample to this argument.
- If you have no more formulas to process, your construction is over, and you have a counterexample to the argument.

Phew! That's a long description of the process of constructing a counterexample. Let's try this to construct a counterexample for another argument, this time for the argument $\Box(p \rightarrow q), p \succ \Box q$. Here is the start of the process:

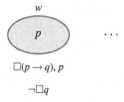

The rule for $\Box$ tells us that since $\Box(p \rightarrow q)$ is true here, $p \rightarrow q$ is true in every world we add. So, we'll add that.

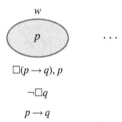

Now, the rule for $p \rightarrow q$ tells us to make a choice, so we'll check if there's anything else to do first—and there is. Since $\neg \Box q$ is true at $w$, we need some world where $q$ is false. Since we don't already have a world where $q$ is false, we'll add a new one:

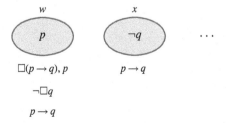

and since $x$ is a new world, we'll add $p \rightarrow q$ to it, like we did to $w$. Now, the only things left to check are the $p \rightarrow q$ at each world. Here, the rule tells us to either make $p$ false or $q$ true.

In $w$, we can't make $p$ false (it's already true!), so we make $q$ true. In $x$, on the other hand, we can't make $q$ true (it's already false!), so we must make $p$ false. This is the picture we end up with:

Here, at world $w$, we have $p \to q$ true since $q$ is true, while at $x$, $p \to q$ is true since $p$ is false. In either case, we have $p \to q$, so $\Box(p \to q)$ is true at both worlds. At $w$, we have $p$, too. But we don't have $\Box q$ at $w$, since although $q$ is *true*, it isn't *necessary*. So, our model provides a counterexample to $\Box(p \to q), p \succ \Box q$ at world $w$. Here, $\Box(p \to q)$ and $p$ are both true, while $\Box q$ is false.

* * *

This logic of possible worlds models has been extensively studied. It has a standard name, S5, which comes from the naming conventions of C. I. Lewis.[78] Lewis defined several systems of modal logic, the most commonly studied of which are S4 and S5. We will use the notation $X \vDash_{S5} A$ for validity defined using possible worlds models, and we will also call this S5 validity. We will also say that $A$ is an S5 tautology if $\vDash_{S5} A$. Here are some examples of S5 valid arguments:

**Fact 1**  1. $\Box A \vDash_{S5} \neg\Diamond\neg A$,  $\neg\Diamond\neg A \vDash_{S5} \Box A$.
2. $\Diamond A \vDash_{S5} \neg\Box\neg A$,  $\neg\Box\neg A \vDash_{S5} \Diamond A$.

*Proof.* Let $\langle W, V \rangle$ be a possible worlds model with $w \in W$. For (1), we argue as follows.

$$
\begin{aligned}
V(\Box A, w) = 1 \quad &\text{iff} \quad \text{for all } u \in W, V(A, u) = 1 \\
&\text{iff} \quad \text{for all } u \in W, V(\neg A, u) = 0 \\
&\text{iff} \quad \text{it is not the case that for some } u \in W, V(\neg A, u) = 1 \\
&\text{iff} \quad V(\Diamond\neg A, w) = 0 \\
&\text{iff} \quad V(\neg\Diamond\neg A, w) = 1
\end{aligned}
$$

Because each line is an "iff," the proof goes in both directions. So, we have shown in any model at all, and in any world in that model, $\Box A$ is true at that world iff $\neg\Diamond\neg A$ is true at that world. For (2), we argue in the same way.

$$
\begin{aligned}
V(\Diamond A, w) = 1 \quad &\text{iff} \quad \text{for some } u \in W, V(A, u) = 1 \\
&\text{iff} \quad \text{it is not the case that for all } u \in W, V(A, u) = 0 \\
&\text{iff} \quad \text{it is not the case that for all } u \in W, V(\neg A, u) = 1 \\
&\text{iff} \quad V(\Box\neg A, w) = 0 \\
&\text{iff} \quad V(\neg\Box\neg A, w) = 1
\end{aligned}
$$

Thus, in the same way, in any model at all, and in any world in that model, $\Diamond A$ is true at that world iff $\neg\Box\neg A$ is true at that world. $\qquad\qquad \Box$

---

78. Clarence Irving Lewis (1883–1964) was an American logician who made important contributions to modal logic and also worked on epistemology and pragmatism. He is famous for his work on modal logic in his *A Survey of Symbolic Logic* (1918) and his *Symbolic Logic* (1932), coauthored with Cooper Langford.

Necessity and possibility also interact in distinctive ways with conjunction and disjunction.

**Fact 2**   1. $\Box(A \land B) \models_{S5} \Box A \land \Box B$, and $\Box A \land \Box B \models_{S5} \Box(A \land B)$.
   2. $\Diamond(A \lor B) \models_{S5} \Diamond A \lor \Diamond B$, and $\Diamond A \lor \Diamond B \models_{S5} \Diamond(A \lor B)$.

*Proof.* Consider a possible worlds model $\langle W, V \rangle$. For (1), suppose we had a counterexample at $w$, that is, we have $V(\Box(A \land B), w) = 1$ but $V(\Box A \land \Box B, w) = 0$. Then, by the conjunction evaluation clause, $V(\Box A, w) = 0$ or $V(\Box B, w) = 0$. Suppose $V(\Box A, w) = 0$, then for some $u \in W$, $V(A, u) = 0$. The assumption that $V(\Box(A \land B), w) = 1$ implies $V(A \land B, x) = 1$, for every $x \in W$, in particular for $u$. Thus, $V(A, u) = 1$, which is a contradiction. The case for $V(\Box B, w) = 0$ is similar. As both cases lead to a contradiction, we conclude that the initial assumption is false. Thus, $\Box(A \land B) \models_{S5} \Box A \land \Box B$.

For the other direction of (1), suppose that $V(\Box(A \land B), w) = 0$ and $V(\Box A \land \Box B, w) = 1$. The second part of the assumption is equivalent to $V(\Box A, w) = 1$ and $V(\Box B, w) = 1$, by the conjunction evaluation clause. The first part of the assumption implies that for some $u \in W$, $V(A \land B, u) = 0$, which implies either $V(A, u) = 0$ or $V(B, u) = 0$. There are two cases. Suppose that $V(A, u) = 0$. This contradicts the fact that $V(\Box A, w) = 1$. The case for $V(B, u) = 0$ is similar. Both cases lead to a contradiction, so the initial assumption is false. Thus, $\Box A \land \Box B \models_{S5} \Box(A \land B)$.

The argument for (2) is similar to the argument for (1). □

One way to put the preceding facts is to say that necessity distributes over conjunction and possibility distributes over disjunction. Necessity does not distribute over disjunction. Since $p \lor \neg p$ is a tautology, it is true in every world of every model. So, $\Box(p \lor \neg p)$ is true in every world of every model. Take a model $\langle W, V \rangle$ with $W = \{w, v\}$, $V(p, w) = 1$, and $V(p, v) = 0$. This model has $V(\neg p, v) = 1$, so both $V(\Box p, w) = 0$ and $V(\Box \neg p, w) = 0$. Similarly, possibility does not distribute over conjunction. One can have a $\Diamond p$ and $\Diamond q$ but not have $\Diamond(p \land q)$. To see this, add to the previous model that $V(q, w) = 0$ and $V(q, v) = 1$.

* * *

The modal operators $\Box$ and $\Diamond$ can be iterated. It makes sense not only to say that $p$ is necessary or that it is possible, but that it is *necessarily* necessary or *possibly* necessary. We can consider formulas such as $p \rightarrow \Box \Diamond p$.

We will evaluate this in the model from the earlier example. Take the possible worlds model $\langle W, V \rangle$ with $W = \{w, x, y\}$ and $V(p, w) = 0$, $V(p, x) = 1$, and $V(p, y) = 1$, depicted like this:

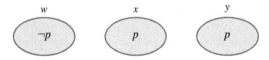

What is $V(p \rightarrow \Box \Diamond p, y)$? Since $V(p, y) = 1$, we need to work out the value of the consequent, $\Box \Diamond p$ at $y$. $V(\Box \Diamond p, y) = 1$ iff for all $u \in W$, $V(\Diamond p, u) = 1$. $V(\Diamond p, u) = 1$ is the case iff for some $z \in W$, $V(p, z) = 1$. As $V(p, y) = 1$, this gives us $V(\Diamond p, u)$, for each $u \in W$. Therefore, $V(\Box \Diamond p, y) = 1$. Thus, $V(p \rightarrow \Box \Diamond p, y) = 1$. In the picture, we have this:

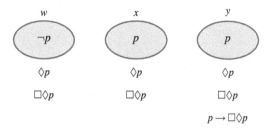

$$p \rightarrow \Box \Diamond p$$

listing at each world the formulas that have figured in our proof and get value 1 at that world, and indeed, $p \rightarrow \Box \Diamond p$ is true at every world in this model, not just at $y$.

* * *

Let's call formulas of the form $\Box B$, $\Diamond B$, $\neg \Box B$, or $\neg \Diamond B$ MODAL FORMULAS. A useful fact about modal formulas in our possible worlds models is that if one is true at any world in a model, it is true at every world in the model.

**Lemma 14**   *Let $\langle W, V \rangle$ be a possible worlds model. Suppose that $A$ is a modal formula and there is a world $w \in W$ such that $V(A, w) = 1$. Then for all worlds $u \in W$, $V(A, u) = 1$.*

*Proof.*   There are four cases, one for each form the modal formula $A$ can have. They are pretty similar, so we will do the $\Box$ and $\Diamond$ cases.

Suppose $A$ is of the form $\Box B$. $V(\Box B, w) = 1$ by assumption. Let $x$ be an arbitrary world in $W$.

$$V(\Box B, w) = 1 \quad \text{iff} \quad \text{for all } u \in W, V(B, u) = 1$$
$$\text{iff} \quad V(\Box B, x) = 1$$

Since the world $x$ is arbitrary, we can conclude that for all worlds $u \in W$, $V(\Box B, u) = 1$.

Suppose $A$ is of the form $\Diamond B$. $V(\Diamond B, w) = 1$ by assumption. Let $x$ be an arbitrary world in $W$.

$$V(\Diamond B, w) = 1 \quad \text{iff} \quad \text{for some } u \in W, V(B, u) = 1$$
$$\text{iff} \quad V(\Diamond B, x) = 1$$

Since the world $x$ is arbitrary, we can conclude that for all worlds $u \in W$, $V(\Diamond B, u) = 1$.

The remaining two cases are similar and are left to the reader.                                            $\Box$

What is valid in S5? It turns out all arguments that are valid according to Boolean models (i.e., all classically valid arguments) are valid in S5. This is because each possible world acts just like a Boolean model, and Boolean models are possible worlds models where $W$ contains just a single world.

**Theorem 20**   *If $X \vDash_{CL} A$, then $X \vDash_{S5} A$.*

We have done enough to show that $\Box A \leftrightarrow \neg \Diamond \neg A$ and $\Diamond A \leftrightarrow \neg \Box \neg A$ are both S5 tautologies. Here are a few more tautologies distinctive of the logic S5.

**Theorem 21**   *The following are S5 tautologies.*

1. $A \rightarrow \Diamond A$
2. $\Box A \rightarrow \Box \Box A$
3. $A \rightarrow \Box \Diamond A$

4. $\Diamond A \to \Box\Diamond A$
5. $\Box(A \to B) \to (\Box A \to \Box B)$

*Proof.* We will prove only the first two here. The others are also straightforward to demonstrate.

For (1), let $\langle W, V \rangle$ be a possible worlds model and suppose that $w \in W$ with $V(A, w) = 1$. This means that there is a world $u$ such that $V(A, u) = 1$, namely, $w$, so by the evaluation clause for $\Diamond$, $V(\Diamond A, w) = 1$. Thus, $V(A \to \Diamond A, w) = 1$, as desired.

For (2), suppose there is a counterexample. So there is a model $\langle W, V \rangle$ and $w \in W$ with $V(\Box A, w) = 1$ and $V(\Box\Box A) = 0$. As $V(\Box\Box A) = 0$, there is some world $u \in W$ such that $V(\Box A, u) = 0$. This is the case iff there is some world $x \in W$ such that $V(A, x) = 0$. This contradicts the assumption that $V(\Box A, w) = 1$, which implies there is no world $x$ such that $V(A, x) = 0$. Therefore, the supposed counterexample does not exist. $\qquad\Box$

There's more we can say about S5 validity. Suppose that $A$ is S5 valid. Is $\Box A$ also S5 valid? Suppose $\langle W, V \rangle$ is a possible worlds model. Since $A$ is S5 valid, $A$ is true in all worlds in the model. This means that $\Box A$ is true at some world of the model. By lemma 14, $\Box A$ is true at all worlds of the model. So, if $A$ is S5 valid, then $\Box A$ is S5 valid as well.

**Theorem 22**    *The following are true about S5 validity:*

1. *If $\models_{S5} A$, then $\models_{S5} \Box A$.*
2. *If $B_1, \ldots, B_n \models_{S5} A$, then $\Box B_1, \ldots, \Box B_n \models_{S5} \Box A$.*
3. *If $M_1, \ldots, M_n \models_{S5} A$, then $M_1, \ldots, M_n \models_{S5} \Box A$, where each of the $M_i$ is a modal formula.*

*Proof.* The proof for (1) was presented informally above.

For (2), suppose that the antecedent is true and the consequent false, so $B_1, \ldots, B_n \models_{S5} A$ but that $\Box B_1, \ldots, \Box B_n \not\models_{S5} \Box A$. There is then a counterexample, a model $\langle W, V \rangle$ and world $w \in W$ with $V(\Box B_i, w) = 1$, for $1 \leq i \leq n$, and $V(\Box A, w) = 0$. Since each premise, $\Box B_i$, is true, each $V(B_i, u) = 1$ for every $u \in W$. Since $B_1, \ldots, B_n \models_{S5} A$, it follows that $V(A, u) = 1$, for each world $u \in W$. By the evaluation clause for $\Box$, this means that $V(\Box A, w) = 1$, but this contradicts $V(\Box A, w) = 0$. Thus, there is no counterexample.

For (3), suppose that the antecedent is true and the consequent is false, so $M_1, \ldots, M_n \models_{S5} A$, but $M_1, \ldots, M_n \not\models_{S5} \Box A$, where each $M_i$ is a modal formula. Then, there is a counterexample $\langle W, V \rangle$ with $w \in W$ such that $V(M_i, w) = 1$, for $1 \leq i \leq n$, and $V(\Box A, w) = 0$. Since $V(\Box A, w) = 0$, there is some world $u \in W$ such that $V(A, u) = 0$. By lemma 14, $V(M_i, u) = 1$, at all $u \in W$ and for $1 \leq i \leq n$. By the assumption that $M_1, \ldots, M_n \models_{S5} A$, it follows that $V(A, u) = 1$, which contradicts the assumption, so there is no counterexample. $\qquad\Box$

## 7.3  Two Applications

We've seen modal models and validity defined using those models. In this section, we're going to look at two different ways these techniques can be used.

### 7.3.1  Strict conditionals and ambiguities

C. I. Lewis did not start with the necessity and possibility operators when he introduced modal logic. Instead, he focused on different forms of *strict conditionals*, or strict implications. His modal logic was designed primarily as a way to analyze conditionality that is more discriminating than the conditional of classical logic. After all, a conditional expressed in ordinary English does not seem to be true when the antecedent is false or when the consequent is true. Yet, that is how $A \rightarrow B$ works in classical logic. For this reason, there is plenty of reason to look for a conditional notion that requires more than the falsity of the antecedent or the truth of the consequent in order to be true. Lewis's strict conditional is one such analysis. Lewis used his strict conditional to define necessity and possibility. In terms of our modal operators, the strict conditional, $\dashv 3$ (read: fishhook), of S5 can be defined like this:

- $A \dashv 3 B$ is defined as $\Box(A \rightarrow B)$, or equivalently $\Box(\neg A \vee B)$.

Since the strict conditional is definable out of our other logical vocabulary, we don't need to add it as a new connective. The truth condition for the strict conditional we obtain from this definition is

- $V(A \dashv 3 B, w) = 1$ iff for all $x \in W$, either $V(A, x) = 0$ or $V(B, x) = 1$.

The values of $A$ and $B$ at a single world may not be enough to determine the value of $A \dashv 3 B$ at that world. If $V(A, w) = 0$ or $V(B, w) = 1$, we will not yet be able to determine $V(A \dashv 3 B, w)$, which requires checking values at other worlds. If $V(A, w) = 1$ and $V(B, w) = 0$, however, then we know that $V(A \dashv 3 B, w) = 0$. This gives some force to the idea that $\dashv 3$ is a *strict* implication, since it requires not only that $A$ (classically) implies $B$ at a single world but that it requires that $A$ (classically) implies $B$ at every world. The strict conditional tells us what would be the case when the antecedent is true, as opposed to the classical conditional, which tells us merely what is the case when the antecedent is true.

As one would expect, the strict conditional is quite unlike the classical conditional. One difference is in the weakening law we saw before. While $p \models_{S5} q \rightarrow p$ holds, $p \not\models_{S5} q \dashv 3 p$. The latter invalidity can be seen with a model $\langle W, V \rangle$ such that $W = \{w, u\}$ with $V(p, w) = 1$, $V(q, u) = 1$, and $V(p, u) = 0$. Then, $V(q \rightarrow p, u) = 0$, so $V(q \dashv 3 p, w) = 0$.

$$* * *$$

The presence of modal operators—and the strict conditional—lets us distinguish two things that one might mean by a sentence like

- If Bug is a cat, then, necessarily, she likes tuna.

There are two ways of rendering this, taking $c$ to represent "Bug is a cat" and $t$ to represent "Bug likes tuna." One is the strict conditional form and the other is the necessitated consequent form.

- $c \dashv 3 t$ (or $\Box(c \rightarrow t)$), and
- $c \rightarrow \Box t$.

These two statements are significantly different in meaning. Take the following model:

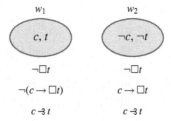

Here, in $w_1$, $c \rightarrow \Box t$ is false, while $c \rightarrow t$ is true. In every world where Bug is a cat, Bug likes tuna. However, it is *not* the case (in world $w_1$ at least) that if Bug is a cat, then the claim that Bug likes tuna is *necessarily* true. For while the statement that Bug likes tuna is true at $w_1$, it is not true at $w_2$, so it is not necessary.

The modal vocabulary provides tools to distinguish scope in English sentences clearly, and modal models give us some way to understand the significance of different readings. In a statement like "if $p$ then necessarily $q$," it means one thing to think of this as a conditional with $p$ in the antecedent and $\Box q$ in the consequent. It is another thing to understand it as a conditional in which the "necessarily" has *wide* scope, over the whole conditional. What does the English "if $p$ then necessarily $q$" *mean*? While we cannot give a definitive answer in this chapter, consideration of alternative circumstances can provide some guidance. If some alternative circumstance in which $p$ is true and $q$ isn't is enough to falsify "if $p$ then necessarily $q$" in *this* circumstance, then we have evidence that we should understand the conditional as something more like $\Box(p \rightarrow q)$ than as $p \rightarrow \Box q$.

### 7.3.2 Propositions

Possible worlds models provide an elegant way to represent *meaning*. They can be used to define *propositions*, which are useful as they can abstract away from a particular formula. It is a reasonable idea that sometimes different sentences can be used to say the same thing. If the job of a declarative sentence is to describe the way the world is, then two sentences $A$ and $B$ that differ in that $A$ is true at a world where $B$ is not must clearly differ in what they mean. Taking a *proposition* expressed by a declarative sentence to be the set of worlds where that sentence is true is a good first stab at modeling that sentence's meaning.[79] Let's work with this idea, using our models.

**Definition 44**    A PROPOSITION *in a possible worlds model* $\langle W, V \rangle$ *is a subset of W.*
    *In a possible worlds model* $\langle W, V \rangle$, *the proposition that A,* $\|A\|$, *is the set of worlds at which A gets value 1, that is,* $\|A\|$ *is the set* $\{w \in W : V(A, w) = 1\}$.

When we want to specify the model $M$ we are talking about, we can write $\|A\|_M$, for the proposition that $A$ in the model $M$.

How do propositions abstract away from formulas? Take $p \wedge q$ and $q \wedge p$. These are different formulas, but, in a sense, they say the same thing. They are true under the same conditions. The propositions $\|p \wedge q\|$ and $\|q \wedge p\|$ are identical in any given possible worlds

---

79. At the end of this section, we will see a reason for considering alternative definitions of propositions.

model, as we can see from the following reasoning:

$$
\begin{aligned}
w \in \|p \wedge q\| \quad &\text{iff} \quad V(p \wedge q, w) = 1 \\
&\text{iff} \quad V(p, w) = 1 \text{ and } V(q, w) = 1 \\
&\text{iff} \quad V(q, w) = 1 \text{ and } V(p, w) = 1 \\
&\text{iff} \quad V(q \wedge p, w) = 1 \\
&\text{iff} \quad w \in \|q \wedge p\|
\end{aligned}
$$

When two formulas $A$ and $B$ are logically equivalent in S5 (i.e., when $A \leftrightarrow B$ is an S5 tautology) $\|A\|$ will be identical to $\|B\|$ in all possible worlds models. In a particular model, formulas that are not logically equivalent may express the same proposition. The following diagram presents an example. In this possible worlds model $\langle W, V \rangle$, $W = \{w, v, u\}$ and $V$ is given by the diagram.

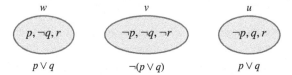

As we can see, $\|p \vee q\| = \{w, u\}$ and $\|r\| = \{w, u\}$, but it is not the case that $(p \vee q) \leftrightarrow r$ is an S5 tautology. A possible worlds model $\langle W, V' \rangle$ where $V'$ matches $V$ except that $V'(r, u) = 0$ and $V'(p, w) = 0$ would provide a counterexample to both directions of the biconditional.

The propositions that are being considered here lack any structure. They are simply sets of worlds. Because of this, the classical logical connectives have some striking connections to set-theoretic operations on propositions, which are summarized as follows for all models:

- $\|A \wedge B\| = \|A\| \cap \|B\|$
- $\|A \vee B\| = \|A\| \cup \|B\|$
- $\|\neg A\| = W - \|A\|$

Using these, one can work out the operation on propositions expressed by the conditional. What about operations for the modal operators? On our modal models, they are very simple.

$$
\|\Box A\| =
\begin{cases}
W & \text{if } \|A\| = W \\
\emptyset & \text{otherwise}
\end{cases}
$$

$$
\|\Diamond A\| =
\begin{cases}
W & \text{if } \|A\| \neq \emptyset \\
\emptyset & \text{otherwise}
\end{cases}
$$

The proposition expressed by a necessary truth in a model is the set of worlds $W$. This means that any two necessary truths express the same proposition. Similarly, any two true possibility claims express the same proposition. A true possibility claim and a true necessity claim, in a model, express the same proposition. These are the same propositions expressed by S5 tautologies in that model. In any two possible worlds models $\langle W, V \rangle$ and $\langle W, V' \rangle$, S5 tautologies express the proposition $W$, while a necessity claim may express different propositions in different models.

Some philosophers and logicians have thought that while unstructured propositions model *some* of the features of propositions and meaning, they abstract too much away to be a comprehensive model of meaning. On the common view that mathematical truths are necessary, it turns out that the claim $2 + 2 = 4$ expresses the same proposition as the first incompleteness theorem, a groundbreaking fact about arithmetic proved by the logician Kurt Gödel.[80] Since you know that $2 + 2 = 4$, you know the proposition expressed by that claim, which is just what the first incompleteness theorem expresses. Some philosophers have wanted more structure in propositions as a way to distinguish different necessary and logical truths. We will not be able to explore these options here, but we do want to register that unstructured propositions, while useful, do not model *everything* there is to say about meaning.[81]

### 7.4  More Modal Models

This final section of the chapter on models for modal logic is a little more advanced than the previous sections. It looks at two different ways to generalize our possible worlds models to draw more modal distinctions than our original models can allow.

#### 7.4.1  Another notion of necessity

The Leibnizian necessity of S5 is not the only notion of necessity studied in modal logic, although it is one of the most commonly discussed in philosophy. We can modify possible worlds models slightly to get another interesting and important kind of necessity. We will call these S4 models, since the logic they validate is known as S4.[82] The key idea in these models is that worlds can differ in not only what they take to be *true* but also what they take to be *necessary* and what they take to be *possible*. This can't happen in our Leibnizian models, because possibility and necessity are a matter of what is true in our worlds, and each world makes reference to the same set of worlds. What makes things possible from the point of view of world $w$ makes things possible from the point of view of world $v$, too.

Well, things don't have to be that way. We could say that some world $v$ is possible from the point of view of world $w_1$ but not possible from the point of view of $w_2$. If $w_1$ is the case, $v$ is possible. If $w_2$ is the case, $v$ is not possible. Such a notion of *relative possibility* seems to be the way to make sense of choice and regret. Imagine you have a choice of only one of two activities for the evening: staying home for dinner with friends or going out to see a band that is playing in Melbourne for one night only. If you have dinner with friends, some things are no longer possible (like going to the concert or later *remembering* going to the concert) that would have been possible had other choices been made.

---

80. Kurt Gödel (1906–1978) was a twentieth-century logician who proved many important results in logic. His incompleteness theorems are his most widely known results. See https://plato.stanford.edu/entries/goedel -incompleteness/ for an overview.

81. For more on the motivations for structured propositions and some of the options for the structure, see Hanks (2009), Jespersen (2012), or King (2019).

82. This name is due to C. I. Lewis. Lewis also defined logics S1, S2, and S3, but they are less prominent than S4 and S5. Mares (2016) explores Lewis's motivations for some of his weaker logics, including his preferred logics, S2 and S3.

This is a key idea that motivates other kinds of models for modal logics. In these models, we keep track of which worlds are possible, from the point of view of other worlds. In a model, we have not only a set of worlds and a valuation function assigning truth values to atoms at worlds but a way of keeping track of which worlds are possible from the point of view of which world. This is done by an *accessibility relation R*. The relation $R$ links $x$ to $y$ if and only if from the world $x$, $y$ is possible. Here is the definition of the models of the logic S4.

**Definition 45 (S4 model)**   *An* S4 MODEL $\langle W, R, V \rangle$ *is a triple consisting of a nonempty set $W$, a binary relation $R$ on $W$ that is reflexive and transitive, and a valuation function $V$ from atoms and worlds to $\{0, 1\}$ such that $V(\bot, w) = 0$, for all $w \in W$.*

To understand this definition, we need to understand what it is for a relation to be reflexive and transitive. We have already seen these notions when we introduced the notion of a partial order, in chapter 1, in definition 1 (page 6).

**Definition 46 (Reflexive, transitive, symmetric relations)**   *A binary relation $R$ on the set $W$ is* REFLEXIVE *iff for all $w \in W$, $wRw$.*
  *A binary relation $R$ on the set $W$ is* TRANSITIVE *iff for all $x, y, z \in W$, if $xRy$ and $yRz$, then $xRz$.*
  *Finally, even though it does not appear in the definition of models for S4, we say that a binary relation $R$ is* SYMMETRIC *iff for all $w, v \in W$, if $wRv$, then $vRw$.*

In an S4 model, the binary relation is not required to be a partial order, but it can be one. The binary relation $R$ is often called the accessibility relation, since it provides a kind of relative accessibility. The S4 models make up a species of a general kind, *Kripke models*. Kripke models are triples $\langle W, R, V \rangle$, where $W$ is a nonempty set of worlds, $R$ is a binary relation on $W$, and $V$ is a valuation function.[83] The binary relation of a Kripke model may obey some conditions, such as reflexivity, but it does not have to. The conditions we have listed above are important, but they are by no means exhaustive. When it comes to conditions on accessibility relations for modal logics, the possibilities are endless. In the next subsection, we will look at some Kripke models whose accessibility relation is reflexive, symmetric, and transitive, and we will see that those Kripke models yield S5.

\* \* \*

Models for S4 can be used to represent different kinds of possibility and necessity. If we think of moments ordered in time, where one moment accesses itself and any *later* moment, then $\Box A$ is true if $A$ is true at *every moment from then on*, and $\Diamond A$ is true at a moment if $A$ is true *for some moment, then or later*. This interpretation of the modal notions is used in *temporal logic*, the logic of time.

We might think of moments in time as progressing along a line. However, models for S4 allow for branching. Another use for models like these is the behavior of a system that can evolve in different ways. For example, we could take worlds to be positions that are possible in a game of chess, where $wRv$ when $v$ is a position that can come after $w$ in some

---

83. A KRIPKE FRAME is a pair $\langle W, R \rangle$ where $W$ is a nonempty set and $R$ is a binary relation on $W$. One obtains a model from a frame by adding a valuation. Much work on modal logic focuses on frames rather than models.

possible play of the game. Here, paths can branch, and as some choices are made, some possibilities are ruled in and others are ruled out.[84]

\* \* \*

The clauses for the Boolean connectives in **S4** models are the same as in possible worlds models, but the clauses for necessity and possibility are changed to incorporate the accessibility relation.

- $V(\Box A, w) = 1$ iff for all $x \in W$, if $wRx$, then $V(A, x) = 1$.
- $V(\Diamond A, w) = 1$ iff for some $x \in W$, both $wRx$ and $V(A, x) = 1$.

As an example of how these clauses are used, let us take a model $\langle W, R, V \rangle$ with $W = \{w, v, u\}$ and $wRw$, $wRv$, $wRu$, $vRv$, and $uRu$.[85] Here, the relation *is* reflexive and transitive. Let $V(p, w) = 1$, $V(p, v) = 1$, and $V(p, u) = 0$. In pictures, the model looks like this:

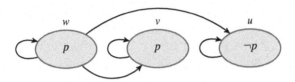

We draw directed links for the relation $R$ using arrows. To work out $V(\Diamond p, w)$, we need to check the value of $p$ at the worlds accessible from $w$. If there is such a world where $p$ is true, $\Diamond p$ is true at $w$. So, since $p$ is true at $w$ and $wRw$, $\Diamond p$ is true at $w$. Since $\neg p$ is true at $u$, $\Diamond \neg p$ is true at $w$ too. When we shift from $w$ to $v$ or to $u$, the possibilities change, since fewer worlds are accessible from $v$ and from $u$. At $v$, $\Diamond p$ is true, since $V(p, v) = 1$ and $vRv$, but $\Diamond \neg p$ is not, as there is no world $x$ such that $vRx$ and $V(p, x) = 0$. At $u$, $\Diamond \neg p$ is true, as $uRu$ and $V(\neg p, u) = 1$, and $\Diamond p$ is not, since there is no world $x$ such that $vRx$ and $V(p, x) = 1$. Necessity formulas also vary from world to world. In the picture, we have:

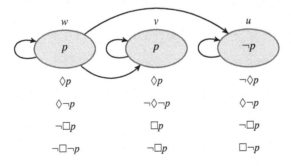

In these models, possibility and necessity vary from world to world. We can construct counterexamples to arguments using **S4** models in the same way as we do for **S5**. The

---

84. Branching time structures are important for *temporal logics* and *logics of agency*. These ideas are developed in many places, including Belnap, Perloff, and Xu (2001) and Horty (2001).

85. An alternative way to specify the binary relation is as a set of pairs of worlds, where $xRy$ iff $\langle x, y \rangle$ is in the set of pairs. The specification for this example would be $R = \{\langle w, w \rangle, \langle w, v \rangle, \langle w, u \rangle, \langle v, v \rangle, \langle u, u \rangle\}$.

definitions of counterexample and validity for an S4 model are similar to the definitions of those concepts for possible worlds models.

**Definition 47 (S4 counterexample, S4 validity)**   *An S4 model $\langle W, R, V \rangle$ is a* COUNTEREX-
AMPLE *to $X \succ A$ iff there is a world $w \in W$ such that $V(X, w) = 1$ and $V(A, w) = 0$.*

   *The argument $X \succ A$ is* VALID *according to S4 models, or S4 valid, iff there is no S4 model $\langle W, R, V \rangle$ that is a counterexample to the argument. We write $X \models_{S4} A$ when $X \succ A$ is S4 valid.*

The logic that is valid on S4 models, namely, S4, is similar to S5, although it is some-what weaker than S5, in the sense that there are some arguments that are valid in S5 that are not valid in S4.[86]

**Theorem 23**   *The following are true about S4:*

1. $\models_{S4} \Box A \rightarrow A$.
2. $\models_{S4} \Box A \rightarrow \Box\Box A$.
3. *If $\Box B_1, \dots, \Box B_n \models_{S4} A$, then $\Box B_1, \dots, \Box B_n \models_{S4} \Box A$.*

*Proof.*   For (1), suppose we have an S4 counterexample $\langle W, R, V \rangle$, so there is a world $w \in W$ such that $V(\Box A \rightarrow A, w) = 0$. Then, $V(\Box A, w) = 1$ and $V(A, w) = 0$. So, for all $x$, if $wRx$, then $V(A, x) = 1$. Since this is an S4 model, $R$ is reflexive, which entails $wRw$, so $V(A, w) = 1$, which is a contradiction.

   For (2), let $\langle W, R, V \rangle$ be an S4 counterexample so there is a world $w \in W$ with $V(\Box A, w) = 1$ but $V(\Box\Box A, w) = 0$. The latter implies that there is a world $x$ such that $wRx$ and $V(\Box A, x) = 0$. This, in turn, implies that there is a world $y$ such that $xRy$ and $V(A, y) = 0$. Since this is an S4 model, $R$ is transitive, so, given that both $wRx$ and $xRy$, it follows that $wRy$. This, together with the assumption that $V(\Box A, w) = 1$, implies $V(A, y) = 1$, which is a contradiction.

   The proof of (3) is similar to the proof of (3) from theorem 22.                      ◻

One place where S4 and S5 diverge is with respect to formulas involving both necessity and possibility.

**Fact 3**   *The following S5 tautologies are not S4 tautologies:*

1. $A \rightarrow \Box\Diamond A$
2. $\Diamond A \rightarrow \Box\Diamond A$

*Proof.*   To construct a countermodel to (i), in particular the instance $p \rightarrow \Box\Diamond p$, we want a model with a world $w$ where $p$ is true but $\Box\Diamond p$ is not. We need a model like this:

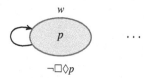

So, we need some world $v$ accessible from $w$ where $\Diamond p$ is not true.

---

86. If we think of consequence relations as pairs $\langle X, A \rangle$, where $X$ is a set of formulas and $A$ is a formula, then the consequence relation for S4, $\models_{S4}$, is a proper subset of the consequence relation for S5, $\models_{S5}$.

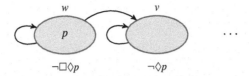

To ensure that $\Diamond p$ is not true at $v$, we need to make sure that $p$ is not true at any world accessible from $v$. But we can do that here, provided that $v$ cannot access $w$. If we make $p$ false at $v$ itself and stop here, we have an S4 model, since $R$ is reflexive and transitive:

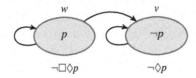

In this model, $p$ is true at $w$, and $\Box\Diamond p$ is false there. While $p$ is true at $w$, it isn't *necessarily* possible there. Our model has this structure: $W = \{w, v\}$ with $wRv, wRw, vRv$, $V(p, w) = 1$ and $V(p, v) = 0$.

The same countermodel will work as a countermodel to (ii), too. Since $V(p, w) = 1$ and $wRw$, we have $V(\Diamond p, w) = 1$ too. As $V(\Diamond p, v) = 0$, we have $V(\Box\Diamond p, w) = 0$, and so $V(\Diamond p \rightarrow \Box\Diamond p, w) = 0$. $\qquad\square$

The counterexample in this proof has the feature that the accessibility relation is not symmetric, since $wRv$ but not $vRw$. Requiring that the relation be symmetric, as well as reflexive and transitive, will ensure that the two formulas above do not have counterexamples. In fact, restricting attention to the class of Kripke models in which the accessibility relation is symmetric, reflexive, and transitive will get us back to S5. In the next section, we will see one of the applications of models in which the accessibility relation is reflexive, transitive, and symmetric.

### 7.4.2 Equivalence relations and epistemic logic
An important concept in logic is that of an equivalence relation.

**Definition 48 (Equivalence relation)** *A binary relation that is reflexive, transitive, and symmetric is called an* EQUIVALENCE RELATION.

We will use equivalence relations to define a kind of Kripke model.

**Definition 49 (Equivalence model)** *An* EQUIVALENCE MODEL, *or S5E model, is a triple* $\langle W, R, V \rangle$ *where $W$ is a nonempty set of worlds, $R$ is a binary relation on $W$ that is an equivalence relation, and $V$ is a function from* Atom *and $W$ to $\{0, 1\}$ such that $V(\bot, w) = 0$, for all $w \in W$.*

The verification conditions for complex formulas in an equivalence model are the same as those in an S4 model. The differences with S4 models emerge in evaluating formulas whose main connective is $\Box$ or $\Diamond$, since the accessibility relation used in those verification conditions will be an equivalence relation.

**Definition 50 (Counterexample, validity)**  *An equivalence model* $\langle W, R, V \rangle$ *is a* COUNTEREX-
AMPLE *to the argument* $X \succ A$ *iff there is a world* $w \in W$ *such that both* $V(X, w) = 1$ *and* $V(A, w) = 0$.

*The argument* $X \succ A$ *is* VALID *according to equivalence models iff there is no equivalence model* $\langle W, V \rangle$ *with a world* $w \in W$ *such that* $V(X, w) = 1$ *and* $V(A, w) = 0$.

*If the argument* $X \succ A$ *is valid according to equivalence models, write* $X \models_{\text{S5E}} A$.

It turns out that in a Kripke model, if the accessibility relation $R$ is symmetric, then $A \rightarrow \Box \Diamond A$ will be true at all worlds in that model. Suppose we have a Kripke model whose accessibility relation $R$ is symmetric. Take an arbitrary world $w \in W$, and suppose $V(A, w) = 1$. Suppose for reductio that $V(\Box \Diamond A) = 0$. Then there is some world $u \in W$ such that $wRu$ and $V(\Diamond A, u) = 0$. This implies that for every world $x \in W$, if $uRx$, then $V(A, x) = 0$. Since $R$ is symmetric and $wRu$, $uRw$, so $V(A, w) = 0$, which is a contradiction.

An equivalence relation $R$ on a set $W$ divides $W$ into sets called EQUIVALENCE CLASSES, which are the sets of all equivalent elements. More precisely, the equivalence class of $w \in W$, $[w]$, is $\{u \in W : wRu\}$. The equivalence classes are DISJOINT, meaning distinct equivalence classes have no members in common, and they are EXHAUSTIVE, meaning every $w \in W$ is in some equivalence class.[87] Let us look at three examples, where $W = \{w_1, w_2, w_3, w_4\}$. In the first example, $w_1Rw_2$, $w_2Rw_1$, $w_3Rw_4$, and $w_4Rw_3$.

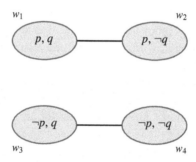

In the diagram, we do not need to indicate a direction on the lines, since accessibility is symmetric. There are two equivalence classes in this model; one is $[w_1]$, which is the set $\{w_1, w_2\}$, and the other is $[w_3]$, which is the set $\{w_3, w_4\}$. In this model, $V(\Box p, w_1) = 1$ and $V(\Diamond p, w_4 = 0)$.

For the second example, we change the accessibility relation.

---

87. In basic question 6, we ask you to prove some facts about equivalence classes.

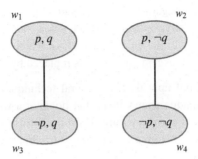

This model also has two equivalence classes, $[w_1]$, which here is $\{w_1, w_3\}$, and $[w_2]$, which is $\{w_2, w_4\}$. In this model, $V(\Box p, w_1) = 0$ but $V(\Diamond p, w_4 = 1)$.

For a third example, we change the accessibility relation again.

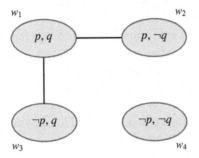

This model also has two equivalence classes, namely, $[w_1]$, which is $\{w_1, w_2, w_3\}$, and $[w_4]$, which is $\{w_4\}$. In this model, $V(\Box p, w_1) = 0$ and $V(\Diamond p, w_4) = 0$.

Possible worlds models and equivalence models validate the same logic, **S5**, in the sense spelled out in the following theorem.

**Theorem 24**   $X \models_{S5} A$ *iff* $X \models_{S5E} A$.

*Proof.*   One direction of this is easy to prove, the right-to-left direction. A possible worlds model can be viewed as an equivalence model by defining $R$ to be the universal relation, where every world relates to every world. A possible worlds counterexample can then be viewed as an equivalence model counterexample. Thus, if an argument has no counterexamples among the class of equivalence models, it has no countermodels among the class of possible worlds models, viewed as equivalence models with a single equivalence class of worlds.

The other direction, from left to right, is more involved, so we only sketch it. Suppose we have an equivalence model counterexample, $\langle W, R, V \rangle$, so there is a world $w \in W$ such that $V(X, w) = 1$ and $V(A, w) = 0$. The proof rests on the observation that the counterexample will not need to appeal to any world outside of the equivalence class $[w]$. Because of this, we can form a possible worlds model $\langle W', V' \rangle$ by setting $W' = \{u \in W : wRu\}$ and setting $V'$ to match $V$ over $W'$, that is, $V'(p, u) = V(p, u)$, for $u \in W'$. We can prove by induction on the construction of the formula $A$ that for every $u \in W'$ that $V'(A, u) = V(A, u)$—that worlds in the smaller model make the same formulas true as those they do in the equivalence relation model. This holds for atoms by the definition of $V'$. The cases for the falsum, conjunction, disjunction, conditional, and negation are immediate, since evaluating these involves no shifting of worlds. For necessity formulas, $V'(\Box A, u) = 1$ iff $V'(A, v) = 1$ for each $v$ in $W'$, that is, for each $v$ where $uRv$ in the equivalence class, which holds iff $V(\Box A, u) = 1$. The same reasoning works for $\Diamond A$, except "for all $v$" is replaced by "for some $v$." $\qquad\Box$

The argument in this proof appeals to a general technique, the use of *bisimulations* between models. This technique shows that what holds at a world in one model can be simulated by what holds at a corresponding (bisimilar) world in the other.[88] It has been developed as a general way to obtain a variety of results about modal logics.

* * *

Any two worlds in an equivalence class of an equivalence model are indiscernible: they make the same modal formulas true. This leads to the application of equivalence models to *epistemic logic*. Epistemic logic interprets the modal operator $\Box$ as saying some agent "knows" a formula. Often, instead of writing "$\Box$," the operator is written "$K_a$," for "the agent $a$ knows that." The accessibility relation $R$ then represents the relation of indiscernibility by an agent, given the agent's information. Two worlds are related by $R$ when they cannot be distinguished by any information the agent has. The truth condition for $\Box$ remains the same as in the S4 models: $\Box A$ is true at a world $w$ if and only if $A$ is true at all of the worlds accessible from $w$. $\Diamond A$ is true at a world if $A$ is true at some world accessible from $w$. The result is that $\Diamond A$ holds at a world just in case the agent $a$ has no information that could rule $A$ out, and $\Box A$ is true at a world just in case the information the agent has ensures that $A$ is true.

To represent the knowledge of multiple agents, we permit the equivalence models to have multiple accessibility relations, one for each agent (e.g., $R_a$ and $R_b$ where the agents under consideration are $a$ and $b$). When there are multiple accessibility relations, each is required to be an equivalence relation, which ensures that we can think of each as representing a relation of indiscernibility. The truth condition for an epistemic modal $K_a$ is similar to that of $\Box$, except that the corresponding accessibility relation is used:

- $K_a A$ is true at a world $w$ iff $A$ is true at all worlds $u$ such that $wR_a u$.

The equivalence models are especially interesting when we have more than one agent. Suppose agents $a$ and $b$ each flip a coin. Agent $a$ knows the result of $a$'s coin flip, but not $b$'s, and vice versa. Let $p$ be the claim that $a$'s coin landed heads, and let $q$ be the claim that $b$'s coin landed heads. Then we have four possible outcomes, $w_1$, $w_2$, $w_3$, and $w_4$.

---

88. Van Benthem (2010, chap. 3) provides a good introduction to bisimulations.

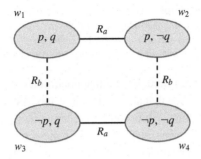

Agent $a$ cannot tell the difference between $w_1$ and $w_2$, or between $w_3$ and $w_4$, but can distinguish the other pairs. The equivalence relation $R_a$, encoding $a$'s information, is indicated by the solid lines. Agent $b$ cannot tell the difference between $w_1$ and $w_3$, or between $w_2$ and $w_4$, but can distinguish the other pairs. So, the relation $R_b$, for $b$'s information, is indicated by the dashed lines.

In the above model, at $w_1$, we have $K_a p$, but $\neg K_a q$ (for all $a$ knows, $q$ could be false) and $\neg K_b \neg q$ (for all $b$ knows, $q$ could be true). Similarly, at $w_1$, we have $K_b q$, but $\neg K_b p$ and $\neg K_b \neg p$.

Where this comes in handy, though, is when we want to reason about the information available to agents about the knowledge of other agents. Consider $K_a K_b q$ at $w_1$. For all $a$ knows, we are either in $w_1$ or $w_2$. At $w_1$, $K_b q$ is true, but at $w_2$, $K_b q$ is false. So, $K_a K_b q$ is false. Agent $a$ does not know that agent $b$ knows $q$. (That makes sense, because if agent $a$ knew that $b$ knew that $q$, then agent $a$ would know that $q$, too.) But in this setup, agent $a$ does know something about agent $b$'s knowledge. Notice that in every world, $K_b q \vee K_b \neg q$ is true. (This is one way to say that agent $b$ knows *whether q or not* and does so at every world.) So, agent $a$, in this setup at least, knows this. There is no circumstance possible in this model for $a$ where $K_b q \vee K_b \neg q$ fails, so we have $K_a(K_b q \vee K_b \neg q)$—$a$ knows that $b$ knows whether or not $q$, but we don't have $K_a q \vee K_a \neg q$—agent $a$ doesn't know whether $q$ or not themselves.

This is one example of how we can use simple **S5E** models to represent information states and the different agents' varying access to information. These **S5E** models have proved very useful, not only in reasoning about agents' knowledge but in the interaction between knowledge and other modal notions, like possibility and necessity, change over time, and different kinds of actions. Epistemic logic has been extended with dynamic concepts to model what happens when agents update their information. The field of *dynamic epistemic logic* is a rich and exciting field of current research (van Ditmarsch, van der Hoek, and Kooi 2007).

### 7.5    Appendix: Intuitionistic Kripke Models

In chapter 6, we showed that intuitionistic logic is not complete for validity defined in terms of Boolean valuations. In section 6.4, we presented Heyting algebras as an alternative sort of valuation for which intuitionistic logic is complete. There is another kind of model, intuitionistic Kripke models, for which intuitionistic logic is complete. These models are just like **S4** models, except with one difference.

**Definition 51 (Intuitionistic Kripke model)**   *An* INTUITIONISTIC KRIPKE MODEL *is a triple*
$\langle W, R, V \rangle$ *where W is a nonempty set of worlds; R is a reflexive, transitive binary relation on W; and
V is a function from pairs of atoms and worlds to* $\{0, 1\}$*, such that* $V(\perp, w) = 0$ *for all* $w \in W$*, obeying
the condition that, for all atoms p and all* $x, y \in W$*, if* $V(p, x) = 1$ *and* $xRy$*, then* $V(p, y) = 1$*.*

These models differ from S4 models with that condition on the valuation $V$. If an atom
is true at a world $x$, and $xRy$, then that atom is true at $y$, too. As you go from world to world
along the $R$ relation, more atoms can become true, but none go from true to false. Kripke
(1963, 97ff.) suggests an intuitive understanding of these models in terms of evidential
situations, where the worlds are stages in a scientific inquiry and $xRy$ holds just in case the
evidence available at $y$ extends the evidence available at $x$.

The truth conditions for complex formulas preserve the property imposed on atoms. The
clauses for $\wedge$ and $\vee$ and $\perp$ are just what you would expect, but the clauses for $\neg$ and $\rightarrow$
are different—they make use of the relation $R$. (Note that there are no clauses for necessity
and possibility, because we are giving models for the language of intuitionistic logic, rather
than S4 or S5.)

$$
\begin{aligned}
V(A \wedge B, w) = 1 \quad &\text{iff} \quad V(A, w) = 1 \text{ and } V(B, w) = 1 \\
V(A \vee B, w) = 1 \quad &\text{iff} \quad V(A, w) = 1 \text{ or } V(B, w) = 1 \\
V(A \rightarrow B, w) = 1 \quad &\text{iff} \quad \text{for all } x \in W, \text{ if } wRx, \text{ and } V(A, x) = 1, \text{ then } V(B, x) = 1 \\
V(\neg A, w) = 1 \quad &\text{iff} \quad \text{for all } x \in W, \text{ if } wRx, \text{ then } V(A, x) = 0 \\
V(\perp, w) = 0 \quad &\phantom{\text{iff}} \quad \text{always}
\end{aligned}
$$

The truth condition for the implication can also be stated as $V(A \rightarrow B, w) = 1$ iff for all
$x \in W$, if $wRx$, then either $V(A, x) = 0$ or $V(B, x) = 1$. This second way of stating the truth
condition brings out its similarity with the strict implication of S4.

Notice how the clauses for negation and implication in intuitionistic Kripke models differ from those in possible worlds models and Kripke models. For $\neg B$ to be true at $w$, $B$ has
to be false at $w$ and any worlds that $w$ bears $R$ to. (Since $R$ is reflexive, $wRw$, so we must
have $V(B, w) = 0$ in order to have $V(\neg B, w) = 1$.) That way, if a negation is true at a world,
it remains true at all the worlds farther along the relation $R$. In contrast, for $V(\neg B, w) = 1$
in possible worlds models, we only need $V(B, w) = 0$. In intuitionistic Kripke models, the
implication is interpreted similarly to the strict conditional, whereas in possible worlds
or S4 models, the implication is interpreted using the standard clause for the classical
conditional.

It is important to note that while $V(A, x) = 1$ or $V(A, x) = 0$ for each formula $A$ and world
$x$, it is not the case that we always have $V(A, x) = 1$ or $V(\neg A, x) = 1$. There are models and
formulas such that $V(A, x) = 0$ and $V(\neg A, x) = 0$, as there must be for excluded middle to
fail. Note, however, that $A$ and $\neg A$ cannot both get value 1 at a single world.

* * *

Intuitionistic Kripke models are S4 models, since their accessibility relations are reflexive and transitive. There are some S4 models that are not intuitionistic Kripke models,
since not all S4 models have valuations that obey the additional condition required by
intuitionistic Kripke models.

The condition on $V$ in the definition of intuitionistic Kripke models is sometimes called
the *heredity condition*. Using the truth conditions above, one can prove a lemma showing

that the preservation feature of heredity extends to all formulas of the language, not just atoms.

**Lemma 15 (Heredity)** *For any intuitionistic Kripke model* $\langle W, R, V \rangle$, *any worlds* $x, y \in W$, *and any formula A, if* $V(A, x) = 1$ *and* $xRy$, *then* $V(A, y) = 1$.

*Proof.* The proof is by induction on the complexity of the formula $A$. Let $\langle W, R, V \rangle$ be an intuitionistic Kripke model and let $x, y \in W$ be such that $xRy$.

The base case is when $A$ is atomic, and that is taken care of by the heredity condition on intuitionistic Kripke models.

Suppose $A$ is $\bot$. Then we never have $V(\bot, x) = 1$, so the conditional, that if $xRy$ and $V(\bot, x) = 1$, then $V(\bot, y) = 1$, is established.

Suppose $A$ is of the form $B \wedge C$. Suppose $V(B \wedge C, x) = 1$, so $V(B, x) = 1$ and $V(C, x) = 1$. Since $B$ is less complex than $B \wedge C$, we can use the inductive hypothesis with the facts that $V(B, x) = 1$ and $xRy$ to conclude $V(B, y) = 1$. Similarly, since $C$ is less complex than $B \wedge C$, we can use the inductive hypothesis to conclude that $V(C, y) = 1$. Therefore, $V(B \wedge C, y) = 1$.

Suppose $A$ is of the form $B \vee C$. Suppose $V(B \vee C, x) = 1$, so $V(B, x) = 1$ or $V(C, x) = 1$. As with the previous case, $B$ and $C$ are both simpler than $B \vee C$, so we can use the inductive hypothesis to conclude that $V(B, y) = 1$ or $V(C, y) = 1$, so $V(B \vee C, y) = 1$.

Suppose $A$ is of the form $\neg B$. Suppose $V(\neg B, x) = 1$, so for all $z$ such that $xRz$, $V(B, z) = 0$. Let $u$ be an arbitrary world such that $yRu$. From the transitivity of $R$ with the assumptions $xRy$ and $yRu$, we conclude $xRu$, so $V(B, u) = 0$. Therefore, since $u$ was an arbitrary world such that $yRu$, $V(\neg B, y) = 1$.

Suppose $A$ is of the form $B \rightarrow C$. Suppose $V(B \rightarrow C, x) = 1$, so, for all $z$ such that $xRz$, if $V(B, z) = 1$, then $V(C, z) = 1$. Suppose $u$ is an arbitrary world such that $yRu$. Suppose, in addition, that $V(B, u) = 1$. Since $R$ is transitive and $xRy$, it follows that $xRu$. By the assumption, $V(C, u) = 1$. Therefore, since $u$ was an arbitrary world such that $yRu$, $V(B \rightarrow C, y) = 1$. $\qquad\square$

The role of heredity in intuitionistic Kripke models is important and worth dwelling on. Recall that for possible worlds models, the proposition $\|A\|$ is defined as the set $\{w \in W : V(A, w) = 1\}$. This definition works for **S4** models as well. If we use this definition in the context of intuitionistic Kripke models, we see that every proposition is *closed upward*, in the sense that if $x \in \|A\|$ and $xRy$, then $y \in \|A\|$. This means that in intuitionistic Kripke models, not every set of worlds is a proposition. We may have a model and a proposition $\|A\|$ but not have any formula whose proposition in that model is $W - \|A\|$. Here is an example to illustrate. Take a model with $W = \{w, v, u\}$ and $V(p, w) = 0$, $V(p, u) = 1$, and $V(p, v) = 0$, with the $R$ relation as in the following diagram.

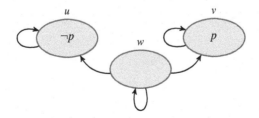

In this model, $\|p\| = \{v\}$ and $\|\neg p\| = \{u\}$. The set $\{w, u\}$, which is $W - \{v\}$, is not a proposition, which is one aspect of the difference between negation in intuitionistic logic and negation in classical logic. The propositions on this model are $\emptyset, \{v\}, \{u\}, \{v, u\}, \{w, v, u\}$. In fact, these are the only propositions available on any intuitionistic Kripke model with the same set $W$ and relation $R$. The remaining propositions can be expressed with formulas: $\|p \to p\| = \{w, v, u\}$, $\|p \vee \neg p\| = \{v, u\}$, and $\|p \wedge \neg p\| = \emptyset$.

The definitions of counterexample and validity are similar to those of the modal logics presented in this chapter.

**Definition 52 (Counterexample, validity)** *An intuitionistic Kripke model $\langle W, R, V \rangle$ is a* COUNTEREXAMPLE *to the argument $X \succ A$ iff there is a world $w \in W$ such that $V(X, w) = 1$ and $V(A, w) = 0$.*

*The argument $X \succ A$ is* VALID *according to intuitionistic Kripke models iff there is no intuitionistic Kripke model $\langle W, R, V \rangle$ with a world $w \in W$ such that $V(X, w) = 1$ and $V(A, w) = 0$.*

*We will write $X \models_{IKM} A$ when $X \succ A$ is valid according to intuitionistic Kripke models.*

The intuitionistic proof system of earlier chapters is sound and complete with respect to intuitionistic Kripke models. Slightly more formally, we have this theorem.

**Theorem 25 (Soundness and completeness for intuitionistic logic)**

$$X \vdash_I A \ \textit{iff} \ X \models_{IKM} A.$$

\* \* \*

The proof of the *soundness* portion of theorem 25 has exactly the same shape as the proof of soundness for classical logic. We prove by induction on the structure of the proof from $X$ to $A$ that whenever $X \vdash_I A$ that $X \models_{IKM} A$ too.

*Proof.* The proof breaks into several cases, one for each rule that could be used in the construction of the proof from $X$ to $A$.

The assumption rule provides the base case. The atomic proof of $A$ is a proof for $A \succ A$. Let $\langle W, R, V \rangle$ be an intuitionistic Kripke model with a world $w$ where $V(A, w) = 1$. Right away, the conclusion is also true at $w$, so we are done. The argument from $A$ to $A$ is truth preserving.

For the inductive hypothesis, we assume that we are given proofs that are truth preserving in intuitionistic Kripke models, and we wish to show that the proofs we make using an inference step are also truth preserving in those models. There are quite a few cases. We will go through many of them here, illustrating all of the techniques we need to use, and leave the rest to you.

First, the $\wedge I$ case. We assume we have proofs $\Pi_1$ for $X \succ A$ and $\Pi_2$ for $Y \succ B$ and that both are truth preserving in our models. Suppose we form a new proof, using the rule $\wedge I$.

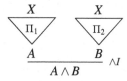

We want to show that $X, Y \models_{IKM} A \wedge B$. Suppose that there is a counterexample in some model, so we have a world $w$ where $V(X, Y, w) = 1$ and $V(A \wedge B, w) = 0$. Then either

$V(A, w) = 0$ or $V(B, w) = 0$. Suppose $v(A) = 0$. By assumption, $V(X, Y, w) = 1$, so $V(X, w) = 1$. From the inductive hypothesis, if $V(X, w) = 1$, then $V(A, w) = 1$, which is a contradiction. Suppose, instead, that $V(B, w) = 0$. By assumption, $V(X, Y, w) = 1$, so $V(Y, w) = 1$. From the inductive hypothesis, if $V(Y, w) = 1$, then $V(B, w) = 1$, which is a contradiction. In both cases, we reached a contradiction, so we conclude that there is no counterexample.

Next, consider the $\vee I$ case. We assume we have a proof $\Pi$ for $X \succ A$. We then form a new proof using the $\vee I$ rule.

$$\frac{\overbrace{\quad\Pi\quad}^{\textstyle X}}{\dfrac{A}{A \vee B}} \vee I$$

We need to show that $X \models_{\text{IKM}} A \vee B$. Assume there is a counterexample, so there is an intuitionistic Kripke model with a world $w$ such that $V(X, w) = 1$ and $V(A \vee B, w) = 0$. By definition, $V(A \vee B, w) = 0$ iff $V(A, w) = 0$ and $V(B, w) = 0$. By the inductive hypothesis, $X \models_{\text{IKM}} A$. Since $V(X, w) = 1$, $V(A, w) = 1$, which is a contradiction. Therefore, there is no counterexample.

Next, consider the $\to I$ case. We assume we have a proof $\Pi$ for $X, A \succ B$. We then form a new proof using the $\to I$ rule.

$$\frac{\overbrace{\quad\Pi\quad}^{\textstyle [A]^1 \quad X}}{\dfrac{B}{A \to B}} \to I^1$$

We need to show that $X \models_{\text{IKM}} A \to B$. Assume there is a counterexample, so there is an intuitionistic Kripke model with a world $w$ such that $V(X, w) = 1$ and $V(A \to B, w) = 0$. By definition, $V(A \to B, w) = 0$ iff there is some world $v$ where $wRv$, and $V(A, v) = 1$ and $V(B, v) = 0$. But then, by the heredity lemma, $V(X, v) = 1$ too, and so, $V(X, A, v) = 1$. By the inductive hypothesis, $X, A \models_{\text{IKM}} B$, so we have $V(B, v) = 1$ too, which is a contradiction. Therefore, there is no counterexample.

Next, we will do the $\to E$ case. We assume we have proofs $\Pi_1$ for $X \succ A \to B$ and $\Pi_2$ for $Y \succ A$ and we form a new proof using the rule $\to E$.

$$\frac{\overbrace{\quad\Pi_1\quad}^{\textstyle X} \qquad \overbrace{\quad\Pi_2\quad}^{\textstyle Y}}{\dfrac{A \to B \qquad A}{B}} \to E$$

We want to show that $X, Y \models_{\text{IKM}} B$. By the inductive hypothesis, $X \models_{\text{IKM}} A \to B$ and $Y \models_{\text{IKM}} A$. Suppose that there is a counterexample to the argument from $X, Y$ to $B$, so we have a model with a world $w$ where $V(X, Y, w) = 1$ and $V(B, w) = 0$. Since $V(X, Y, w) = 1$, $V(X, w) = 1$, and it then follows from the inductive hypothesis that $V(A \to B, w) = 1$. Similarly, $V(Y, w) = 1$, so it follows from the inductive hypothesis that $V(A, w) = 1$. But since $V(A \to B, w) = 1$ and $V(A, w) = 1$, we must have $v(B, w) = 1$ (since $wRw$), which contradicts the assumption that $V(B, w) = 0$. Therefore, there is no counterexample to the argument from $X, Y$ to $B$.

As our last example, we'll do the $\neg E$ case. We assume we have proofs $\Pi_1$ for $X \succ \neg A$ and $\Pi_2$ for $Y \succ A$, and we form a new proof using the rule $\neg E$.

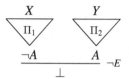

We want to show that $X, Y \models_{\mathsf{IKM}} \perp$. By the inductive hypothesis, $X \models_{\mathsf{IKM}} \neg A$ and $Y \models_{\mathsf{IKM}} A$. Suppose that there is a counterexample to the argument from $X, Y$ to $\perp$, so there is some model and some world $w$ where $V(X, Y, w) = 1$ and $V(\perp, w) = 0$. Since $V(X, Y, w) = 1$, $V(X, w) = 1$. It then follows from the inductive hypothesis that $V(\neg A, w) = 1$. Similarly, $V(Y, w) = 1$, and so it follows from the inductive hypothesis that $V(A, w) = 1$. But $V(\neg A, w) = 1$ iff for all $u$ such that $wRu$, $V(A, w) = 0$, so, in particular, $V(A, w) = 0$, as $wRw$. We then have $V(A, w) = 1$ and $V(A, w) = 0$, which yields a contradiction. Therefore, there is no counterexample.

There are several cases left to do before the *soundness theorem* is exhaustively demonstrated. We are going to leave those cases for the reader to do to test their understanding, with the easiest cases being $\wedge E$ and $\perp E$ and the hardest being $\vee E$. Despite there being some cases left to cover, we will consider soundness proved. □

To prove completeness, as we did in the last chapter for classical logic, we show that if we have no proof from $X$ to $A$ (this time no *intuitionistic* proof), then we show that $X \not\models_{\mathsf{IKM}} A$, by constructing a model with a world $w$ where each member of $X$ is true and where $A$ is not. Here, we reuse the results from the previous chapter, and since those results were tagged with a **WARNING LABEL**, the rest of this section bears that label, too. Please skip the rest of this section until you are very confident with the rest of this chapter, and you have also mastered the completeness proof from the previous chapter.

$* * *$

The first step in our completeness proof is to appeal to lemma 12 (see page 100) to conclude that whenever $X \not\vdash_I A$, then there is a maximal $A$-avoiding set (according to $I$) $X'$ extending $X$. By lemma 9, we know that the formulas in $X'$ act rather like those true at a world in an intuitionistic Kripke model.

1. $\perp \notin X'$.
2. $B \wedge C \in X'$ if and only if $B \in X'$ and $C \in X'$.
3. $B \vee C \in X'$ if and only if $B \in X'$ or $C \in X'$.
4. If $\neg B \in X'$, then $B \notin X'$.
5. If $B \to C \in X'$, then either $B \notin X'$ or $C \in X'$.

The conditions for $\perp$, $\wedge$, and $\vee$ are exactly the conditions for truth in a world. The conditions for $\neg$ and $\to$ fall short of the conditions for Boolean models, just as the corresponding truth conditions in an intuitionistic Kripke model differ from the Boolean truth conditions. To construct our Kripke model, we need not just *one* $I$-maximal set $X'$ but a whole family of them. For if $\neg B \notin X'$, then it does not necessarily mean that $B \in X'$, but we can show that there is *some* set (call it $X''$) extending $X'$ where $B \in X''$. How do we know that there is such an $X''$? Since $X' \not\vdash_1 \neg B$, we must have $X', B \not\vdash_1 \bot$ (since from $X', B \vdash_1 \bot$, we would have $X' \vdash_1 \neg B$, by $\neg I$), so there is a maximal $\bot$-avoiding set $X''$ extending $X' \cup \{B\}$. That is the $X''$ we are after. The same sort of reasoning applies to conditionals, too. If $X' \not\vdash_1 B \to C$, then $X', B \not\vdash_1 C$ too, and so, there is an I-maximal $C$-avoiding set extending $X' \cup \{B\}$.

The fact that we can move from set to set—from world to world in the model that we are constructing—means that we can find one model that will count as a counterexample to *any* argument at all.

**Definition 53 (The canonical intuitionistic Kripke model)** *The* CANONICAL INTUITION-ISTIC KRIPKE MODEL *is defined as* $\langle M, \subseteq, V \rangle$ *where*

$$M = \{X \subseteq \text{Form} : \textit{for some formula } A, X \textit{ is an I-maximal } A\textit{-avoiding set of formulas}\},$$

*the set of all sets $X$ of I-maximal $A$-avoiding formulas, for any formula $A$. $V$ is defined by setting $V(p, X) = 1$ iff $p \in X$, and the order relation on the model is the subset relation between sets of formulas.*

It is worth reflecting for a moment on the sets of formulas in $M$. In the terminology of the previous chapter, they are I-theories. The C-maximal $A$-avoiding sets of formulas are all in there as well, since every C-maximal $A$-avoiding set of formulas is also I-maximal. There are, however, some I-maximal sets that are not C-maximal sets, as there must be to provide worlds at which classically valid, but intuitionistically invalid, formulas fail, such as $p \lor \neg p$. There is an I-maximal $p \lor \neg p$-avoiding set, but there is no such C-maximal set, since every C-theory will contain $p \lor \neg p$.

The accessibility relation of the canonical model clearly satisfies the conditions for an intuitionistic Kripke model, since the relation $\subseteq$ is reflexive and transitive. The first thing for us to prove is that in this model, for any formula $A$ at all, $V(A, X) = 1$ iff $A \in X$.

**Lemma 16** *In the canonical intuitionistic Kripke model, for any world $X$ and any formula $A$, $V(A, X) = 1$ iff $A \in X$.*

*Proof.* The proof is by induction on the construction of the formula $A$. The base case is given by the definition of $V$: $V(p, X) = 1$ iff $p \in X$. To prove the cases for the connectives, we appeal to lemma 9. For $\bot$, we know that $\bot \notin X$. For the conjunction case, $V(B \land C, X) = 1$ iff $V(B, X) = 1$ and $V(C, X) = 1$. By the inductive hypothesis, this holds iff $B \in X$ and $C \in X$. By lemma 9, this holds iff $B \land C \in X$, as desired. Similarly, $V(B \lor C, X) = 1$ iff $V(B, X) = 1$ or $V(C, X) = 1$. By the inductive hypothesis, this holds iff $B \in X$ or $C \in X$. By lemma 9, this holds iff $B \lor C \in X$.

For negation and the conditional, we must do a little more work. First, $V(\neg B, X) = 1$ iff for each $Y \in M$ where $X \subseteq Y$, $V(B, Y) = 0$. By hypothesis, this means that for each $Y \in M$ where $X \subseteq Y$, we have $B \notin Y$. This holds iff $\neg B \in X$, as we will see. From right to left, if $\neg B \in X$, then whenever $X \subseteq Y$ and $Y$ is a maximal $C$-avoiding set for some formula $C$, we cannot have $B \in Y$, since $B, \neg B \vdash_1 C$. From left to right, let's suppose that $\neg B \notin X$. So, $X \not\vdash_1 \neg B$, and so, $X, B \not\vdash_1 \bot$. So, there is some I-maximal $\bot$-avoiding set $Y$ where $B \in Y$ and $X \subseteq Y$, as desired. So, we have $V(\neg B, X) = 1$ iff for each $Y \in M$ where $X \subseteq Y$, $V(B, Y) = 0$.

For the conditional, $V(B \to C, X) = 1$ iff for each $Y \in M$ where $X \subseteq Y$, if $V(B, Y) = 1$, then $V(C, Y) = 1$ too. By hypothesis, this means that for each $Y \in M$ where $X \subseteq Y$, if $B \in Y$, then $C \in Y$. This holds iff $B \to C \in X$, as we will see. From right to left, if $B \to C \in X$, then whenever $X \subseteq Y$ and $Y$ is some $C$-avoiding set for some formula $C$, we cannot have $B \in Y$ and $C \notin Y$, since $B \to C, B \vdash_I C$. From left to right, let's suppose that $B \to C \notin X$. So, $X \nvdash_I B \to C$, and so, $X, B \nvdash_I C$. So, there is some I-maximal $C$-avoiding set $Y$ where $B \in Y$ (and $C \notin Y$) and $X \subseteq Y$, as desired. So, we have $V(B \to C, X) = 1$ iff for each $Y \in M$ where $X \subseteq Y$, if $V(B, Y) = 1$, then $V(C, Y) = 1$ too.

$\square$

So, the canonical model has a world for every I-maximal $A$-avoiding set, for whatever formula $A$ we choose. The worlds are ordered by the subset relation, and the formulas true at a world $X$ are those formulas that are *in* the set. It then satisfies the additional condition on the valuation, so it is an intuitionistic Kripke model. This model is so comprehensive that inside it somewhere, we can find a counterexample to any argument that has no intuitionistic proof, since if $X \nvdash_I A$, there is some I-maximal $A$-avoiding set $X'$ extending $X$, and this $X'$ is a world in our canonical model. With this we can declare the completeness result, showing that proofs in intuitionistic logic are complete for validity as defined by intuitionistic Kripke models, *proved*.

### 7.6   Key Concepts and Skills

☐ You should be able to determine the truth or falsity of modal formulas in worlds in a given possible worlds model.

☐ You should be able to generate simple possible worlds models as counterexamples to arguments in S5.

☐ You should be able to distinguish the different ways to understand the scope of modal formulas, especially in the interaction between necessity and conditionals.

☐ You should be able to model propositions in sets of worlds and determine the propositions corresponding to formulas in a given possible worlds model.

☐ You should be able to determine the truth or falsity of modal formulas in worlds in a given model for S4 or for S5E.

☐ You should be able to use equivalence models for S5E to interpret the information available to agents in simple situations.

A POSSIBLE WORLDS MODEL is a pair $\langle W, V \rangle$, where $W$ is a nonempty set of *possible worlds* and $V$ is a function that gives a truth value (from $\{0, 1\}$) to each choice of an atomic formula and a world, with $V(\bot, w) = 0$ for all $w \in W$.

$$V(A \wedge B, w) = 1 \quad \text{iff} \quad V(A, w) = 1 \text{ and } V(B, w) = 1$$
$$V(A \vee B, w) = 1 \quad \text{iff} \quad V(A, w) = 1 \text{ or } V(B, w) = 1$$
$$V(A \rightarrow B, w) = 1 \quad \text{iff} \quad V(A, w) = 0 \text{ or } V(B, w) = 1$$
$$V(\neg A, w) = 1 \quad \text{iff} \quad V(A, w) = 0$$
$$V(\Box A, w) = 1 \quad \text{iff} \quad V(A, v) = 1 \text{ for all } v \in W$$
$$V(\Diamond A, w) = 1 \quad \text{iff} \quad V(A, v) = 1 \text{ for some } v \in W$$
$$V(A \prec B, w) = 1 \quad \text{iff} \quad \text{for all } x \in W, \text{either } V(A, x) = 0 \text{ or } V(B, x) = 1$$

An S4 MODEL $\langle W, R, V \rangle$ is a triple consisting of a nonempty set $W$, a binary relation $R$ on $W$ that is reflexive and transitive, and a valuation function $V$ from atoms and worlds to $\{0, 1\}$ such that $V(\bot, w) = 0$, for all $w \in W$.

- $V(\Box A, w) = 1$ iff for all $x \in W$, if $wRx$, then $V(A, x) = 1$.
- $V(\Diamond A, w) = 1$ iff for some $x \in W$, both $wRx$ and $V(A, x) = 1$.

An EQUIVALENCE MODEL, or S5E model, is a triple $\langle W, R, V \rangle$ where $W$ is a nonempty set of worlds, $R$ is a binary relation on $W$ that is an equivalence relation, and $V$ is a function from Atom and worlds to $\{0, 1\}$, with $V(\bot, w) = 0$ for all $w \in W$.

- $K_a A$ is true at a world $w$ iff $A$ is true at all worlds $u$ such that $wR_a u$.

## 7.7 Questions for You

### Basic Questions

1. Take the possible worlds model with worlds $W = \{w, x, y\}$ and where $V(p, w) = V(p, x) = 1$, $V(p, y) = 0$, and $V(q, w) = V(q, y) = 1$ and $V(q, x) = 0$ and $V(r, w) = 1$ and $V(r, x) = V(r, y) = 0$. Draw this model in a diagram, and then determine the worlds in the model where the following formulas are true and the worlds where they are false:
   i. $\Box(p \vee q)$
   ii. $\Diamond(p \wedge q)$
   iii. $p \rightarrow \Box p$
   iv. $\Diamond q \rightarrow q$
   v. $\Box(p \rightarrow q)$

2. For this question, we will use the possible worlds model from question 1. Write every set of worlds (there should be eight of them), and see if you can find a formula $A$, whose *proposition* in the model is that set. So, for example, for the set $\{w, x\}$, we can choose, $p$, since $\|p\| = \{w, x\}$. But we can choose $p \vee r$ also, since $\|p \vee r\| = \{w, x\}$ too. Find formulas for the seven other sets of worlds.

3. Construct possible worlds model counterexamples to these arguments, if there are any. If there aren't, try to explain as clearly as possible why the argument is valid.

    i. $\Diamond p, \Diamond q \succ \Diamond(p \wedge q)$

    ii. $\Diamond p, \Box q \succ \Diamond(p \wedge q)$

    iii. $\Diamond(p \to q) \succ \Diamond p \to \Diamond q$

    iv. $\Box(p \to q) \succ \Diamond p \to \Box q$

    v. $\Box(p \to q) \succ \Diamond p \to \Diamond q$

4. Consider the sentence "It will rain today, and it's possible that it will hail, too." Find two different ways to represent the structure of this sentence, making the ambiguous scope clear, and then find a model that shows the difference between the two ways to understand that sentence.

5. Use the countermodel generation procedure from section 7.2 to find counterexamples to these two arguments.

    i. $\Diamond p, \Diamond q \succ \Box(p \vee q)$

    ii. $\Box(p \vee q) \succ \Box p \vee \Box q$

Verify that the models you obtain are counterexamples. Can you find counterexamples with fewer worlds?

6. Prove the following for any equivalence model $\langle W, R, V \rangle$.

- For every world $w \in W$, there is some equivalence class $X$ such that $w \in X$.
- For any two worlds $w, u \in W$, $u \in [w]$ if and only if $w \in [u]$.
- For any worlds $w, u, v \in W$, if $w \in [u]$ and $w \in [v]$, then $[u] = [v]$. Does it follow that $u = v$?

7. For this problem, you will construct some S4 models. You can focus on just the propositional atoms $p$ and $q$. For each model that you construct, determine the truth values of the formulas $\Box p$ and $\Diamond q$ at each world.

    i. Construct two distinct S4 models with two worlds.

    ii. Construct two distinct S4 models with three worlds.

    iii. Construct one S4 model that has a world that bears $R$ to every world and one S4 model that has no world that bears $R$ to every world.

    iv. Construct one S4 model that is a tree and one that is not a tree. Both models should have four worlds.

8. For this problem, you will construct some equivalence models. You can focus on just the propositional atoms $p$ and $q$. For each model that you construct, determine the truth values of the formulas $\Box p$ and $\Diamond q$ at each world.

    i. Construct an equivalence model with three worlds.

    ii. Construct an equivalence model in which there are two worlds that do not bear the $R$ relation to each other.

9. Construct an S4 model that is not an equivalence model and then create an equivalence model out of this by adding additional connections to the $R$ relation to obtain a symmetric relation. How many equivalence classes are in this new model? Can you use this procedure to generate an equivalence model from a new S4 model that will give you a different number of equivalence classes?

10. Models can be infinite. The natural numbers, $0, 1, 2, 3, \ldots$, along with their standard ordering $\leq$, can be used to define many infinite models. Consider the S4 model $\langle W, R, V \rangle$ where

- $W = \{w_n : n \text{ is a natural number}\}$,
- $w_i R w_j$ iff $i \leq j$,
- $V(p, w_k) = 1$ iff $k$ is an even number, and
- $V(q, w_k) = 1$ iff $5 \leq k$.

First, verify that $R$ is reflexive and transitive, so that this is an **S4** model. Then, evaluate the following formulas at $w_0$:

- $\Box p$
- $\Box \neg p$
- $\Box \Diamond p$
- $\Diamond \Box p$

- $\Box \Diamond q$
- $\Diamond \Box q$
- $\Box (p \lor q)$

11. Verify the following facts about **S5** and **S4** validity.

    i. $\Diamond \Diamond A \models_{\mathsf{S4}} \Diamond A$

    ii. $\Diamond \Box A \models_{\mathsf{S5}} \Box A$

    iii. $\Diamond \Box A \not\models_{\mathsf{S4}} \Box A$

    iv. $\Diamond \Box A \models_{\mathsf{S5}} \Box \Diamond A$

    v. $\Diamond \Box A \not\models_{\mathsf{S4}} \Box \Diamond A$

    vi. $\models_{\mathsf{S5}} \Box(\Box A \to B) \lor \Box(\Box B \to A)$

    vii. $\not\models_{\mathsf{S4}} \Box(\Box A \to B) \lor \Box(\Box B \to A)$

### Challenge Questions

1. Construct an equivalence model with two agents $a$ and $b$ to invalidate the argument from $K_a K_b p$ to $K_b K_a p$. In other words, construct a model to show how it could be for $a$ to know that $b$ knows that $p$ without $b$ knowing that $a$ knows that $p$.

    In the model you make, with a world where $K_a K_b p$ is true and $K_b K_a p$ is false, is $K_a \neg K_b K_a p$ true too? That is, does $a$ know that $b$ doesn't know that $a$ knows that $p$—or *not*? Could this have gone either way?

2. In section 7.3.1, we introduced the strict conditional, $A \prec B$, and gave its truth conditions. We showed how to define it in terms of $\Box$ and the classical, truth-functional connectives. Suppose that you have the classical connectives ($\to, \land, \lor, \neg, \bot$) and the strict conditional ($\prec$) but no necessity or possibility operators. Define necessity using the strict conditional and the classical connectives and show that your definition has the same truth conditions as the primitive operator introduced in section 7.1. Once you've done that, define possibility using the strict conditional and the classical connectives and show that the connective you defined has the same truth conditions as the primitive possibility operator.

3. To get a feel for intuitionistic Kripke models, it is useful to come up with counterexamples to a few arguments that are classically valid but not provable intuitionistically. Construct intuitionistic Kripke model counterexamples to the following arguments:

- $\succ p \lor \neg p$
- $\neg \neg p \succ p$
- $(p \to q) \to p \succ p$
- $\neg(p \to q) \succ p$

- $\neg(p \wedge q) \succ \neg p \vee \neg q$

4. Define a translation function $m : \mathsf{Form} \mapsto \mathsf{MForm}$ as follows.[89]

   - $m(p) = \Box p$
   - $m(\neg B) = \Box \neg m(B)$
   - $m(B \wedge C) = m(B) \wedge m(C)$
   - $m(B \vee C) = m(B) \vee m(C)$
   - $m(B \to C) = \Box(m(B) \to m(C))$

   Show that $\models_{\mathsf{IKM}} A$ iff $\models_{\mathsf{S4}} m(A)$.

5. This is an open-ended discussion question. In this chapter, the logic of knowledge is S5. Consequently, the knowledge operator obeys the following principles.

   - (Positive introspection) $\models_{\mathsf{S5E}} KA \to KKA$.
   - (Negative introspection) $\models_{\mathsf{S5E}} \neg KA \to K \neg KA$.
   - (Logical omniscience) If $A \models_{\mathsf{S5}} B$, then $\models_{\mathsf{S5E}} KA \to KB$.

   Knowledge in this epistemic logic is idealized. Are these plausible idealizations? Can you come up with reasons for doubting any of these principles? If so, how can you modify the definition of a model to avoid these principles? Can you supply reasons that these are plausible or good idealizations? Are there any other consequences of using S5 for our epistemic logic that you think are suspicious?

---

89. The translation is due to J. C. C. McKinsey and Alfred Tarski (1948).

# 8 Actuality & 2D Logic

In the possible worlds models of the previous chapter, the modal operators $\square$ and $\lozenge$ shift the world at which a formula is evaluated. When we evaluate $\square p$ at $w$, for example, it is not enough to check the value of $p$ at $w$; we have to check its value at other worlds, too. In simple possible worlds models, this means that $p$ must be true at every world. In more sophisticated models, this means that $p$ has to be true at all worlds accessible from $w$. In these models, to evaluate a formula, we need to check not just the state of things *here* but also in worlds accessible from here.

Crossley and Humberstone (1977) pointed out that there are sentences of English that seem to require us to not only check worlds accessible from here but also to backtrack. Here is an example. Consider the different ways we could understand this sentence:

It is possible for every red thing to be shiny.

One understanding is straightforward: there is a world (an accessible world) where all of the red things are shiny. But another way to understand it is to be talking about all the red things and taking *them* to be shiny. In other words, you could understand it like this:

It is possible for everything that is *actually* red to be shiny.

Here, the "actually" marks the point at which we backtrack. We are looking for some world where lots of things are shiny. Which things? Not the things that are red in *that* world (the world we've "gone to"), but the things that are red in *this* world (the world we've "come from").

To understand all of the details of this statement's structure is beyond our tools so far, since it uses predicates, "is red," and quantifiers, "everything." (For how quantifiers and predicates behave, we wait until the next part of the text.) However, the question is pressing for us to understand what the word "actually" is doing in claims like this one. Crossley and Humberstone formalize "actually" with another modal operator, $\mathbb{A}$. For a claim of the form $\mathbb{A}p$ to be true, evaluated with respect to some world or other (whether a possible alternative of the actual world, or at the actual world itself), we need $p$ to be true at the actual world of the model.

To make sense of this way of evaluating formulas, we need to have some means of keeping track of which world is actual. We need to move from the conceptual resources of the set of possible worlds as a kind of *map* of logical space (where any location on the map is a place we might be) to a map equipped with a "you are here" marker, where there is only one place we *might* be (the place we are, the spot marked out) and the other locations are places we could go, or places we might *have been*, had things gone differently.

### 8.1 Actuality Models

So, let's add this "you are here" marker to our models. The update to our models is straightforward.

**Definition 54 (Actuality models)** *An* ACTUALITY MODEL *is a triple* $\langle W, g, V \rangle$ *where W is a set of worlds, $g \in W$, and V is a valuation function from pairs of atoms and worlds to $\{0, 1\}$ such that $V(\bot, w) = 0$, for all $w \in W$.*

The valuation is extended to complex formulas using the truth conditions for possible worlds models, with one addition. The truth condition for actuality is

- $V(\mathbb{A}A, w) = 1$ iff $V(A, g) = 1$.

In the other notation, the condition is

- $w \Vdash_V \mathbb{A}A$ iff $g \Vdash_V A$.

On its own, the actuality operator does not appear to do much. Evaluating it at the actual world is trivial.

**Lemma 17** *For all actuality models $\langle W, g, V \rangle$, $V(\mathbb{A}A, g) = V(A, g)$.*

*Proof.* This is immediate from the truth condition, since the world index does not change as it is already $g$. □

Evaluating an actuality formula at a *nonactual* world is more substantive. Take the actuality model $\langle W, g, V \rangle$ with $W = \{g, w\}$, $V(p, g) = 0$, and $V(p, w) = 1$. Then, $V(p \wedge \neg \mathbb{A}p, w) = 1$. Here is how we can represent actuality models. The *actual* world in the model is marked by a double boundary.

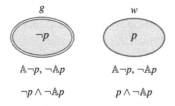

Within the scope of a modal operator, the actuality operator shifts the location of evaluation back to the actual world. Note that $p \wedge \neg p$ is never true at any world, so $\Diamond(p \wedge \neg p)$ is also true nowhere. However, we *can* have $\Diamond(p \wedge \neg \mathbb{A}p)$, as we do in the actual world (and all worlds) in our model.

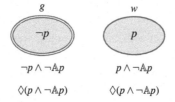

It turns out that even at other nonactual worlds, iterated actuality operators, such as $\mathbb{A}\mathbb{A}B$, do not do anything.

**Lemma 18**  *For all actuality models $\langle W, g, V \rangle$, for all worlds $w \in W$, $V(\mathbb{A}\mathbb{A}A, w) = V(\mathbb{A}A, w)$.*

*Proof.*  This follows from lemma 17.                                                              $\square$

The interaction of the actuality operator and negation is strikingly different from the interaction of negation and necessity.

**Lemma 19**  *For all actuality models $\langle W, g, V \rangle$ and all worlds $w \in W$, $V(\neg\mathbb{A}A, w) = V(\mathbb{A}\neg A, w)$.*

*Proof.*

$$
\begin{aligned}
V(\neg\mathbb{A}A, w) = 1 \quad &\text{iff} \quad V(\mathbb{A}A, w) = 0 \\
&\text{iff} \quad V(A, g) = 0 \\
&\text{iff} \quad V(\neg A, g) = 1 \\
&\text{iff} \quad V(\mathbb{A}\neg A, w) = 1
\end{aligned}
$$

$\square$

This means that $V(\neg\mathbb{A}\neg A, w) = V(\mathbb{A}A, w)$. This is not what we saw with $\square$. (What is a model with $V(\neg\square\neg A, w) \neq V(\square A, w)$?)

In addition, if an actuality formula is true, it is true everywhere.

**Lemma 20**  *For all actuality models $\langle W, g, V \rangle$, for all worlds $w, u \in W$, if $V(\mathbb{A}A, w) = 1$, then $V(\mathbb{A}A, u) = 1$.*

*Proof.*  Suppose $V(\mathbb{A}A, w) = 1$. Then, $V(A, g) = 1$. So it follows that $V(\mathbb{A}A, u) = 1$ too.    $\square$

We might call this the *rigidity* of actuality. What is actual doesn't vary from world to world. It is a surprising feature of the actuality models, since it is really plausible that there are a lot of things that are actually the case that are contingently true. We will return to this issue below.

## 8.2   Validity

As you would expect, actuality models can be used to define counterexamples and validity. The concepts we will define are for the logic we call **S5A**, following Crossley and Humberstone: it is **S5** with actuality.

**Definition 55 (Counterexamples, validity)**  *A* COUNTEREXAMPLE *to the argument $X \succ A$ is an actuality model $\langle W, g, V \rangle$ and a world $w \in W$, such that $V(X, w) = 1$ and $V(A, w) = 0$.*

*An argument $X \succ A$ is* **S5A** VALID *iff there are no counterexamples to it, or equivalently, if for all actuality models $\langle W, g, V \rangle$, for all $w \in W$, if $V(X, w) = 1$, then $V(A, w) = 1$.*

*If the argument $X \succ A$ is* **S5A** *valid, you write $X \models_{\text{S5A}} A$. A formula $A$ is an* **S5A** *tautology iff $\models_{\text{S5A}} A$.*

Now that we have a definition of validity and tautology, we can look at the logic **S5A**. What are some tautologies? All of the **S5** tautologies are still tautologies here, since every possible worlds model is an actuality model, once we choose a world to be the actual world. There are some distinctive tautologies that use the new actuality operator.

**Theorem 26**   *These formulas are all S5A tautologies.*

1. $\mathbb{A}(\mathbb{A}A \to A)$
2. $\neg\mathbb{A}\neg A \leftrightarrow \mathbb{A}A$
3. $\Box A \to \mathbb{A}A$
4. $\mathbb{A}A \to \Box\mathbb{A}A$
5. $\mathbb{A}(A \to B) \to (\mathbb{A}A \to \mathbb{A}B)$

*Proof.*   We will provide proofs of (1) and (4) here.[90] Let $\langle W, g, V \rangle$ be an actuality model, with $w \in W$. For (1), suppose $V(\mathbb{A}(\mathbb{A}A \to A), w) = 0$. Then, $V(\mathbb{A}A \to A, g) = 0$. It follows that $V(\mathbb{A}A, g) = 1$ and $V(A, g) = 0$. This contradicts lemma 17, $V(\mathbb{A}A, g) = V(A, g)$. Therefore, the supposition was false, and so $V(\mathbb{A}(\mathbb{A}A \to A), w) = 1$.

For (4), suppose $V(\mathbb{A}A, w, g) = 1$. Then, by lemma 20, $V(\mathbb{A}A, u) = 1$, for every world $u \in W$. So, $V(\Box\mathbb{A}A, w) = 1$. Therefore, $V(\mathbb{A}A \to \Box\mathbb{A}A, w) = 1$.                    □

Much as we had the rule of necessitation (Nec) for $\Box$, we have it for $\mathbb{A}$ in **S5A**, too.

**Theorem 27**   *If $\models_{\text{S5A}} A$, then $\models_{\text{S5A}} \mathbb{A}A$ too.*

*Proof.*   If $\models_{\text{S5A}} A$, then in any model $\langle W, g, V \rangle$ and in any world $w$ in that model, $V(A, w) = 1$. It follows that $V(A, g) = 1$ in that model (since $g$ is a world in that model), and hence, $V(\mathbb{A}A, w) = 1$, so it follows that $\mathbb{A}A$ is true at any world in any model, so $\models_{\text{S5A}} \mathbb{A}A$, too.    □

Actuality models provide one way to incorporate actuality into possible worlds models. In the next section, we turn to another way to do so.

### 8.3   Double Indexing

Crossley and Humberstone remark that the rigidity tautology, $\mathbb{A}A \to \Box\mathbb{A}A$, is surprising. There is some sense in which what is actual could have been different, had a different world been actual. Instead of thinking of a model as a set of worlds with a valuation function and a single world marked out as actual (the kind of map we might find fixed in a location, to be read only in one place), we can have models that can be interpreted from the point of view of any of the worlds. So, if we have three worlds, $w$, $u$, and $v$, we view the three worlds in three different ways, from the point of view of different worlds being *actual*.

---

90. The rest will be left as a basic question at the end of the chapter.

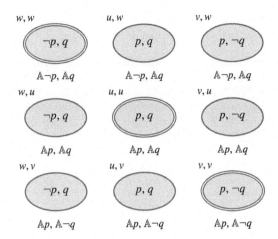

In this picture, each row involves a different choice of a world as actual. The labels in the diagram are a pair where the first part is the name of that world and the second names the world taken to be actual. In the first row, $w$ is actual and $u$ and $v$ are nonactual alternatives, so the $v$ world of that row is labeled $v, w$. In the second row, $u$ is actual, and $w$ and $v$ are nonactual alternatives, so the $v$ world of that row is labeled $v, u$. In the third row, $v$ is actual and is consequently labeled $v, v$. The atoms like $p$ and $q$ have their truth values at worlds, independently of what is taken to be actual. But statements like $\mathbb{A}p$ or $\mathbb{A}q$ vary from row to row, since the selection of actual world varies from row to row.

Once we interpret the language like this, a formula no longer has a truth value relative only to a choice of a world. Is $\mathbb{A}\neg p$ true at $w$? Well, it is if $w$ is actual, but it isn't if $u$ or $v$ is actual. Formulas have their truth value relative to a *pair* of worlds: the world being evaluated (which amounts to choosing the column in this diagram) and the world taken to be actual (which amounts to choosing the row).

Given the valuation $V$ that assigns to each atom a truth value at each world, we define a *pair* evaluation, assigning formulas a truth value at each pair of worlds, like this:

**Definition 56 (Double indexed models)** *A* DOUBLE INDEXED MODEL *is given by a pair* $\langle W, V \rangle$ *consisting of a nonempty set $W$ of worlds and a valuation function $V$ assigning a truth value from $\{0, 1\}$ to each pair consisting of an atom and a world, where $V(\bot, w, v)$ for all $w, v \in W$. Given these data, we can define a function assigning a truth value to each formula and a* pair *of worlds, like this:*

- $V(p, w, v) = 1$ *iff* $V(p, w) = 1$.
- $V(A \land B, w, v) = 1$ *iff* $V(A, w, v) = 1$ *and* $V(B, w, v) = 1$.
- $V(A \lor B, w, v) = 1$ *iff* $V(A, w, v) = 1$ *or* $V(B, w, v) = 1$.
- $V(A \to B, w, v) = 1$ *iff* $V(A, w, v) = 0$ *or* $V(B, w, v) = 1$.
- $V(\neg A, w, v) = 1$ *iff* $V(A, w, v) = 0$.
- $V(\bot, w, v) = 1$ *never.*
- $V(\Box A, w, v) = 1$ *iff* $V(A, u, v) = 1$ *for each world* $u \in W$.
- $V(\Diamond A, w, v) = 1$ *iff* $V(A, u, v) = 1$ *for some world* $u \in W$.
- $V(\mathbb{A}A, w, v) = 1$ *iff* $V(A, v, v) = 1$.

*We understand $V(A, w, v) = 1$ as saying that the formula A is true* at *the world w, given that the world v is actual.*

Double indexed models can do everything that actuality models can do, without forcing the choice of an actual world. *Any* world can be treated as actual.

**Definition 57 (Double indexed counterexamples, validity)**   *A double indexed model* $\langle W, V \rangle$ *is a* COUNTEREXAMPLE *to an argument* $X \succ A$ *iff there are worlds* $w, v \in W$ *such that* $V(X, w, v) = 1$ *and* $V(A, w, v) = 0$.

*An argument* $X \succ A$ *is* DOUBLE INDEXED VALID *iff there are no counterexamples to the argument. Or, equivalently, for all models* $\langle W, V \rangle$, *for all* $w, v \in W$, *if* $V(X, w, v) = 1$, *then* $V(A, w, v) = 1$. *If the argument* $X \succ A$ *is double indexed valid, then we write* $X \models_{DI} A$.

*Double indexed validity is sometimes also called* GENERAL VALIDITY, *because we consider any pair of points to be a possible counterexample, whether actual or not.*

Once we have double indexed models, a new option is open to us: we can naturally define other kinds of necessity. The necessity of $\square$ is one kind of necessity, but there is another. This kind can be obtained by introducing an operator that changes the *actual world* position of the pair of worlds. This operator is called "fixedly" by Davies and Humberstone (1980). It has the following truth conditions.

- $V(\mathbb{F}A, w, v) = 1$ iff $V(A, w, u) = 1$ for each world $u \in W$.

The language including $\mathbb{F}$ allows us to express much more than we could without it. Consider again the rigidity of actuality. If $\mathbb{A}A$ is true (with respect to a pair of worlds), it is necessarily true (with respect to that pair of worlds), that is, $\mathbb{A}A \to \square \mathbb{A}A$ is a tautology. To make the example concrete, if it happens to be raining right now (if it's actually raining), then it is *necessary* that it's actually raining. What is actually happening is necessary.

On the other hand, it seems that what is actual is, in some obvious and natural sense, contingent. It didn't have to be raining. Had things gone differently, from the point of view of that circumstance, what is actual would have been *that* rather than what is (as a matter of fact) actual. Something might actually be true, but had a different world been actual, different things would have been true—and hence, different things would have been *actually* true. Although $\mathbb{A}A \to \square \mathbb{A}A$ is a tautology, $\mathbb{A}A$ is not settled to be true in such a way as to make it impossible to consider what it would take for $\mathbb{A}A$ to be false or, in the case where $A$ is false, what it would take for $\mathbb{A}A$ to be true.

The reasoning in cases like these involves some way to vary the world considered as actual. This is not the behavior of $\square$, but this is what $\mathbb{F}$ does. Although $\mathbb{A}A \to \square \mathbb{A}A$ is a tautology, $\mathbb{A}A \to \mathbb{F}\mathbb{A}A$ is not a tautology. Here is why: consider a circumstance where $\mathbb{A}p$ is true at $w$ (so $p$ is true at the pair $\langle w, w \rangle$, and hence $\mathbb{A}p$ is true there), but there is another possible world $v$ where $p$ is false. Since $p$ is false at world $v$, it is false at $v$ if $v$ is considered as actual, and hence, so is $\mathbb{A}p$, considered at the pair $\langle v, v \rangle$. So, from the point of view of $w$, $\mathbb{A}p$ is true, while $\mathbb{F}\mathbb{A}p$ is false, for if we shift the *actual* world index to $v$, at the pair $\langle w, v \rangle$, $\mathbb{A}p$ is false, since $p$ is false at $\langle v, v \rangle$. In a diagram, we have this:

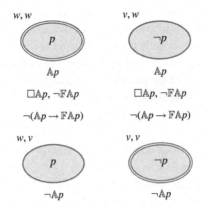

In the $w$ row, $p$ is true (at $w$) but could have been false (as shown in $v$). In both worlds in this row, $\mathbb{A}p$ is true, so it is also necessary at both of these worlds. However, in the $v$ row, $p$ is false at $v$, so $\mathbb{A}p$ is false in both worlds in this row. So, $\mathbb{F}\mathbb{A}p$ is false back at world $w$ in row $w$, because $\mathbb{A}p$ is false when we vary the world considered as actual to $v$. Although $\mathbb{A}p$ is true at $w$ when $w$ is considered as actual, this truth is not *fixed*. If it is evaluated at $w$ from the point of view of another world considered as actual, the statement is false.

It is worth examining the statement $\mathbb{F}\mathbb{A}A$ and what it takes for this to be true at a pair of worlds in a model.

$$V(\mathbb{F}\mathbb{A}A, w, v) = 1 \quad \text{iff} \quad V(\mathbb{A}A, w, u) = 1 \text{ for each } u \in W$$
$$\text{iff} \quad V(A, u, u) = 1 \text{ for each } u \in W$$

A formula is FIXEDLY ACTUALLY true iff it is true at each world—when that world is taken to be actual. It is another kind of necessity. Necessity expresses truth across a row. Fixity expresses truth down a column, and fixed actuality expresses truth down the diagonal of the model. The diagonal captures an important notion, which we will discuss more below.

The actuality operator moves from a world, the world of evaluation, to the world considered actual. There is a natural operation that shifts the world considered actual instead:[91]

- $V(\dagger A, w, v) = 1$ iff $V(A, w, w) = 1$.

The dagger operator lets us evaluate what would be the case if we take the world of evaluation as actual. In terms of the diagrams, the dagger operator shifts along a column, whereas the actuality operator shifts along a row. Just as the combination of fixedly and actually, $\mathbb{F}\mathbb{A}$, is evaluated along the diagonal, so is the combination of the box and dagger, $\Box\dagger$. The diagonal proposition, or diagonal necessity, is an important notion, and we can introduce a primitive connective, $\mathbb{D}$, to express it.

- $V(\mathbb{D}A, w, v)$ iff $V(A, u, u) = 1$, for each $u \in W$.[92]

---

91. The dagger was introduced by Robert Stalnaker (1978).
92. There is not a standard notation for this operator. We have opted for a mnemonic for "diagonal."

To see why the diagonal proposition is called that, recall the previous diagram. $\mathbb{D}A$ is true at a point in the diagram iff $A$ is true on all of the pairs along the diagonal, the worlds that are considered as actual. The formula $\mathbb{D}A$ is evaluated at the pairs $\langle w, w \rangle$, that is, where a world is considered as actual and is used as the world of evaluation, all of which lie along the diagonal of the grid. Each of the points along the diagonal is from the point of view of the world $w$, considering $w$ as actual.[93]

* * *

The $\mathbb{D}$ operator provides a good account of the kind of *necessity* possessed by a claim like $A \to \mathbb{A}A$. There is an obvious sense in which $A \to \mathbb{A}A$ *has to* be true. But it is not necessary. I don't need to know anything about the world to ensure that if $A$ holds, then $\mathbb{A}A$ holds. However, $A \to \mathbb{A}A$ need not be *necessarily* true. Neither need it be *fixedly* true. However, in every model, it is fixedly actually true. Consider the model from before:

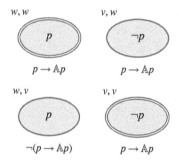

Here, $p \to \mathbb{A}p$ is true at $\langle w, w \rangle$, and at $\langle v, v \rangle$, $\mathbb{D}(p \to \mathbb{A}p)$ is true. However, $p \to \mathbb{A}p$ is not true everywhere in the $v$ row, so $\square(p \to \mathbb{A}p)$ is false at $\langle v, v \rangle$, and $p \to \mathbb{A}p$ is not true everywhere in the $w$ column, so $\mathbb{F}(p \to \mathbb{A}p)$ is not true at $\langle w, w \rangle$. The kind of "necessity" borne by $p \to \mathbb{A}p$ is expressed well by the $\mathbb{D}$ operator. Whenever the claim is expressed, no matter from *which* world, if it is taken as a claim about the world from which it is produced, it expresses a truth. We need know nothing about the world in order to ensure that it's true. It is not guaranteed to be true when considering nonactual worlds, either as worlds we are evaluating as alternative possibilities (shifting along the row) or if we are considering this possibility as viewed from other candidates for the actual world (shifting along the column). However, if we consider alternative possibilities *as other candidates for being actual*, it remains true.

## 8.4  Real-World Validity

There is another definition of validity that one can consider for double indexed models, which is inspired by our attention to the $\mathbb{D}$ operator. Rather than focus on truth in *all* pairs of worlds of a model, one focuses just on the diagonal points of the model.

---

93. Laura Schroeter (2003) provides an overview and discussion of the philosophical significance of the diagonal proposition and related notions.

**Definition 58 (Real-world counterexample, validity)** *A double indexed model* $\langle W, V \rangle$ *is a* REAL-WORLD COUNTEREXAMPLE *to an argument* $X \succ A$ *iff there is a world* $w \in W$ *such that* $V(X, w, w) = 1$ *and* $V(A, w, w) = 0$.

*An argument* $X \succ A$ *is* REAL-WORLD VALID *iff there are no real-world counterexamples to the argument. Equivalently, for all double indexed models* $\langle W, V \rangle$, *for all worlds* $w \in W$, *if* $V(X, w, w) = 1$, *then* $V(A, w, w) = 1$. *If the argument* $X \succ A$ *is real-world valid, then we write* $X \models_{RW} A$.

Real-world validity contrasts with general validity. The two concepts of validity are similar, although they differ with respect to arguments involving actuality formulas. For example, $\models_{RW} \mathbb{A}A \leftrightarrow A$ but $\not\models_{DI} \mathbb{A}A \leftrightarrow A$. The formula $\mathbb{A}A \leftrightarrow A$ is real-world valid, because any counterexample must be at the actual world of the model. At the actual world, however, the value assigned to $\mathbb{A}A$ and the value assigned to $A$ are the same.

∗ ∗ ∗

There is an ongoing debate over which of the two concepts of validity, general and real-world, is correct. Nelson and Zalta (2012) have pointed out that on real-world validity, there are contingent logical truths. An example is $\mathbb{A}p \leftrightarrow p$. This is real-world valid. A formula is contingent in an actuality model if there is some world of the model at which it is true and some world at which it is false. Since the formula $\mathbb{A}p \leftrightarrow p$ is true at the actual world of every model, we need to come up with a model and a world at which it is false. A model with $W = \{w, g\}$ such that $V(p, w, g) \neq V(p, g, g)$ will work to show that $\mathbb{A}p \leftrightarrow p$ is not true at all worlds of all actuality models. In terms of real-world validity, the contingency of some logical truths means that the rule of necessitation, if $\models_{RW} A$, then $\models_{RW} \Box A$, does not generally hold. This is striking since necessitation is a rule that holds in a wide range of modal logics, logics that are known as *normal* modal logics, which includes S4 and S5.

∗ ∗ ∗

We will end this chapter by linking these models and the concepts we have modeled with them to three key notions that have been central to philosophy through much of its history: *necessity*, *a priori knowability*, and *analyticity*.[94] These notions are easy to run together, though many philosophers have taken it to be important to distinguish them.

[NECESSITY] A statement is necessarily true if and only if it is true in all possible worlds. This is usually understood as a kind of metaphysical notion. Another way to understand necessity is truth under all counterfactual assumptions. A statement is necessarily true if and only if it would still have been true had things gone differently.

[A PRIORI KNOWABILITY] A statement is a priori knowable if and only if it can be known to be true without any appeal to the facts of how the world is. This is fundamentally an *epistemic* notion, to do with our knowledge and the way it is obtained. For something to be a priori knowable, it is not required that it be learned without appeal to matters of fact, only that it is *learnable* without such appeal. (I can learn some complex mathematical fact by testimony or by some other empirical means. If it is a priori knowable, it is at

---

94. Schroeter (2017) provides an overview of two-dimensional semantics with many connections to philosophical issues. Two-dimensional semantics has many applications in semantics, as illustrated by Fusco (2015), among others.

least possible that I can learn it without such means—for example, by going through a proof of that fact.)

[ANALYTICITY] A statement is analytically true if and only if it is true in virtue of the meanings of the words or the concepts involved in that statement. This is a semantic or a conceptual feature of the statement involved.

It is easy to confuse these notions. It is tempting, too, to attempt to define one or two of these notions in terms of the remaining one. One useful thing about the models discussed in this chapter is the way they can give us tools for understanding the connections between these notions.

We have already seen that truth across a row of a double indexed model is intended as a model for necessity. $\Box$ is designed to express necessity. If a claim is a priori knowable, it is plausible that it be true across the diagonal of a double indexed model. I can know a priori that $p \rightarrow \mathbb{A}p$, for example, no matter what the world says about $p$. In any world, considered as actual, $p \rightarrow \mathbb{A}p$ is true. This looks like the kind of fact I could come to know, a priori. I can know a priori that the world I am in is the actual world. That is what it is to be actual, from my point of view—it is to be the world that I am in, in just the same way that to be "here," when I talk about being here, is to be at the location where I am when I express that claim. So, truth down the diagonal is at least a first attempt at a model for a priori knowability.

If something is analytically true—if it is true in virtue of the meanings of the words—it is plausible to think that it must be true at every pair of points in the model, if the language is interpreted in such a way as to respect the meanings of the concepts involved, in just the same sort of way that $\wedge$, $\vee$, $\neg$, and so on, are interpreted across the points of the model. If a model for a language is a structure designed to interpret the language, in such a way as to respect the semantics of the expressions in the language, then the analytic truths seem to be those preserved in all of the points in the model.

We can use double indexed models to help us understand ways that these notions can come apart. We have already seen that if $p$ is true, then $\mathbb{A}p$ is necessary. However, if $p$ is true, it doesn't follow that $\mathbb{D}\mathbb{A}p$. While $\mathbb{A}p$ might be necessary, it need not be a priori knowable, or analytically true. On the other hand, it is intuitively a priori knowable that $p \rightarrow \mathbb{A}p$, while this is not necessary.

Claims like $p \rightarrow \mathbb{A}p$ are rather esoteric, and they are not the most interesting examples of claims that distinguish necessity and a priori knowability. We cannot discuss other examples in depth here, but it is worth considering one, due to Saul Kripke (1972). Water is (basically) $H_2O$. This was a scientific discovery about the constitution of water. Water had this constitution before we discovered it, and this discovery tells us something about what kind of substance water is. It is not only true that water has this structure. It is plausible that this is necessary, as argued by Hilary Putnam (1975). Had things gone differently, and had the lakes and rivers and oceans of earth been filled with things of a different chemical constitution (say, XYZ), that wouldn't have been *water*, even if the inhabitants of the planet called that substance "water." We could imagine a different possible world, with a planet we might call "Twin Earth," where the inhabitants use the language in just the same way that we do, where the oceans and lakes and rivers are filled with XYZ and they call that substance "water." In fact, before we discovered the chemical constitution of water, as far as we knew, the world we inhabited might have been *that* world, rather than this one. If that is the case, it makes sense to think that Water is $H_2O$ is false in that world, when that world is considered as actual. It is true in that world when considered as an alternate possibility to our (actual) world. On this view, the claim that Water is $H_2O$ is not a priori knowable, and it is not analytically true.

## 8.5  Key Concepts and Skills

☐ You should be able to evaluate formulas in **S5A** models and construct simple counterexamples to invalid formulas.

☐ You should understand the difference between fixity ($\mathbb{F}$) and necessity ($\square$), as well as fixed actuality ($\mathbb{D}$) in double indexed models. (The key definitions of these models are presented in the box on the next page.)

☐ You should be able to evaluate formulas in double indexed models and use these models to construct counterexamples to general validity and real-world validity.

☐ You should be able to use double indexed models to clarify or model the difference between necessity, a priori knowability, and analyticity.

## 8.6  Questions for You

### Basic Questions

1. Consider the following actuality model, with worlds $W = \{g, x, y\}$, and $V(p, g) = V(p, x) = 1$, $V(p, y) = 0$ and $V(q, g) = 0$, $V(q, x) = V(q, y) = 1$. Draw a diagram for this model, and evaluate the formulas $\mathbb{A}p$, $\neg p \wedge \mathbb{A}p$, $q \wedge \mathbb{A}\neg q$, $\Diamond \mathbb{A}q$, and $\mathbb{A}\Diamond q$ at each world in the model.

2. Which of these arguments are **S5A** valid? For those that aren't valid, provide an actuality model as a counterexample. For those that are valid, explain why it is valid.

    i. $\Diamond(p \wedge \mathbb{A}q) \succ \mathbb{A}p \wedge \Diamond q$

    ii. $\Diamond(p \wedge \mathbb{A}q) \succ \Diamond p \wedge \mathbb{A}q$

    iii. $\Diamond p \wedge \mathbb{A}q \succ \Diamond(p \wedge \mathbb{A}q)$

    iv. $\Diamond \mathbb{A}p \succ \square \mathbb{A}p$

    v. $\square p, \mathbb{A}q \succ \square(p \wedge \mathbb{A}q)$

3. Consider the following double indexed model, with worlds $W = \{w, x, y\}$, and $V(p, w) = V(p, x) = 1$, $V(p, y) = 0$ and $V(q, w) = 0$, $V(q, x) = V(q, y) = 1$. Draw a diagram for this model, and evaluate the formulas $\mathbb{A}p$, $\mathbb{F}q$, $\dagger p$, $\dagger(p \vee q)$, $\mathbb{D}p$ at each world in the model.

4. One feature of double indexed models that is important to understand is that *atoms* are true or false at *worlds*, independently of the second index in a pair. So, if $p$ is true at $\langle w, v \rangle$, it is true at $\langle w, u \rangle$ too. An atom has the same truth value up and down its column. The same doesn't work for other formulas. $\mathbb{A}p$ need not have the same value at $\langle w, v \rangle$ as at $\langle w, u \rangle$. Use this feature of atoms to explain why both of these facts are correct:

    i. $p \models_{\text{DI}} \mathbb{F}p$.

    ii. $A \not\models_{\text{DI}} \mathbb{F}A$, and in fact, $A \not\models_{\text{RW}} \mathbb{F}A$.

An ACTUALITY MODEL is a triple $\langle W, g, V \rangle$ where $W$ is a nonempty set of worlds, $g \in W$, and $V$ is a valuation function from pairs of formulas and worlds to $\{0, 1\}$ such that $V(\bot, w) = 0$, for all $w \in W$.

- $V(\mathbb{A}A, w) = 1$ iff $V(A, g) = 1$.

A DOUBLE INDEXED MODEL is given by a pair $\langle W, V \rangle$ consisting of a nonempty set $W$ of worlds and a valuation function $V$ assigning a truth value from $\{0, 1\}$ to each pair consisting of an atom and a world, such that $V(\bot, w, v) = 0$ for all $w, v \in W$. We define a function assigning a truth value to each formula and a *pair* of worlds, like this:

- $V(p, w, v) = 1$ iff $V(p, w) = 1$.
- $V(A \wedge B, w, v) = 1$ iff $V(A, w, v) = 1$ and $V(B, w, v) = 1$.
- $V(A \vee B, w, v) = 1$ iff $V(A, w, v) = 1$ or $V(B, w, v) = 1$.
- $V(A \rightarrow B, w, v) = 1$ iff $V(A, w, v) = 0$ or $V(B, w, v) = 1$.
- $V(\neg A, w, v) = 1$ iff $V(A, w, v) = 0$.
- $V(\bot, w, v) = 1$ never.
- $V(\Box A, w, v) = 1$ iff $V(A, u, v) = 1$ for each world $u \in W$.
- $V(\Diamond A, w, v) = 1$ iff $V(A, u, v) = 1$ for some world $u \in W$.
- $V(\mathbb{A}A, w, v) = 1$ iff $V(A, v, v) = 1$.
- $V(\dagger A, w, v) = 1$ iff $V(A, w, w) = 1$.
- $V(\mathbb{F}A, w, v) = 1$ iff $V(A, w, u) = 1$ for each world $u \in W$.
- $V(\mathbb{D}A, w, v)$ iff $V(A, u, u) = 1$, for each $u \in W$.

A double indexed model $\langle W, V \rangle$ is a DOUBLE INDEXED COUNTEREXAMPLE to an argument $X \succ A$ iff there are worlds $w, v \in W$ such that $V(X, w, v) = 1$ and $V(A, w, v) = 0$. A double indexed model $\langle W, V \rangle$ is a REAL-WORLD COUNTEREX-AMPLE to an argument $X \succ A$ iff there is a world $w \in W$ such that $V(X, w, w) = 1$ and $V(A, w, w) = 0$.

---

Can you find an example of a formula for $A$ (made up out of atoms, using the operators) where $A$ is true at a pair of worlds and $\mathbb{F}A$ is not?

5. Complete the proof of theorem 26 by showing that (2), (3), and (5) are **S5A** tautologies.
6. Prove that $\mathbb{A}A_1, \dots, \mathbb{A}A_n \models_{\text{DI}} \mathbb{A}B$ iff $A_1, \dots, A_n \models_{\text{RW}} B$.

## Challenge Questions

1. Let's introduce the singular connective $\mathbb{N}$, meaning nonactually, and evaluate it on the actuality models with the clause

   - $V(\mathbb{N}A, w)$ iff for every $u \in W$ such that $u \neq g$, $V(A, u) = 1$.

   Which of the following are tautologies: $\mathbb{N}A \rightarrow \mathbb{N}\mathbb{N}A$, $A \rightarrow \mathbb{N}A$, $A \rightarrow \mathbb{N}\neg\mathbb{N}\neg A$, $\mathbb{N}(\mathbb{N}A \rightarrow B) \vee \mathbb{N}(\mathbb{N}B \rightarrow A)$? How does this connective interact with the other modal connectives? To answer this, verify the validity of $\Box A \succ \mathbb{N}A$ and $\mathbb{A}A, \mathbb{N}A \succ \Box A$ and then identify some further tautologies and valid arguments involving $\mathbb{N}$ and the other modal connectives.

2. Lois Lane doesn't know that Clark Kent is Superman. But Clark Kent *is* Superman. In fact, it is consistent to say that it is *necessary* that Clark Kent is Superman and that Lois is ignorant of some necessary truth. What Lois takes to be the case (that Clark Kent isn't Superman) is a *consistent scenario*, but nonetheless, it is *impossible*.

   Use a double indexed model with at least two possible worlds to explain this scenario. Start with one possible world, the world as the Superman/Lois Lane stories describe. In this world, there is one person, the Kryptonian, Kal-El, who goes by the name Superman when dressed up in his outfit and acting out the superhero role, and who also goes by the name Clark Kent when working as a reporter. Then, there is another possible world, where this Kal-El decided to *not* play the Superman role at all and is happy with his life as Clark Kent. But in *this* world, there is someone *else* who goes by the name "*Superman*." In this second world, the Superman of the original world (Kal-El) is called "Clark Kent," but he is not, in this world, the person everyone calls "Superman."

   Now, using these two worlds as the basis for a double indexed model, consider the right way to evaluate the statement "Clark Kent is Superman" at each *pair* of worlds (at world $w$ from the point of view of world $v$ as actual). Is there a sense in which "Clark Kent is Superman" is necessary, in some pairs of worlds? Is there a sense in which it can fail to be true at other pairs? Given a model like this, how would you represent the world *as Lois takes it to be*?

3. Is the distinction between what is *actual* and what is merely *possible* purely relative, or is the distinction a difference in what is truly *real*? Consider the analogy with "now" and moments in time. Is there a difference in reality between the present time and moments in the past or the future, or is there no such difference in existence? Or consider the analogy with "here" and locations in space. Is there a difference in reality between the location that is *here* and the locations that are elsewhere, or is this difference not a distinction in what is truly real?[95]

---

95. Adriane Rini and Max Cresswell (2012) explore these questions at length in their book, *The World-Time Parallel: Tense and Modality in Logic and Metaphysics.*

# 9 Modal Natural Deduction

In this chapter, we will turn from models for modal logics to think about proofs for modal notions. We have, for two chapters, looked at how to define counterexamples to arguments using □, ◊, and other modal concepts. This raises a natural question: what can we say about *proofs* using modal notions? It is one thing to prove something from premises. How could you ever prove that something is *necessary*? It is one thing to prove a conclusion from the assumption that something is *true*. What could it be to prove a conclusion from the assumption that something is merely *possible*? What are the rules (if there are any) for reasoning using modal concepts?

We can't answer all of these questions in this chapter,[96] but we will give some answers concerning □ and ◊. In this chapter, we will look at natural deduction proof systems for the logics S4 and S5. The natural deduction systems for these logics are similar. The system for S4 is a bit simpler than that for S5, so we will start there.

## 9.1 Natural Deduction for S4

We introduced the logic S4 in chapter 7. Its models are triples of the form $\langle W, R, V \rangle$, where $W$ is a nonempty set of worlds; $R$ is a reflexive and transitive binary relation on $W$, of relative accessibility (where $wRv$ holds if and only if from the point of view of world $w$, world $v$ is a possibility); and $V$ is a valuation assigning a truth value to each atom at each world. The connectives $\wedge$, $\vee$, $\rightarrow$, $\neg$ are interpreted in the Boolean manner of classical logic at each world. $\square A$ is true at a world $w$ iff $A$ is true at *every* world accessible from $w$, while $\lozenge A$ is true at a world iff $A$ is true at *some* world accessible from $w$. An argument $X \succ A$ is said to be S4 *valid* iff in any model, in any world where each member of $X$ is true, so is $A$.

Now, since the propositional connectives $\wedge$, $\vee$, $\rightarrow$, $\neg$ are interpreted using the usual Boolean two-valued valuations (modified only by evaluating formulas at more than one world), the proof rules for these connectives will be the rules for classical natural deduction. We have the usual rules for the connectives, together with *DNE*. The additions for S4 are the rules for the necessity and possibility operators.

* * *

---

96. We have no space to give proof systems for actuality, fixity, or the dagger operator, frankly because there is little agreement on how to best do this. The proof theory for double indexed notions is an active topic for current research (Restall 2012).

What could rules for these modal operators look like? We are looking for introduction and elimination rules for $\Box$ and $\Diamond$. At the very least, it seems straightforward to specify at least one thing we could infer from $\Box A$ — $A$. And there is one way we could infer $\Diamond A$ — from $A$. So, we could use these rules to eliminate a $\Box$ and to introduce a $\Diamond$.

$$\frac{\Box A}{A} \; \Box E \qquad \frac{A}{\Diamond A} \; \Diamond I$$

The conclusions of these rules depend on the same assumptions as the premises of the rules.

The introduction rule for $\Box$ and the elimination rule for $\Diamond$ require more care. Let's start with the necessity operator. When should one be able to introduce a necessity operator? The rule

$$\frac{A}{\Box A} \; \Box I$$

clearly won't work, since it will make it all too easy to conclude necessity statements. For example, suppose that it is raining. From the unrestricted rule $\Box I$, it would then follow that necessarily, it is raining. This conclusion does *not* follow. The weather is a contingent matter, if anything is. However, while the unrestricted rule does not work, as it is not sound for **S4** validity, it is on the right track.

We do not want to be able to introduce a necessity quite that freely, but we do want to be able to introduce a necessity operator under some circumstances. For example, we would like to be able to derive the tautology $\Box A \rightarrow \Box\Box A$. The problem with the unrestricted rule $\Box I$ was that it permitted the introduction of a necessity operator when an assumption was contingent. If the information from which we infer our conclusion is contingent, this is not strong enough grounds to support the claim that the conclusion is *necessary*. But if the premises themselves are necessary, the grounds will be strong enough. So, we will keep the rule $\Box I$ but add the condition that all of the assumptions on which the premise depends are themselves necessary. One way to do this is to require that the premises are of the form $\Box B$ or $\neg\Diamond B$. In this case, we have inferred $A$ from assumptions that not only hold in *this* circumstance but would still hold had other things been the case. We call such formulas *necessitives*.

**Definition 59 (Necessitive formulas)** *A formula is a* NECESSITIVE *if and only if it has the form* $\Box B$ *or* $\neg\Diamond B$.

The restriction on assumptions is called a "side condition" for the inference. We can specify the rule like this:

where the condition (that the assumptions on which $A$ depends—if any—are necessitives) is written to the side of the rule of inference, and this is why it is called a "side condition." The conclusion of $\Box I$ depends on the same assumptions that the premise of the rule depends on. The rule with this side condition is what we need to construct proofs for $\Box$ formulas in a way that is well suited to **S4** validity.

Here is an example proof, using this rule:

$$\frac{\dfrac{[\Box A]^1}{\Box\Box A}\ \Box I}{\Box A \to \Box\Box A}\ {\to} I^1$$

In this proof, we started with the assumption $\Box A$. This is a proof from $\Box A$ to $\Box A$, so we have proved $\Box A$ from a necessitive formula, and so, the side condition for $\Box I$ is satisfied, and we can infer $\Box\Box A$. Then we discharge the assumption in the ${\to}I$ step to derive the conclusion $\Box A \to \Box\Box A$. We couldn't do the same thing to infer $\Box A$ from $A$, since the assumption $A$ is not a necessitive. So, our rule is not so liberal as to make mistakes like that.

To illustrate the rule again, here is a second proof, for the argument $\Box(A \to B), \Box A \succ \Box B$.

$$\frac{\dfrac{\Box(A \to B)}{A \to B}\ \Box E \qquad \dfrac{\Box A}{A}\ \Box E}{\dfrac{B}{\Box B}\ \Box I}\ {\to} E$$

Here, the final inference is a legitimate application of $\Box I$ because the two assumptions that the premise, $B$, depends on both have $\Box$ as their main operators, and so, are necessitives.

$$* * *$$

One rule remains, $\Diamond E$. What follows from a possibility claim? If it is possible that it is raining, it certainly does not follow that it is, in fact, raining. But if we can infer something (say $C$) from the claim that it *is* raining, then we could at least infer $\Diamond C$ from the claim that it is possible that it is raining. So, our first attempt at the rule will follow the form of $\lor E$, with $C$ being the minor premise.

$$\frac{\Diamond A \qquad \overset{\displaystyle [A]^1 \quad X}{\underset{\displaystyle C}{\nabla_\Pi}}}{\Diamond C}\ \Diamond E^{?1} \qquad (X \text{ contains only necessitives.})$$

To work out what follows from the possibility of $A$, one assumes $A$ to reach an intermediate conclusion, $C$, and then concludes that $C$ is possible. Assumptions other than $A$ can be used in $\Pi$, but they must be necessitives too.

[**S4** SIDE CONDITION] All assumptions in $X$ upon which the premise of an application of the rule $\Box I$ depends and the assumptions in $X$ upon which the minor premise of an application of $\Diamond E$ depends are necessitives.

The idea behind this side condition is that the auxiliary assumptions that one can appeal to in working out the consequences of a possibility claim must themselves be necessary.[97]

The rule $\Diamond E$ is useful as we have stated it, although it is not quite strong enough to give us all of S4 validity. The problem is that the rules so far do not permit us to derive certain connections between necessity and possibility. The more general rule for $\Diamond E$ has this shape, where the occurrence of $\bot$ concluding $\Pi$ is the minor premise:

$$\frac{\Diamond A \qquad \underset{\Diamond E^1}{\bot}}{\bot}$$

$$\begin{array}{c} [A]^1 \quad X \\ \diagdown \underset{\Pi}{\phantom{x}} \diagup \end{array}$$

(*X* contains only necessitives.)

We say that the conclusion of the rule depends on the assumptions that the major premise $\Diamond A$ depends on as well as the assumptions that the minor premise $\bot$ depends on, apart from the occurrences of *A* discharged by $\Diamond E$. This rule also requires the S4 side condition that all of the undischarged assumptions of $\Pi$ must be necessitives. This rule is plausible. After all, if we can show that *A* is contradictory, then *A* cannot *possibly* be true. We will then take the proof system for S4 to be our system for classical logic plus the rules $\Box I$, $\Box E$, $\Diamond I$, and $\Diamond E$, where proofs are required to obey the S4 side condition. We will write $X \vdash_{S4} A$ when there is a proof in our S4 proof system for $X \succ A$.

It is not obvious that this form of $\Diamond E$ rule is actually more *general* than $\Diamond E^?$. In this form of the possibility elimination rule, we do not weaken the conclusion from *C* to $\Diamond C$, but the inference requires a proof to a contradiction. It turns out (as we will see) that if we use $\Diamond E$, we can have all of the effect of $\Diamond E^?$, without needing to adopt it as a separate rule. We will show that $\Diamond E^?$ is present as a derived rule in our natural deduction system. To show this, we will first show that the proof system for S4 permits us to derive all four De Morgan connections between $\Box$ and $\Diamond$.[98]

**Theorem 28**   *The following are provable in* S4:

1. $\Box A \vdash_{S4} \neg\Diamond\neg A$
2. $\neg\Diamond\neg A \vdash_{S4} \Box A$
3. $\Diamond A \vdash_{S4} \neg\Box\neg A$
4. $\neg\Box\neg A \vdash_{S4} \Diamond A$

*Proof.*   Here are the proofs. First, for (1).

$$\frac{[\Diamond\neg A]^2 \qquad \dfrac{[\neg A]^1 \quad \dfrac{\dfrac{\Box A}{A}\Box E}{\bot}\neg E}{\bot}\Diamond E^1}{\dfrac{\bot}{\neg\Diamond\neg A}\neg I^2}$$

---

97. The side condition might strike you as overly restrictive, since $\Box p \land \Box q$ will not qualify as a necessitive, despite being the conjunction of two necessitives. In the challenge questions, we will ask you to consider some alternative formulations of the modal rules $\Box I$ and $\Diamond E$.

98. The classical equivalences of $\neg(A \land B)$ with $\neg A \lor \neg B$ and of $\neg(A \lor B)$ with $\neg A \land \neg B$ are also known as De Morgan equivalences. They are named after the British mathematician Augustus De Morgan (1806–1871).

For (2):

$$\cfrac{\cfrac{\neg\Diamond\neg A \qquad \cfrac{[\neg A]^1}{\Diamond\neg A}\;\Diamond I}{\cfrac{\bot}{\cfrac{\neg\neg A}{\cfrac{A}{\Box A}\;\Box I}\;DNE}\;\neg I^1}\;\neg E}{}$$

For (3):

$$\cfrac{\Diamond A \qquad \cfrac{\cfrac{[\Box\neg A]^2}{\neg A}\;\Box E \qquad [A]^1}{\bot}\;\neg E}{\cfrac{\bot}{\neg\Box\neg A}\;\neg I^2}\;\Diamond E^1$$

For (4):

$$\cfrac{\neg\Box\neg A \qquad \cfrac{\cfrac{[\neg\Diamond A]^2 \qquad \cfrac{[A]^1}{\Diamond A}\;\Diamond I}{\cfrac{\bot}{\cfrac{\neg A}{\Box\neg A}\;\Box I}\;\neg I^1}\;\neg E}{}}{\cfrac{\bot}{\cfrac{\neg\neg\Diamond A}{\Diamond A}\;DNE}\;\neg I^2}\;\neg E$$

$\Box$

We can now show that $\Diamond E^?$ is derivable if we have $\Diamond E$. Given a proof $\Pi$ of $C$ from $A$ (where the other premises are necessitive), we can extend the proof like this:

$$\cfrac{\Diamond A \qquad \cfrac{[\neg\Diamond C]^2 \qquad \cfrac{\cfrac{\cfrac{[A]^1}{\Pi}}{C}}{\Diamond C}\;\Diamond I}{\bot}\;\neg E}{\cfrac{\bot}{\cfrac{\neg\neg\Diamond C}{\Diamond C}\;DNE}\;\neg I^2}\;\Diamond E^1$$

This shows how we can get from a proof $\Pi$ from $A$ (and other necessitive assumptions) to $C$ and the side assumption of $\Diamond A$ to the conclusion $\Diamond C$. (Note that the assumption of $\neg\Diamond C$, which is also an assumption in the proof of $\bot$ to which we applied the $\Diamond E$ rule, does not interfere with the application of that rule, as the assumption is also a necessitive formula.)

* * *

The issue of normalization for these modal proofs is complicated. The logic for $\wedge$, $\vee$, $\rightarrow$, and $\neg$ is classical, rather than intuitionistic, so the proofs do not normalize in the way that intuitionistic proofs do. However, we can at the very least reduce *immediate* detours, where a $\Box$ or $\Diamond$ formula is introduced and then immediately eliminated. We can reduce $\Box$ detours like this:

$$
\cfrac{\cfrac{\cfrac{\Pi}{A}}{\Box A}\ {}_{\Box I}}{A}\ {}_{\Box E} \qquad \rightsquigarrow \qquad \cfrac{\Pi}{A}
$$

And we can reduce $\Diamond$ detours like this:

$$
\cfrac{\cfrac{\cfrac{\Pi_1}{A}}{\Diamond A}\ {}_{\Diamond I} \qquad \cfrac{[A]^1 \atop \Pi_2}{\bot}}{\bot}\ {}_{\Diamond E^1} \qquad \rightsquigarrow \qquad \begin{matrix}\Pi_1 \\ A \\ \Pi_2 \\ \bot \end{matrix}
$$

However, combining these rules with *DNE* (and with the conditions to permute rules) proves much more complicated than this direct case. Dag Prawitz, who first presented these rules for $\Box$, observed that the conditions on the $\Box I$ rule are overly restrictive. He provided two generalizations, one of which was meant to permit normalization. His proof of normalization was flawed, as noted by Medeiros (2006). She presented a new rule, a kind of *generalized* introduction rule, to replace $\Box I$.[99]

$$
\cfrac{\Box A_1\ \cdots\ \Box A_n \qquad \cfrac{[\Box A_1]^{j_1}\ \cdots\ [\Box A_n]^{j_n}}{\underset{\displaystyle B}{\nabla \Pi_1}}}{\Box B}\ {}_{\Box IM^{j_1,\,\cdots,\,j_n}}
$$

The rule $\Box IM$ has the side condition that the $B$ premise of the rule can depend on no other assumptions than $\Box A_1, \ldots, \Box A_n$, all of which are discharged by the rule. This rule is different from ones we have seen before in that it does not have a fixed number of premises. The spirit of the rule is similar to that of the $\Box I$ rule we use: in order to introduce a necessity operator, one must have obtained the premise by appeal only to assumptions that are themselves necessary.

We will close this section, noting that the proof system we have introduced is sound and complete for S4 validity as given by possible worlds models.

**Theorem 29**  $X \vdash_{\mathsf{S4}} A$ *if and only if* $X \vDash_{\mathsf{S4}} A$.

We do not have the space to go through this proof (or rather, *two* proofs) in detail. However, here is a sketch of how the argument goes. For the interested reader, the details can be found in section 9.4.

For the left-to-right direction (soundness), we show that if we have a proof for $X \succ A$, then no model $\langle W, R, V \rangle$ provides a counterexample. We show this by an induction argument on the structure of the proof. For this, the modal rules are the rules that require most close attention. $\Box E$ and $\Diamond I$ are straightforward. If $\Box A$ is true at a world, then so is $A$. If $A$ is true at a world, so is $\Diamond A$. For $\Box I$, we need to show that assuming that the proof

---

99. Recall that we saw generalized elimination rules in the challenge questions for chapter 2.

$\Pi$, from premises $X$, of the form $\Box B$ or $\neg\Diamond B$, to conclusion $A$, has no counterexamples in our model, then neither does the argument from those premises to $\Box A$. This must be the case, since if there were a world $w$ where $\Box A$ were false but the elements of $X$ were true, then there must be a world $v$ where $wRv$ and at which $A$ is false. But the members of $X$ have the form $\Box B$ or $\neg\Diamond B$, and these formulas have the property that if they are true at $w$ and $wRv$, then they are true at $v$ too. And this means that $v$ would be a counterexample to the argument from $X$ to $A$ too, which (by hypothesis) has no counterexample. So, proofs constructed using $\Box I$ are sound. The same sort of reasoning applies for $\Diamond E$, and this is how soundness is proved.

For the right-to-left direction (completeness), we show that if there is no proof for $X \succ A$, then there is some counterexample for that argument. The standard technique for this argument is what is called a *canonical model* construction. We construct a model whose worlds are consistent complete sets of formulas, where the members of a set are the formulas true at that world. Accessibility between worlds is defined by setting $wRv$ if and only if whenever $\Box A$ is in $w$, then $A$ is in $v$. We show, then, that indeed this structure satisfies the constraints of being a model, and that if an argument $X \succ A$ has no proof, then there is some world in this model where the members of $X$ are true and $A$ is false. This is enough to show completeness.

## 9.2  Natural Deduction for S5

Adjusting the natural deduction system to provide S5 rather than S4 is not difficult. The rules $\Box I$, $\Box E$, $\Diamond I$, and $\Diamond E$ remain the same, but the side conditions on $\Box I$ and $\Diamond E$ change. Recall the definition of a *modal formula*: a formula is a modal formula iff it has the form $\Box B$, $\Diamond B$, $\neg\Box B$, or $\neg\Diamond B$. These formulas have the property in S5 of expressing necessary claims. Not only are $\Box B$ and $\neg\Diamond B$ always necessary, if true, but so are $\Diamond B$ and $\neg\Box B$.

[S5 SIDE CONDITION] All assumptions in $X$ upon which the premise of an application of the rule $\Box I$ and the assumptions in $X$ upon which the minor premise of an application of $\Diamond E$ depend are modal formulas.

This liberalizes the condition from S4, which requires all additional assumptions of those rules to be necessitives. The S5 condition lets possibility formulas into the mix as well. We can see how this condition delivers the S5 tautology, $\Diamond A \to \Box\Diamond A$.

$$\cfrac{\cfrac{[\Diamond A]^1}{\Box\Diamond A}\;\Box I}{\Diamond A \to \Box\Diamond A}\;\to I^1$$

The $\Box I$ step would not be allowed with the S4 side condition.

We will define the proof system for S5 similarly to that of S4. We add to classical logic the rules $\Box I$, $\Box E$, $\Diamond I$, and $\Diamond E$ and require the S5 side condition. We write $X \vdash_{S5} A$ when there is a proof in this proof system for $X \succ A$.

Let us look at how to derive some of the corresponding facts about S5 validity for S5 provability.

**Theorem 30**  *The following are true about S5 provability:*

1. *If* $\vdash_{S5} A$, *then* $\vdash_{S5} \Box A$.
2. *If* $B_1, \ldots, B_n \vdash_{S5} A$, *then* $\Box B_1, \ldots, \Box B_n \vdash_{S5} \Box A$.
3. *If* $M_1, \ldots, M_n \vdash_{S5} A$, *then* $M_1, \ldots, M_n \vdash_{S5} \Box A$, *where each of the $M_i$ is a modal formula.*

*Proof.* For (1), take the proof $\Pi$ of $A$ and extend it with $\Box I$.

This is a proper application of the rule, since there are no assumptions upon which $A$ depends.

For (2), suppose $\Pi$ is a proof of $B_1, \ldots, B_n \vdash_{S5} A$.

$$
\begin{array}{ccc}
B_1 & \cdots & B_n \\
\end{array}
$$
$$\Pi$$
$$A$$

Extend it on both ends, as follows:

$$
\frac{\Box B_1}{B_1} \Box E \quad \cdots \quad \frac{\Box B_n}{B_n} \Box E
$$
$$\Pi$$
$$\frac{A}{\Box A} \Box I$$

The application of the $\Box I$ rule at the end is a proper application because all of the assumptions on which $A$ depends, the displayed $\Box B_i$s, are all modal formulas.

For (3), the proof is basically the same as (1), although this time, there are assumptions in the proof. By supposition, all of the assumptions are of the appropriate form for the final application of $\Box I$ to be proper.                                                                        $\Box$

This natural deduction system is sound and complete for S5 validity.

**Theorem 31**   $X \vdash_{S5} A$ *if and only if* $X \models_{S5} A$.

We will not present the details of the proof here. The argument has the same shape as the proof of theorem 29 (see page 172).

## 9.3   Features of S5

In this section, we will go over a few features of the proof system for S5. First, we will note that all S4 proofs are S5 proofs.

**Fact 4**   *If* $X \vdash_{S4} A$, *then* $X \vdash_{S5} A$.

*Proof.* The rules for the S4 proof system are those of the S5 proof system, except that the S4 side condition is stricter than the S5 side condition. Whenever an application of $\Box I$ or $\Diamond E$ satisfies the S4 side condition, it will satisfy the S5 side condition, so any S4 proof is also an S5 proof.                                                                                          $\Box$

Next, we will look at the *compositionality* of proofs. In a previous chapter, we saw that for intuitionistic and classical logic, we could combine proofs, so that if $\Pi_1$ is a proof of $A \vdash_I B$ and $\Pi_2$ is a proof of $B \vdash_I C$, then there is a straightforward way of chaining $\Pi_1$ and $\Pi_2$ together to obtain a proof $\Pi_3$ of $A \vdash_I C$. Proofs can be *composed*. Composing two intuitionistic proofs will always result in an intuitionistic proof, and composing two classical proofs results in a classical proof. What about S5 proofs (or S4 proofs)? It is not the case that simple composition will always work. To show this, let $\Pi_1$ and $\Pi_2$ be

$$\frac{q \wedge \Box p}{\Box p} \wedge E \qquad \frac{\Box p}{\Box \Box p} \Box I$$

and the composition, $\Pi_3$, is the following:

$$\frac{\dfrac{q \wedge \Box p}{\Box p} \wedge E}{\Box \Box p} \Box I$$

The problem is that $\Pi_3$ is not an S5 proof. The assumption does not have the correct form for the final $\Box I$ step. The problem of being unable to compose proofs was part of the reason that Prawitz proposed alternative forms of $\Box I$, a relative of which we saw with $\Box IM$. Despite the fact that the proofs do not compose, in general, in S5, it is still the case that if $A \vdash_{S5} B$ and $B \vdash_{S5} C$, then $A \vdash_{S5} C$. We know this because the proof system is sound and complete for possible worlds validity, and it is the case that if $A \vDash_{S5} B$ and $B \vDash_{S5} C$, then $A \vDash_{S5} C$.

We can allow a "lax" form of compositionality for S5 proofs. Suppose $\Pi_1$ is a proof of $A \vdash_{S5} B$ and $\Pi_2$ is a proof of $B \vdash_{S5} C$.

$$\cfrac{\cfrac{[B]^1 \\ \Pi_2 \\ C}{B \to C} \to I^1 \quad \cfrac{A \\ \Pi_1 \\ B}{}}{C} \to E$$

No matter how $\Pi_2$ is constructed, this proof will work as a proof of $A \vdash_{S5} C$ because there are no side conditions on $\to E$. The cost of this approach is the introduction of more non-normality, as we have introduced a detour. This is a lax form of compositionality, since we have had to do some extra work to combine two proofs (adding the conditional as a kind of "interface"), so it is not straightforward composition of proofs.

* * *

Next we will look at the interaction between modal notions and logical equivalence. Let's say that two formulas are LOGICALLY EQUIVALENT in S5 iff $\vdash_{S5} A \leftrightarrow B$. If two formulas are logically equivalent, are their necessitations also logically equivalent? While this is not immediately obvious, this turns out to be true.

**Theorem 32**  *If $\vdash_{S5} A \leftrightarrow B$, then $\vdash_{S5} \Box A \leftrightarrow \Box B$.*

*Proof.* Let $\Pi$ be an **S5** proof of $\succ A \leftrightarrow B$. We then construct a proof as follows:

$$
\cfrac{\cfrac{\begin{array}{c} \Pi \\ A \leftrightarrow B \end{array} \quad \cfrac{[\Box A]^1}{A}\,\Box E}{\cfrac{\cfrac{B}{\Box B}\,\Box I}{\Box A \to \Box B}\,{\to}I^1}\,{\leftrightarrow}E \qquad \cfrac{\begin{array}{c} \Pi \\ A \leftrightarrow B \end{array} \quad \cfrac{[\Box B]^2}{B}\,\Box E}{\cfrac{\cfrac{A}{\Box A}\,\Box I}{\Box B \to \Box A}\,{\to}I^2}\,{\leftrightarrow}E}{\Box A \leftrightarrow \Box B}\,{\leftrightarrow}I
$$

$\Box$

Notice that at the two $\Box I$ steps, the only open premises above them are necessitives. The steps then satisfy the side condition.

The proof demonstrates that the necessity of **S5** is *congruential*.

**Definition 60 (Congruentiality)**   *An operator $\bigcirc$ in the logic L is said to be* CONGRUENTIAL *if and only if whenever* $\vdash_L A \leftrightarrow B$, *we also have* $\vdash_L \bigcirc A \leftrightarrow \bigcirc B$.

We have shown that $\Box$ is congruential in **S5**. Our proof can be modified to show something stronger, namely, that the necessity is *monotonic*.[100]

**Definition 61 (Monotonicity)**   *An operator $\bigcirc$ in the logic L is said to be* MONOTONIC *if and only if whenever* $\vdash_L A \to B$, *we also have* $\vdash_L \bigcirc A \to \bigcirc B$.

Showing that $\Box$ is monotonic in **S5** is achieved by a minor modification to the subproof ending in the left premise of $\leftrightarrow I$. We place the proof of $A \leftrightarrow B$ with a proof of $A \to B$, leaving the rest of the subproof the same.

\* \* \*

It is important to understand that in this proof of congruentiality, we assume that $\vdash_{S5} A \leftrightarrow B$ rather than merely granting $A \leftrightarrow B$ as an assumption in the proof. It is not true that $A \leftrightarrow B \vdash_{S5} \Box A \leftrightarrow \Box B$. A possible worlds countermodel can demonstrate this.[101] Take the argument $p \leftrightarrow q \succ \Box p \leftrightarrow \Box q$, and let $W = \{w, u\}$ with $V(p, w) = V(q, w) = 1$, $V(p, u) = 1$, and $V(q, u) = 0$, like this:

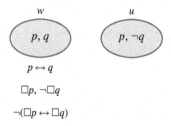

$$p \leftrightarrow q$$

$$\Box p, \neg\Box q$$

$$\neg(\Box p \leftrightarrow \Box q)$$

---

100. Congruentiality as defined here is what Humberstone (2011, 485) defines as $\leftrightarrow$-congruentiality. He defines a connective as being congruential or monotonic with respect to a consequence relation, rather than with respect to a biconditional or conditional connective.

101. The truth condition for the biconditional says that $V(A \leftrightarrow B, w) = 1$ iff $V(A, w) = V(B, w)$.

Here, $V(p \leftrightarrow q, w) = 1$, $V(\Box p, w) = 1$, but $V(\Box q, w) = 0$, which implies $V(\Box p \leftrightarrow \Box q, w) = 0$, so we have a counterexample to the argument from $p \leftrightarrow q$ to $\Box p \leftrightarrow \Box q$.

∗ ∗ ∗

This reasoning, about the congruentiality of $\Box$, can be generalized to other ways we can modify a formula, beyond sticking a box in front of it. The concept of a *formula context* is useful as for more general discussions of how to modify formulas.

**Definition 62 (Formula context)** *A* FORMULA CONTEXT *is a formula $C(p)$ that has some occurrences of an atom $p$. A substitution of the formula $B$ for $p$ in that context is the formula $C(B)$, obtained by replacing all copies of $p$ in $C$ with the formula $B$.*

For example, if the formula context $C(p)$ is

$$q \wedge (\underline{p} \vee (\neg r \vee \underline{p}))$$

with the occurrences of $p$ underlined, and $B$ is $s \to s$, then the substitution $C(B)$, with the newly substituted formulas underlined, is

$$q \wedge ((\underline{s \to s}) \vee (\neg r \vee \underline{(s \to s)})).$$

If $B$ is, instead, $p \wedge p$, then the substitution $C(B)$ is

$$q \wedge ((\underline{p \wedge p}) \vee (\neg r \vee \underline{(p \wedge p)})).$$

We will say that a context $C(p)$ is EXTENSIONAL in the logic $\mathsf{L}$ if and only if $A \leftrightarrow B \vdash_\mathsf{L} C(A) \leftrightarrow C(B)$. It is not hard to show that negation, conjunction, disjunction, and the conditional create extensional contexts, and in fact, if $C(p)$ is *any* complex formula made up from atoms using the connectives $\wedge$, $\vee$, $\neg$, and $\to$, then $C(p)$ is extensional.

A context $C(p)$ is said to be INTENSIONAL in the logic $\mathsf{L}$ if and only if (1) it is not extensional, and (2) $\Box(A \leftrightarrow B) \vdash_\mathsf{L} C(A) \leftrightarrow C(B)$. An *intensional context* is a context where substitution of (classically) equivalent formulas is invalid, but the substitution of *necessarily* equivalent formulas is valid.

The counterexample discussed above tells us that the necessity of $\mathsf{S5}$ creates an intensional context. $\Box(A \leftrightarrow B)$ is a strict equivalence, because it is (or is equivalent to) a biconditional using strict implication, that is, $(A \mathbin{\rightarrow\mkern-9mu\shortmid} B) \wedge (B \mathbin{\rightarrow\mkern-9mu\shortmid} A)$. More generally, it is possible to show that strict equivalents can be substituted in *any* formula context in $\mathsf{S5}$:

$$\Box(A \leftrightarrow B) \vdash_\mathsf{S5} C(A) \leftrightarrow C(B).$$

∗ ∗ ∗

The final important feature of $\mathsf{S5}$ that we will discuss requires one more definition.

**Definition 63 (Modality, positive modality)** *A* MODALITY *is a nonempty sequence of operators $\bigcirc_1, \ldots, \bigcirc_n$ where for $1 \leq i \leq n$, $\bigcirc_i$ is either $\neg, \Diamond,$ or $\Box$.*

*A* POSITIVE MODALITY *is a nonempty sequence of operators $\bigcirc_1, \ldots, \bigcirc_n$ where for $1 \leq i \leq n$, $\bigcirc_i$ is either $\Diamond$ or $\Box$.*

If we have a modality with some negations in the sequence, then we can use theorem 28 to push the negations all the way to the right of the sequence. Pairs of negations can be eliminated, so that will leave at most one negation at the right, leaving the remainder of the sequence of operators, if any, a positive modality. For example, if we have the modality

$$\neg \Diamond \Box \Diamond \neg \Box \neg,$$

we can use equivalences to obtain the equivalent modality,

$$\Box \Diamond \Box \Box \neg,$$

where

$$\Box \Diamond \Box \Box$$

is a positive modality.

In **S5**, all positive modalities $\bigcirc_1, \ldots, \bigcirc_n$ are logically equivalent to the $\bigcirc_n$. To prove this, we will show the following.

**Theorem 33**   *The following are true:*

1. $\vdash_{\mathsf{S5}} \Box A \leftrightarrow \Box\Box A$
2. $\vdash_{\mathsf{S5}} \Diamond A \leftrightarrow \Diamond\Diamond A$
3. $\vdash_{\mathsf{S5}} \Diamond A \leftrightarrow \Box\Diamond A$
4. $\vdash_{\mathsf{S5}} \Box A \leftrightarrow \Diamond\Box A$

*Proof.*   Some parts of this claim are easy to prove. (1) has been proved already, as has (3). The left-to-right directions of (2) and (4) follow from $\Diamond I$, so that leaves two more conditionals to prove.

$$\cfrac{[\Diamond\Diamond A]^3 \qquad \cfrac{\cfrac{\cfrac{[\Diamond A]^1 \quad [\neg\Diamond A]^2}{\bot}\neg E}{\bot}\Diamond E^1}{\cfrac{\cfrac{\bot}{\neg\neg\Diamond A}\neg I^2}{\Diamond A}DNE}}{\Diamond\Diamond A \to \Diamond A}\to I^3$$

$$\cfrac{[\Diamond\Box A]^3 \qquad \cfrac{\cfrac{\cfrac{[\Box A]^1 \quad [\neg\Box A]^2}{\bot}\neg E}{\bot}\Diamond E^1}{\cfrac{\cfrac{\bot}{\neg\neg\Box A}\neg I^2}{\Box A}DNE}}{\Diamond\Box A \to \Box A}\to I^3$$

$\Box$

These logical equivalences can be used to reduce all positive modalities $\bigcirc_1, \ldots, \bigcirc_n$ to whichever of $\Box$ and $\Diamond$ is $\bigcirc_n$.

Inspecting the proofs involved in all the equivalences reveals that they are particular to **S5**. Some of them use the **S5** side condition that is not already covered by the **S4** side condition, so some of the logical equivalences will not be available in **S4**.

## 9.4 Soundness and Completeness

This last section of this chapter will show how we can prove that the natural deduction systems for S4 and S5 are actually sound and complete for the models introduced in chapter 7. This material is more advanced than the rest of the chapter, so it bears a WARNING LABEL . Proceed further only when you are confident with the material in the rest of this chapter and when you have already read and understood the completeness proofs for classical logic and for Kripke models for intuitionistic logic. If you have scaled those mountains, you have everything you need to tackle this material.

$$* * *$$

Our target is the following theorem.

**Theorem 34 (Soundness and completeness for S4 and S5)** $X \vdash_{S4} A$ *if and only if* $X \models_{S4} A$, *and* $X \vdash_{S5} A$ *if and only if* $X \models_{S5} A$.

As with the proofs of our soundness and completeness theorems for classical logic and for intuitionistic logic, the proofs come in two parts. To show soundness, we prove, by induction on the construction of a proof, that any proof we can construct has no counterexample in a model. Then, for completeness, we show that if we have an argument for which there is *no* proof, we can construct a counterexample. This construction makes a model out of sets of formulas, in just the same way as we did for intuitionistic Kripke models in chapter 7. We can construct a canonical model for S4, which is a counterexample for *every* invalid argument, and we can do the same for S5. So, let's tackle these proofs one at a time. First, let us prove soundness.

*Proof.* Our aim is to show, by induction on the structure of a proof $\Pi$ from $X$ to $A$, that if the $\Box I$ and $\Diamond E$ steps in this proof satisfy the S5 side conditions, then the argument $X \succ A$ has no counterexample in any S5 model, and if the $\Box I$ and $\Diamond E$ steps in this proof satisfy the stronger S4 side conditions, then the argument $X \succ A$ has no counterexample in any S4 model, either.

For models for S5, we will consider the larger class of equivalence relation models, since any universal model is also an equivalence relation model in which the accessibility relation is universal. So, as we reason about S4 and S5, we will work with an arbitrary model $\langle W, R, V \rangle$ where $R$ is reflexive and transitive, and if we are focusing especially on S5, we will make the further assumption that $R$ is symmetric.

For our proof from $X$ to $A$, we wish to show that for any world $w$ in $W$ where each member of $X$ is true, so is $A$. For the nonmodal rules (the rules for $\bot$, $\wedge$, $\vee$, $\rightarrow$, $\neg$, and *DNE*), the reasoning is exactly the same as it is in the soundness proof for classical logic, since truth at a world acts, for these connectives, in the same way as truth in a Boolean valuation. The new work for S4 and S5 is in the modal rules.

So, suppose that our proof ends in a $\square E$ step, like this:

and, by the inductive hypothesis, we know that the argument $X \succ \square A$ has no counterexamples. Suppose that there is a counterexample to $X \succ A$, so for some model, $V(X, w) = 1$ but $V(A, w) = 0$. From the assumption, it follows that $V(\square A, w) = 1$, and as $R$ is reflexive, $V(A, w) = 1$, which is a contradiction.

The $\lozenge I$ case is similar. Suppose our proof ends in a $\lozenge I$ step, like this:

so, by the inductive hypothesis, we know that argument $X \succ A$ has no counterexample. It follows that the argument $X \succ \lozenge A$ has no counterexamples, for any world at which $A$ is true also has $\lozenge A$ true (again, since $R$ is reflexive).

The more interesting rules for us to deal with are the $\square I$ and $\lozenge E$ rules, which use the modal side conditions. Let's suppose that our proof ends in a $\square I$ step, like this:

$$
\begin{array}{c}
X \\
\nabla \Pi_3 \\
A \\
\hline
\square A
\end{array} \ \square I
$$

where the argument $X \succ A$ has no counterexample and where the premise of $\square I$ satisfies the S4 side condition: each formula in $X$ is a necessitive. If the argument $X \succ A$ has no counterexample at any world in any model, we know that at every world where each formula in $X$ is true, so is $A$. So, could there be a counterexample to the whole argument? Could we have a world $w$ where each member of $X$ is true but $\square A$ is false? If there were such a world, we would have a $v$ where $wRv$ and where $A$ is false at $v$. Now, since the members of $X$ are necessitive formulas (those of the form $\square B$ or $\neg \lozenge B$), these have the property of being preserved along the relation $R$, since $R$ is transitive. If $\square B$ is true at $w$ and $wRv$, then we must have $\square B$ true at $v$ too. (Since whenever $vRx$, since $wRv$ and $vRx$ we have $wRx$, and so, since $\square B$ holds at $w$, we have $B$ at $x$. So, since $B$ is true at all worlds $x$ accessible from $v$, we have $\square B$ at $v$. The same reasoning holds for negated diamond formulas.) So, each formula in $X$ is true at $w$, and, since each such formula is a necessitive, it is also true at $v$. So, if $w$ is a counterexample to the argument from $X$ to $\square A$, then the world $v$ (where $A$ fails) would be a counterexample to the argument from $X$ to $A$. But our hypothesis is that there is no counterexample to *that* argument, so there can be no counterexample to the whole argument either.

If we are building an S5 proof, then we can no longer appeal to the assumption that each formula in $X$ is a necessitive. We have only the weaker assumption that each formula is *modal*. However, we can appeal to the stronger assumption that the model in an assumed counterexample is an equivalence model, so the relation $R$ is an equivalence relation—in addition to being reflexive and transitive, it is symmetric. In that case, we have the stronger property that all *modal* formulas true at a world $w$ are true at any world $v$ where $wRv$, since if $\Diamond D$ holds at $w$, there is some $v'$ where $wRv'$ and $D$ holds at $v'$, and thus, $\Diamond D$ holds at $v$ too. The reason is that by symmetry, from $wRv$, we also have $vRw$, and by transitivity, $wRv'$ gives $vRv'$, and so, $D$ holding at $v'$ ensures that $\Diamond D$ holds at $v$ too. (Similar reasoning applies to $\neg\Box$ formulas.) So, in this case, we can still be sure that the formulas in $X$ are true not only at $w$ but also at any world accessible from $w$, and so, any counterexample to the argument from $X$ to $\Box A$ would mean that there would have to be a counterexample to the earlier argument, from $X$ to $A$. If $X \succ A$ has no counterexample in S5E models, neither does $X \succ \Box A$, when the formulas in $X$ are modal.

To complete the verification of soundness, we need just check the $\Diamond E$ rule. So, suppose our proof is extended like this:

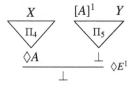

where the assumptions $Y$ are (for the moment) necessitives, and the shorter arguments $\Pi_4$ and $\Pi_5$ have no counterexamples at any world in an S4 model. Could we have a counterexample to the whole argument at any world? Suppose that we did. That is, we have a world $w$ where the formulas in $X$ and in $Y$ are true. Since the argument from $X$ to $\Diamond A$ has no counterexamples in our model, we must have $\Diamond A$ true at our world $w$, which means that there is some world $v$ where $wRv$ and where $A$ is true. Since $wRv$ and the formulas in $Y$ are all necessitives that are true at $w$, they are also true at $v$. So notice that the formulas in $Y \cup \{A\}$ are all true at $v$. But the argument $Y, A \succ \bot$ is valid—it has no counterexamples, so we could have no world where the formulas in $Y \cup \{A\}$ are all true. So, our assumption that the whole argument has a counterexample cannot be true. The whole argument is also S4 valid.

If we weaken the assumption on this inference to the S5 side condition, to the effect that the formulas in $Y$ are merely modal formulas, then similar reasoning applies, provided that the model in question is an equivalence model.

This completes the examination of all the rules of our proof system. Any S4 proof gives rise to an S4 valid argument, and any S5 proof similarly gives rise to an S5 valid argument. Our proof systems are sound. □

To check the completeness half of the soundness and completeness theorem, we will do the same thing we did for the completeness proof for intuitionistic logic. We will construct a *canonical model* for S4 and one for S5. Each model has the form $\langle M, R, V \rangle$ where $M$ is the set of all sets $X$ of maximal $C$-avoiding formulas, for any formula $C$ (defined using S4 proof for the S4 model and S5 proof for the S5 model). Each $V$ is defined by setting $V(p, X) = 1$ iff $p \in X$, and the accessibility relation on both models is defined like this: $XRY$ if and only if for each formula $A$, if $\Box A \in X$ then $A \in Y$.

**Definition 64 (The canonical S4 and S5 models)** *The* CANONICAL S4 MODEL *is* $\langle M_{S4}, R_{S4}, V_{S4} \rangle$ *where*

- $M_{S4} = \{X \subseteq \mathsf{MForm} : for\ some\ formula\ A,\ X\ is\ an\ S4\ maximal\ A\text{-}avoiding\ set\ of\ formulas\}$,
- $XR_{S4}Y$ *if and only if for each formula A, if* $\Box A \in X$ *then* $A \in Y$, *and*
- $V_{S4}(p, X) = 1$ *iff* $p \in X$.

*The* CANONICAL S5 MODEL *is* $\langle M_{S5}, R_{S5}, V_{S5} \rangle$ *where*

- $M_{S5} = \{X \subseteq \mathsf{MForm} : for\ some\ formula\ A,\ X\ is\ an\ S5\ maximal\ A\text{-}avoiding\ set\ of\ formulas\}$,
- $XR_{S5}Y$ *if and only if for each formula A, if* $\Box A \in X$ *then* $A \in Y$, *and*
- $V_{S5}(p, X) = 1$ *iff* $p \in X$.

As noted, the definitions of the two canonical models have a common form. Some of the proofs used to verify that the canonical S4 model has certain properties work for showing that the canonical S5 model has those properties. When an argument will work for either model, we will omit the subscripts.

The first thing to check is that these models are indeed models for S4 and S5, respectively: that is, that the relation $R_{S4}$ so defined is reflexive and transitive, and $R_{S5}$ is reflexive, transitive, and symmetric.

**Lemma 21** *The relation* $R_{S4}$ *in the canonical S4 model is reflexive and transitive, and the relation* $R_{S5}$ *in the canonical S5 model is an equivalence relation.*

*Proof.* $R$ is reflexive, since for any maximal $C$-avoiding set $X$, if $\Box A \in X$, then since $\Box A \vdash_{S4} A$ (and also, $\Box A \vdash_{S5} A$), we must have $A \in X$ too.

$R$ is transitive, since if we have maximal formula-avoiding sets $X$, $Y$, and $Z$ where $XRY$ and $YRZ$, we will be able to show that $XRZ$ as follows: suppose $\Box A \in X$. We'll show that $A \in Z$ in this way. Since $\Box A \vdash_{S4} \Box\Box A$, we have $\Box\Box A \in X$, and so, $\Box A \in Y$, and as a result, $A \in Z$, as desired.

Finally, in the S5 canonical model, $R_{S5}$ is symmetric. If $XR_{S5}Y$, we can show that $YR_{S5}X$ like this. Suppose that $\Box A \in Y$. We must have $A \in X$ since if $\neg A \in X$ (recall, since we have *DNE* in the logic, each maximal $C$-avoiding set is consistent and complete: if $A \notin X$, then $\neg A \in X$). Then, since $\neg A \vdash_{S5} \Box\Diamond\neg A$, we must have $\Diamond\neg A \in Y$, but this would mean that $\bot \in Y$ since $\Diamond\neg A, \Box A \vdash_{S5} \bot$, and this is impossible. So, if $\Box A \in Y$, then $A \in X$, for any $A$, and so, $YR_{S5}X$. The relation $R_{S5}$ is symmetric in the S5 canonical model. $\qquad \square$

The next thing to prove for our canonical model is that the truths at a world (at a set of formulas) are exactly the members of that world.

**Lemma 22** *In the canonical S4 and S5 models, for any world X and any formula A,* $V(A, X) = 1$ *iff* $A \in X$.

*Proof.* The proof is by induction on the construction of the formula $A$. The cases for the nonmodal connectives are straightforward: we have done them before when we proved lemmas 9 and 10. The hard case involves verifying that the result holds for *necessity* and *possibility* formulas.

For necessity, we need to show that for any maximal $C$-avoiding set $X$, $\Box A \in X$ if and only if for each $Y$ where $XRY$, $A \in Y$. The left-to-right direction is given by the definition of $R$. That is, if $\Box A \in X$, and if $XRY$, then $A \in Y$. For the reverse, if $\Box A \notin X$, we want to show that there is some $Y$ where $XRY$ where $A \notin Y$. This is not too difficult to show. Consider the set $X^{\square}$, which is defined as the set of all formulas of the form $\square B$ that are in $X$. We have $X^{\square} \nvdash A$, since if we *had* $X^{\square} \vdash A$, we could extend that proof to a proof of $\square A$ with one step of $\square I$. So, since $X^{\square} \nvdash A$, there is some maximal $A$-avoiding set $Y$ where $X^{\square} \subseteq Y$. We have $A \notin Y$. We also have $XRY$, since whenever $\square B \in X$, $\square B \in X^{\square} \subseteq Y$, and so, $B \in Y$ too. So, we have the $Y$ we wanted.

For possibility, we need to show that for any maximal $C$-avoiding set $X$, $\Diamond A \in X$ if and only if there is some $Y$ where $XRY$ and $A \in Y$. We know that if $\Diamond A \notin X$, then $\square \neg A \in X$, and so, for any $Y$ where $XRY$, $\neg A \in Y$, and so, $A \notin Y$. It remains to be shown that if $\Diamond A \in X$, then there is some $Y$ where $XRY$ and $A \in Y$. As with the reasoning for $\square$, consider the set $X^{\square}$. We know that $X^{\square}, A \nvdash \bot$, since if we had $X^{\square}, A \vdash \bot$, we could extend this proof by a step of $\Diamond E$ to show that $X^{\square}, \Diamond A \vdash \bot$, but this conflicts with the fact that $\Diamond A \in X$ and the fact that $X$ excludes $C$ (and, hence, excludes $\bot$). So, since $X^{\square}, A \nvdash \bot$, we can find a maximal $\bot$-avoiding set $Y$ where $X^{\square} \cup \{A\} \subseteq Y$. It follows that $XRY$ and $A \in Y$, as desired. $\qquad \square$

So, we have constructed our canonical $\mathsf{L}$ model (where $\mathsf{L}$ is either $\mathsf{S4}$ or $\mathsf{S5}$) and shown that the formulas true at a world $w$ in the canonical $\mathsf{L}$ model are exactly the formulas in $w$, viewed as a set. Since the sets in the model are the $\mathsf{L}$-maximal $C$-avoiding sets (for each formula $C$), whenever $X \nvdash_{\mathsf{L}} A$, there is some world in our model where each member of $X$ holds, and $A$ fails. This model shows that $X \nvDash_{\mathsf{L}} A$. This finishes the demonstration of the *completeness* part of the soundness and completeness theorem.

$$* * *$$

Well, we have shown that anything we cannot prove in $\mathsf{S5}$ has a counterexample in the canonical $\mathsf{S5}$ model. This model is an equivalence relation model, not a possible worlds model. To show that there is a counterexample in an equivalence relation model, we can appeal to the proof of theorem 24 (see page 139), to show that we can view each equivalence class of the equivalence relation model as a universal model of its own. In this way, the canonical model supplies counterexamples to every invalid argument, either as one equivalence relation model or as a supply of a whole family of possible worlds models.

### 9.5 Key Concepts and Skills

☐ You should understand the idea of a rule with *side conditions*, the motivation for the $\mathsf{S4}$ and $\mathsf{S5}$ conditions in the modal rules, and you should be able to read proofs and check that instances of the modal rules satisfy or violate the side conditions.

☐ You should be able to construct simple proofs in the proof systems for $\mathsf{S4}$ and $\mathsf{S5}$.

☐ You need to understand what it means for an operator to be *congruential* and why the modal operators $\square$ and $\Diamond$ *are* congruential in $\mathsf{S4}$ and $\mathsf{S5}$.

☐ You need to understand the concept of a *formula context* and the difference between *intensional* and *extensional* contexts.

$$\begin{array}{c} X \\ \bigtriangledown\;\Pi \\ \dfrac{A}{\Box A}\;\Box I \end{array} \qquad \dfrac{\Box A}{A}\;\Box E \qquad \dfrac{A}{\Diamond A}\;\Diamond I \qquad \begin{array}{c} [A]^1\quad X \\ \Pi\;\bigtriangledown \\ \dfrac{\Diamond A\qquad\qquad\bot}{\bot}\;\Diamond E^1 \end{array}$$

The S4 SIDE CONDITION requires that all assumptions in $X$ upon which the premise of an application of the rule $\Box I$ depends and the assumptions in $X$ upon which the minor premise of an application of $\Diamond E$ depends are necessitives.

The S5 SIDE CONDITION requires that all assumptions in $X$ upon which the premise of an application of the rule $\Box I$ depends and the assumptions in $X$ upon which the minor premise of an application of $\Diamond E$ depends are modal formulas.

### 9.6 Questions for You

**Basic Questions**

1. Read these proofs. Check and see whether the S4 side condition (or the S5 side condition) is satisfied by the applications of $\Box I$ and $\Diamond E$ inferences.

$$\dfrac{\dfrac{\Box(p\to q)}{p\to q}\;\Box E \qquad \dfrac{[\Box p]^1}{p}\;\Box E}{\dfrac{\dfrac{q}{\Box q}\;\Box I}{\dfrac{\Box p\to\Box q}{\Box(\Box p\to\Box q)}\;\Box I}\;{\to}I^1}\;{\to}E$$

$$\dfrac{\dfrac{\Box(p\to\Diamond q)}{p\to\Diamond q}\;\Box E \qquad [p]^1}{\dfrac{\dfrac{\Diamond q}{\Box\Diamond q}\;\Box I}{p\to\Box\Diamond q}\;{\to}I^1}\;{\to}E$$

$$\dfrac{\Diamond(\Box p\wedge\Diamond q)\qquad\qquad \dfrac{\dfrac{\dfrac{[\Box p\wedge\Diamond q]^4}{\Diamond q}\;\wedge E \qquad \dfrac{\dfrac{[\Box p]^3}{p}\;\Box E\quad[q]^1}{\dfrac{p\wedge q}{\Diamond(p\wedge q)}\;\Diamond I}\;\wedge I\quad[\neg\Diamond(p\wedge q)]^5}{\dfrac{\bot}{\dfrac{\neg\neg\Diamond(p\wedge q)}{\dfrac{\Diamond(p\wedge q)}{\Box p\to\Diamond(p\wedge q)}\;{\to}I^3}\;DNE}\;\neg I^2}\;\Diamond E^1}{\dfrac{\dfrac{\dfrac{[\Box p\wedge\Diamond q]^4}{\Box p}\;\wedge E}{\Diamond(p\wedge q)}\;{\to}E\qquad\qquad[\neg\Diamond(p\wedge q)]^5}{\bot}\;\neg E}\;\Diamond E^4}{\dfrac{\dfrac{\bot}{\neg\neg\Diamond(p\wedge q)}\;\neg I^5}{\Diamond(p\wedge q)}\;DNE}$$

2. Find S4 proofs for the following arguments:

    i. $\Box p, \Box q \succ \Box(p\wedge q)$

ii. $\Diamond\Diamond p, \neg\Diamond p \succ \bot$

iii. $\Diamond\Diamond p \succ \Diamond p$

iv. $\Diamond p, \Box q \succ \Diamond(p \wedge q)$

v. $\Box(p \rightarrow q), \Diamond p \succ \Diamond q$

3. Find **S5** proofs for the following arguments:

i. $\Diamond\Box p \succ p$

ii. $\Box(p \vee q), \neg\Diamond p \succ \Box q$

iii. $p \succ \Box\Diamond p$

iv. $\Diamond(\Box p \wedge \neg q), \Diamond(\Box q \wedge \neg p) \succ \bot$

v. $\succ \Box(\Box p \rightarrow q) \vee \Box(\Box q \rightarrow p)$

Note that (v) is difficult. It is recommended that you use your proof for (iv) as well as some derived rules to obtain a proof for (v).

### Challenge Questions

1. In section 7.3.1, we introduced the connective of strict implication, $\dashv$3, which was interpreted in possible worlds models by the clause

   • $V(A \dashv3 B, w) = 1$ iff for all $x \in W$, either $V(A, x) = 0$ or $V(B, x) = 1$.

   What are appropriate introduction and elimination rules for this connective? Once you formulate your rules, show that they are sound with respect to **S5** validity.

2. In **S4**, more positive modalities are distinct than in **S5**. In fact, in **S4**, the following modalities are all distinct:

$$\Diamond \quad \Box \quad \Box\Diamond \quad \Diamond\Box \quad \Box\Diamond\Box \quad \Diamond\Box\Diamond$$

   We cannot prove the equivalences (3) and (4) from theorem 33 in **S4**. However, we *can* prove these:

   i. $\vdash_{\mathbf{S4}} \Box\Diamond\Box\Diamond A \leftrightarrow \Box\Diamond A$

   ii. $\vdash_{\mathbf{S4}} \Diamond\Box\Diamond\Box A \leftrightarrow \Diamond\Box A$

   First, show how it follows from these equivalences (and (1) and (2) from theorem 33) that every positive modality is equivalent to one of the six listed above.

   Then (and this part is rather difficult), find **S4** proofs for the two equivalences (i) and (ii).

3. Construct enough models to show that the six modalities listed above are not equivalent in **S4**. You do not need to generate a *huge* number of models for this. For example, first find an **S4** model where $\Diamond p$ is true at a world and *all* of the other modalities of $p$ are false at that world, and then the same sort of reasoning can give you a model where $\Box p$ is false at a world and all of the other modalities of $p$ are *true* at that world. Then look at what it takes for each of $\Box\Diamond p$, $\Diamond\Box p$, $\Box\Diamond\Box p$, and $\Diamond\Box\Diamond p$ to be true at worlds in an **S4** model and the relationships between these conditions.

4. Explain why $C(p)$ is an extensional context, if $C$ involves only the connectives $\wedge$, $\vee$, $\rightarrow$, $\neg$. Then explain why $C(p)$ is either an intensional or an extensional context when $C$ contains any of the connectives of the language of **S5**. (You can prove this result by induction on the construction of the context $C(p)$.)

5. Imagine extending our language with one-place or two-place connectives that have these meanings

   *… but …    … because …    I said that …*

   Would you expect each of these to create extensional or intensional contexts?

6. How do we come to learn the concepts of necessity and possibility? Do we learn something like the truth conditions ($\Box p$ is true iff $p$ is true in all accessible worlds), or do we learn something like the inference rules (you can deduce $\Box p$ only when you have proved $p$ from necessary truths)? Try to list as many reasons for and against either account of how we come to grasp these concepts. How do you weigh up these considerations?

7. The modal rules used in this chapter, in particular, the side conditions on $\Box I$ and $\Diamond E$ for S5, were restrictive. One might think that the rules were too restrictive, since they rule out the premise of a $\Box I$ step depending on an assumption of $\Box p \land \Box q$. We can loosen the side conditions a bit, following essentially what Prawitz (1965, chap. 6) did when he set out his proof systems for modal logics. Say that a formula $A$ is *modally closed* iff every occurrence of an atom, apart from $\bot$, in $A$ is in the scope of either $\Box$ or $\Diamond$. For the extended S5 side condition, change "modal formula" in the S5 side condition to "modally closed formula." Call this new side condition the S5$^+$ side condition.

   First, provide proofs of some S5 valid arguments using the S5$^+$ side condition. Next, show that every proof satisfying the S5 side condition satisfies the S5$^+$ side condition. Finally, show that not every proof satisfying the S5$^+$ side condition satisfies the S5 side condition by supplying a proof satisfying the one but not the other.

8. Another alternative rule, also supplied by Prawitz (1965, chap. 6), changes the rule $\Diamond E$. The new rule, $P\Diamond E$, is the following, where $B$ is required to be a necessitive.

   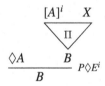

   The S4 side condition is still used. First, provide new proofs for the arguments in basic question 2 using this rule, rather than $\Diamond E$. Second, provide a proof of $\Diamond\bot \succ \bot$ using this rule. Finally, can you think of any philosophical reasons for favoring the rule $P\Diamond E$ over $\Diamond E$ or vice versa?

9. Give introduction and elimination rules for the actuality operator. What sort of side conditions do these rules need? Do the side conditions on the rules for $\Box$ and $\Diamond$ need adjustment? Prove that your rules are sound with respect to one concept of validity defined in the previous chapter.

# III PREDICATE LOGIC

# 10 Proofs for Predicate Logic

Think back to the pair of arguments we considered back at the beginning of chapter 1:

|     | *All footballers are bipeds.* |     | *All footballers are bipeds.* |
| --- | --- | --- | --- |
| (1) | *Sócrates is a footballer.* | (2) | *Sócrates is a biped.* |
|     | So, *Sócrates is a biped.* |     | So, *Sócrates is a footballer.* |

We used this pair of arguments to motivate the two different approaches to formal logic: *proof theory*, for analyzing the structure of good arguments, and *model theory*, for a systematic way of presenting counterexamples to bad arguments. Now, you might have noticed that while we have spent nine chapters on proofs and models for logical concepts like the propositional connectives and modal operators, we have not said anything about how to analyze arguments like (1) and (2). We have said nothing about the *quantifiers* (expressions like "*All*" in the first premise of (1) and (2)) or anything about the *names* or *predicates* that appear in those sentences. In fact, as far as the surface structure of the sentences in (1) and (2) goes, *none* of the connectives or modal operators of the first two parts of this book appear. Yet, it should be obvious that there is logical structure in (1) that makes the step from its premises to its conclusion a good one, and likewise, we can explain why the step in argument (2) is prone to counterexamples. To do that, we need to expand our toolbox once more, and just as we did in the first two parts of the book, we start by expanding our *language*.

## 10.1 Syntax

The formal languages studied in the first two parts of this book have been *propositional*. They are propositional in two senses. First, the sentences in our languages have all been *declaratives*, sentences that express propositions rather than make requests, or ask questions, and so on.[102] But our languages have been propositional in an even deeper sense than this. In a second sense, they have been propositional because in our languages, the components of formulas—if they have components—have been *more formulas* and the connectives that join them together. We have not peered into the structure of declaratives to find anything other than a proposition. The *atomic* formulas in our languages are just propositional atoms, such as *p*. The atoms represent simple declarative sentences, such as "Melbourne is east of Adelaide." While propositional logic is useful for many things, it

---

102. Look back to section 1.2 for the differences between declaratives and other sentence types.

does not let us represent the shared structure between "Melbourne is east of Adelaide" and "Melbourne is east of Perth." Both are represented by propositional atoms, $p$ and $q$. Those two sentences have something in common, namely, "Melbourne is east of ____." We can capture these commonalities by distinguishing more structure inside atoms. So, the time has come to split the atom to see what we can find inside.

* * *

Let's start by looking at some of the sentences that have featured in our arguments, to see if we can find some of the structure that is doing the work in leading from the premises to the conclusion. In addition to argument (1) above, recall this proof:

$$\frac{\displaystyle\text{Sydney is east of Melbourne} \quad \frac{\text{Melbourne is east of Adelaide} \quad \text{Adelaide is east of Perth}}{\text{Melbourne is east of Perth}}\ (1)}{\text{Sydney is east of Perth}}\ (2)$$

Here, steps (1) and (2) seem good, and what is crucial in these steps is the structure of the premises and the conclusion. If we were to try to isolate that structure, we might come up with something like this:

$$\frac{a \text{ is east of } b \quad b \text{ is east of } c}{a \text{ is east of } c}$$

Both step (1) and step (2) have this shape, for different choices to plug in as values of $a$, $b$, and $c$. Now, what sorts of things go in for $a$, for $b$, and for $c$? These are not the sorts of things we used for $p$, $q$, and $r$ before. Whatever they are, they aren't themselves *propositions*. In the case of our argument, they are city names, though a moment's thought should convince you that you don't need to name cities for this argument shape to work. The *general* feature of $a$, $b$, and $c$ in this case is that they are *names*. They name things. A name is not a sentence. If I were just to say "Adelaide," and leave it at that, you'd be hard-pressed to say that I was saying something true or false.[103] We can name things in order to point them out, or to describe them (like when I say "Adelaide is nicer than people think") or compare them (like when I say "Adelaide is east of Perth"). When we describe or compare, we are forming a judgment or making a statement. So let's look at how we do this.

Consider the "frame" into which we slotted the two names to form a declarative sentence. That frame has this shape

____ *is east of* ____

where we have space to substitute two names, in order to form a declarative sentence. Here, the order of the two names is important, because when I say that Adelaide is east of Perth, that is not the same as saying that Perth is east of Adelaide. What underlies the structure of the reasoning in our proof is the fact that we have substituted different names into the same frame. We call phrases like "___ *is east of* ___" *predicates*. In this case, we call it a *two-place* predicate, because it requires two names to be supplied to form a sentence.

---

103. Unless you had asked a question and I was giving you "Adelaide" as an answer.

Not all predicates are two-place. When I say that Sócrates is a footballer, or that he is a biped, these sentences feature the name "Sócrates," and the rest:

_____ *is a footballer*   _____ *is a biped*

are predicates. Both of these predicates require only *one* name to be supplied to form a sentence.

So, we have discovered two components of declarative sentences—we have split the atom to find at least two components inside it: *names* and *predicates*.

\* \* \*

Names and predicates don't exhaust the features of our sentences that play an important role in distinguishing good and bad arguments. The first premise of both arguments:

*All footballers are bipeds*

features no names, and it combines the two predicates "___ *is a footballer*" and "___ *is a biped*" to form a sentence all its own. The crucial glue is what logicians call a *quantifier*.

There are different ways[104] to analyze the structure of claims involving quantifiers. Our approach is simple, though it could be argued that it's not the *best* fit with the surface structure of English or of any natural language. The approach we will take has two virtues: (a) Our approach is the current orthodoxy in logic. When you read other logic texts, or see how logic is used in philosophy, mathematics, computer science, or elsewhere, you will see a language just like this. (b) We will use the raw materials from the language of propositional logic and make as small an addition to it as possible, in order to incorporate quantifiers.

The key idea is that when we say that all footballers are bipeds, we are making a claim something like this:

*If a is a footballer, a is a biped*

except we aren't saying it for some specific object we're calling *a*. We're making the claim for *all* objects whatsoever. For anything at all, if *it* is a footballer, *it* is a biped. Here, the "*it*" plays a crucial role. Instead of picking things out by name, we *quantify over* them, and instead of referring to objects by name, the pronoun (here, "*it*") picks out objects one by one, as supplied by the quantifier (here, "*for anything at all*"). So, in mangled English, we think of "*All footballers are bipeds*" as having this shape:

*For anything at all, if it is a footballer, then it is a biped*

where "*for anything at all*" is a quantifier, "*if ... then ...*" is the conditional, "*... is a footballer*" and "*... is a biped*" are predicates, and the remaining item, the "*it*" is what we call a *variable*. In the formal language we will define below, the structure is written like this,

$$\forall x(Fx \to Bx)$$

---

104. Look at the challenge questions at the end of this chapter to explore an alternative approach to the syntax and semantics of quantifiers.

and we can formalize our argument (1) like this:

$$\frac{\forall x(Fx \rightarrow Bx) \qquad Fa}{Ba}$$

This is the structure we will analyze in these our final two chapters. Atomic formulas in our formal language will be made up from *predicates*, *names*, and *variables*, and we will combine these atomic formulas using our connectives ($\land$, $\lor$, $\neg$, and $\rightarrow$) and the quantifiers ($\forall$ for *universal* generality and its dual $\exists$ for *existential* claims).[105] You might wonder what happened to the modal operators of the last three chapters. Combining $\Box$ and $\Diamond$ with quantifiers introduces a range of complex philosophical and logical issues. The best way to treat modal predicate logic is itself a question of ongoing research, so we leave modal operators aside for the moment. We will return to them in the final chapter.

\* \* \*

So, in predicate logic, our atomic formulas have two parts, a *predicate* and some number of *terms*. Predicates are used to formalize verb phrases, like "is east of," "sees," "is a footballer," or "travels to." For *predicates*, we will use uppercase letters, $F$, $G$, $H$, and so on. Each predicate takes a specified number of terms for arguments, called its *arity*. A predicate with $n$ slots is an $n$-ary, or an $n$-place, predicate.

A term is either a *name*, like "Sócrates" or "Adelaide" or "5," or a *variable*. We will use lowercase letters from the beginning of the alphabet, such as $a, b, c, d, \ldots$, for our names, and lowercase letters from the end of the alphabet, like $x, y, z, \ldots$, for our variables. As with our atomic formulas from the first part of the book, if we run out of letters, we can subscript them with numbers, so $x_1$, $y_5$, and $z_{23}$ are variables, and $a_2$, $b_3$, and $c_{3\,088}$ are names. Terms are intended to pick out *things*, rather than describe them. Sometimes, we will use the letters $t$ and $s$ (and $t_n$ or $s_m$ where $n$ and $m$ are numbers) to stand for terms where we do not care whether the term is a name or a variable.

These will be all the terms that we will use for now. For some purposes in logic, one may want to introduce *function symbols*, which generate new terms from old terms, like "+," which generates a term for a number, when supplied with two other terms for numbers. For example, "2 + 3" is a different term from "5," even though they both designate the same number. As another example, "the birthday of $t$" names the birthday of anyone picked out by the term $t$. We might even be interested in term-forming operators, which generate terms from descriptions or other formulas not restricted to terms.[106] "The first person to land on the moon" could be treated as a term, which is constructed out of predicates. These techniques introduce additional complications, so we won't consider them further in these two chapters.

\* \* \*

Now that we have distinguished predicate and term structure, we can introduce ways to form *general* statements to say that everything, nothing, or something is a certain way. Suppose we are looking at a frog enclosure at a zoo that contains exactly two frogs, Adriane

---

105. *Dual*? Here is a hint about what that might mean. $\exists$ stands to $\forall$ as $\lor$ stands to $\land$ and as $\Diamond$ stands to $\Box$.
106. See Kalish and Montague (1964) for more on term-forming operators.

and Chris. We can say that Adriane is a green frog, symbolized as $Ga \land Fa$, and that Chris is a red frog, symbolized as $Rc \land Fc$. To say that all frogs are in the water, in this scenario, we could use $Wa \land Wc$, to symbolize that Adriane is in the water and Chris is in the water, but providing a conjunct for each frog would get really tedious if there are a lot of frogs. A bigger problem arises if we have not named all the frogs, because then we wouldn't say what we want, namely, something about *all frogs*. (Never mind the difficulties that arise when there are infinitely many frogs!) Instead, we can formalize the statement using the *universal quantifier* ($\forall$) as

$$\forall x(Fx \to Wx).$$

If we grant this, then for any thing at all, we are committed to the conditional claim that if it is a frog, it is in the water. So, we're committed to $Fa \to Wa$ (if Adriane is a frog, she's in the water), $Fc \to Wc$ (if Chris is a frog, she's in the water), and so on, but we are also committed *generally*. If we learn that something else is a frog—and if we still grant $\forall x(Fx \to Wx)$—then we are committed to the claim that it is in the water too.

Now, suppose that a frog is on a rock, rather than in the water, but you are not sure which frog it is. In the zoo scenario, you could express the statement that there is a frog on the rock with $Oa \lor Oc$, but, as before, this approach would take a while if there are a *lot* of frogs, and it would be hard or impossible to do this if you have not named them all. Instead, we can formalize the statement "there is a frog on the rock" using the *existential quantifier* ($\exists$) as

$$\exists x(Fx \land Ox).$$

This claim says that there is something that is both a frog and is on the rock, and we can agree that it's true, even if we don't know which frog it is. In this chapter, we will see how we can use quantified expressions in our reasoning, how we can combine quantifiers to form very rich and expressive statements, and how we can infer *to* quantified expressions and what we can infer *from* them. Specifically, it is the way that quantifiers can be *combined* that distinguishes modern predicate logic from the syllogistic logic of Aristotle. For centuries, we have known how to reason with claims like "All footballers are bipeds" and "Some bipeds are injured." It requires a conceptual advance from Aristotle's logic to explain the structure of claims like "Every football team has some player who kicked all of the goals in at least one of their matches." Understanding claims with this kind of structure requires attention to predicates, names, variables, quantifiers, and scope, and making these ideas clear and precise is the goal of this chapter. So, with that, a definition:

**Definition 65 (Terms, formulas)** *The set of* TERMS *of the language is defined as follows.*

- *Every variable, $x, y, z, x_1, \ldots$, is a term.*
- *Every name, $a, b, c, a_1, \ldots$, is a term.*
- *Nothing else is a term.*

   *The set of* FORMULAS *of predicate logic is defined inductively as follows.*

- *If $F$ is an n-ary predicate and $t_1, \ldots, t_n$ are terms, then $Ft_1 \cdots t_n$ is a formula. In addition, $\bot$ is a formula, as before.*
- *If $A$ and $B$ are formulas, then $(A \land B)$, $(A \lor B)$, $(A \to B)$, and $\neg A$ are formulas,*
- *If $A$ is a formula and $x$ is a variable, $\forall x A$ and $\exists x A$ are formulas.*
- *Nothing else is a formula.*

The first clause here, saying that an *n*-ary predicate requires *n* terms to make a formula, allows for *n* to be any finite number, including the case where *n* is *zero*. What is a predicate that requires no terms to be supplied to make a formula? It's an atomic proposition without any terms inside. Does a natural language like English have any sentences like this? Maybe, maybe not. While "It is raining" contains an "it," it isn't clear that this "it" functions like a name or a variable. (If that "it" names something, what does it name? What is it that is raining when we say that it's raining? If it's raining and it's sunny, is it the same *thing* that is raining and is sunny? If you think questions like these shouldn't be asked or answered, maybe you think that "it's raining" and "it's sunny" are better treated as atomic formulas containing no terms at all.) If we use zero-place predicates in our language, we will continue the convention of writing them with lowercase letters, like *p*, *q*, and *r*.

* * *

As was our practice in previous chapters, we can leave out the outermost parentheses in conjunctions, disjunctions, and conditionals when they occur as the main connective. But, these parentheses can be important. Here is why: if *F* is a one-place predicate, then *Fx* is a formula. If *p* is a zero-place predicate (a propositional atom), then $Fx \rightarrow p$ is a formula, and if we wanted to say that this *conditional* is true *no matter what I chose for x*, then we would universally quantify it: we would get $\forall x(Fx \rightarrow p)$. This says that *for anything you choose at all*, if it has feature *F*, then *p* is true. Here, the parentheses make a difference in what the statements say. For if we leave them out and write $\forall xFx \rightarrow p$, we get something very different.[107] $\forall xFx \rightarrow p$ is a conditional statement with the universally quantified formula $\forall xFx$ as its antecedent, and *p* as its consequent. If I say $\forall xFx \rightarrow p$, then I am saying that *p* is true if *everything* has feature *F*. That is a much stronger condition for showing that *p* holds than $\forall x(Fx \rightarrow p)$. To show *p* using $\forall x(Fx \rightarrow p)$, we could simply find *something* with feature *F*, to conclude *p*. Just as with the formulas in propositional logic, the parentheses matter.

With $\forall x(Fx \rightarrow p)$ and $\forall xFx \rightarrow p$, we have already seen some examples of formulas in this language. Let's look at a few other examples before continuing on with the language. If *G* is another one-place predicate, then we can construct formulas like

$$\exists xFx \wedge \exists xGx \qquad \exists x(Fx \wedge Gx)$$

using the rules of construction. *Fx* and *Gx* are both atomic formulas, so $\exists xFx$ and $\exists xGx$ are both formulas, and we conjoin them to form $\exists xFx \wedge \exists xGx$. To translate this back into almost-English, we would say something like "something is *F* and something is *G*." The fact that we used the same variable in $\exists xFx$ as in $\exists xGx$ does not mean that we are saying the same thing is both *F* and *G*, any more than the fact that we used the same word "something" in the English sentence would mean that. Think of it like this: someone could say "something is *F*" (or $\exists xFx$), and someone *else* could say "something is *G*" (or $\exists xGx$) where there is no connection between the two statements.[108] We, then, could conjoin their

---

107. There is a similarity to the situation in modal logic from chapter 7, namely, the difference between $\Box(q \rightarrow p)$ and $\Box q \rightarrow p$.

108. There is, again, a similarity to modal logic. The claim $\Diamond p \wedge \Diamond q$ says something different about worlds than the claim $\Diamond(p \wedge q)$.

statements, agreeing with both of them, to say $\exists xFx \land \exists xGx$, taking them at their word. There is no requirement that the same object satisfy both $F$ and $G$. If we wanted to say *that*, we have a different formula for it: $\exists x(Fx \land Gx)$. This says that something is both $F$ and $G$. It says that there is some object bearing both those features.

Just as $\exists xFx$ is a formula, so is $\exists yFy$. We could have used a different variable to make the same kind of statement. Here, the use of the different variable does not mean that we have a different object in mind. If someone says $\exists xFx$ and someone else says $\exists yFy$, we could agree with both, by saying $\exists xFx \land \exists yFy$. But in saying this, we are saying nothing more than $\exists xFx$, in the same way that we could agree that "something is $F$" and "at least one thing is $F$" are both true and are both different ways of saying the same thing.

Different variables in *those* sorts of cases give us different ways of saying the same thing. There are some things that we *need* to use more than one variable to say. If we wanted to say that for anything at all, there is something *larger* than it, we need to use two quantifiers. Let $L$ be a two-place predicate where $Lxy$ says that $x$ is larger than $y$. Then, when we say that for anything at all, there is something larger than it, the quantifiers work like this. Translating things halfway, we could write

$$\forall x(\text{there is something larger than } x).$$

But how do I say that there is something larger than $x$? $Lyx$ says that $y$ is larger than $x$, so I can say that there is something larger than $x$ by saying $\exists yLyx$. That is, there is something (call it $y$) that does the job of being larger than $x$. Putting these pieces together, our formula is this:

$$\forall x\exists yLyx.$$

This says that for any $x$ we choose at all, there is something we can choose that is larger than that. Here the variables matter, because we have two different quantifiers, and they operate on different parts of the predicate $L$. If we swapped the $y$ and $x$ around in $Lxy$, we would get

$$\forall x\exists yLxy,$$

which says that for everything, there is something that it is larger than. (So, for anything, there is something smaller than it.) This says something very different.

Just as the order of variables in the predicate makes a difference, so does the order of the quantifiers. If we were to say

$$\exists y\forall xLyx,$$

which flips the quantifiers around from our original statement $\forall x\exists yLyx$, then we get a different statement again, which says that there is something that is larger than *everything*. This is a different statement, not least because the only way it could be true is that if something—this thing that is larger than everything—is larger than *itself*. Remember, if we say that $\forall xLyx$, this is saying that *anything at all* can stand in place of $x$, and in particular, whatever we meant by $y$ is a candidate.

* * *

The distinctive behavior of quantifiers depends on the behavior of the variables. Quantifiers use variables by *binding* them. Any variables in a formula not bound by a quantifier, like the $x$ in $Fx$ or the $y$ in $\exists x(Lxy \land Lyx)$, are said to be *free*. Variables can be free in a part of a formula and bound elsewhere, like the $x$s in $Fx \land \exists xGx$. So, let's make these notions precise in another definition.

**Definition 66 (Free and bound variables, and scope)**  *The set $fv(A)$ of* FREE VARIABLES *in a formula A is defined like this:*

- *If A is an atomic formula, $Ft_1 \cdots t_n$, then $fv(A)$ is the set of variables among the $t_i$, $1 \le i \le n$.*
- *If A is a propositional atom p or $\bot$, then $fv(A) = \emptyset$.*
- *$fv(A \land B) = fv(A \lor B) = fv(A \to B) = fv(A) \cup fv(B)$. (That is, the free variables in a conjunction, disjunction, or conditional of A and B are the variables free in A together with the variables free in B.)*
- *$fv(\neg A) = fv(A)$. (That is, the variables free in $\neg A$ are the variables free in A.)*
- *$fv(\exists xA) = fv(A) \backslash \{x\}$ and $fv(\forall xA) = fv(A) \backslash \{x\}$. (The variables free in $\exists xA$ and in $\forall xA$ are the variables free in A, except for x, which is bound by the quantifier in each case.)*

*The variables in the set $fv(A)$ are said to be* FREE *in the formula A.*

*The* SCOPE *of the initial quantifier $\forall xA$ is A, and the scope of the initial quantifier $\exists xA$ is A.*

*If x is free in A, then x is* BOUND *by the quantifier in $\forall xA$ and in $\exists xA$.*

*A term t is said to be* FREE FOR *the variable x in A iff no free occurrence of x in A is in the scope of a quantifier that binds a variable in t.*

**Definition 67 (Open and closed formulas)**  *A formula A is an* OPEN FORMULA *iff $fv(A) \ne \emptyset$. (That is, an open formula contains some free variables.) A formula A is a* CLOSED FORMULA *iff A is not an open formula. (That is, a closed formula contains no free variables.)*

Before continuing, let's apply these definitions to an example. Consider the formula

$$\forall x(\exists yLxy \to \exists yLyx)$$

This is a closed formula, because the variables it contains ($x$ and $y$) all occur within the scope of quantifiers that bind those variables ($\forall x$ and the two occurrences of $\exists y$). Its set of free variables is empty. If we remove the outermost quantifier, to consider the formula

$$\exists yLxy \to \exists yLyx,$$

then the two occurrences of $x$ are *free*, and the two occurrences of $y$ in the predicates are *bound* by their respective quantifiers. The set of free variables for this formula is simply $\{x\}$. The first existential quantifier has the formula $Lxy$ in its scope, while the second has the formula $Lyx$ in its scope. These two formulas have the same set of free variables, $\{x, y\}$.

\* \* \*

An important concept for predicate logic is *substitution*, which is an important component of how we move from a quantified expression to an *instance* of that expression. We want to explain the relationship between a statement like $\forall x(Fx \to Bx)$ (all footballers are bipeds) and $Fa \to Ba$ (if Sócrates is a footballer, he is a biped). Here, we have not only removed the universal quantifier when we have gone from $\forall x(Fx \to Bx)$ to $Fa \to Ba$. We have also *substituted* the name $a$ for the variable $x$. This is the key feature of variables as they occur in formulas: they allow for substitution.

In simple instances like this one, it is straightforward to see how substitution works, but there are more complex cases to consider. For example, if I were to substitute $a$ for $x$ in $Fx \land \exists xGx$, the result should be $Fa \land \exists xGx$, because the $x$ in $G$ is bound by a quantifier and is not available for substitution. So, to be careful, we define substitution in a systematic way, and as usual, it will be an inductive definition, using the definitions of the terms and formulas themselves.

**Definition 68 (Substitution)** *If $t$ is free for $x$ in $A$, then the* SUBSTITUTION OF $t$ FOR $x$ IN A FORMULA $A$, $(A)[x/t]$, *and the* SUBSTITUTION OF $t$ FOR $x$ IN A TERM $u$, $u[x/t]$, *are defined like this:*

- *If $u$ is $x$, then $u[x/t] = t$.*
- *If $u$ is not $x$, then $u[x/t] = u$.*
- *If $A$ is $Fs_1 \cdots s_n$, then $(Fs_1 \cdots s_n)[x/t] = Fs_1[x/t] \cdots s_n[x/t]$.*
- *If $A$ is $p$ or $\bot$, then $(A)[x/t] = A$.*
- $(A \wedge B)[x/t] = A[x/t] \wedge B[x/t]$.
- $(A \vee B)[x/t] = A[x/t] \vee B[x/t]$.
- $(A \rightarrow B)[x/t] = A[x/t] \rightarrow B[x/t]$.
- $(\neg A)[x/t] = \neg(A[x/t])$.
- $(\forall x A)[x/t] = \forall x A$.
- $(\forall y A)[x/t] = \forall y((A)[x/t])$.
- $(\exists x A)[x/t] = \exists x A$.
- $(\exists y A)[x/t] = \exists y((A)[x/t])$.

Here are some examples, so you can see how the definition works:

- $x[x/z] = z$
- $x[y/z] = x$

If we substitute a term for a variable *in that variable itself*, the result is the term we substitute. But if we substitute a term for a variable *in some other term* (whether a different variable or a name), the substitution makes no difference.

- $Fax[x/b] = Fab$
- $(Fax \wedge Gxy)[x/c] = Fax[x/c] \wedge Gxy[x/c] = Fac \wedge Gcy$

In formulas without quantifiers, the substitution operation simply dives down to the atomic formulas and applies the substitution to each term in each atomic formula. So, the variable being substituted for is swapped out for the term selected to replace it. When quantifiers are around, more care is needed.

- $(\forall x Gxc)[x/a] = \forall x Gxc$
- $(\forall x(Fxy \wedge Gyz))[y/b] = \forall x(Fxb \wedge Gbz)$
- $(\forall x(Fxy \wedge \exists y Gyz))[y/b] = \forall x(Fxb \wedge \exists y Gyz)$

If we substitute a term for a variable in an expression where that variable is *bound*, then the substitution makes no difference (as in the first example here), because there is nothing that gets replaced. If the variable is not bound, the replacement occurs (as in the second). In a formula, a substitution can occur in some places, but not in others, if the variable occurs free in some parts of the formula and is bound elsewhere, as in the last example.

* * *

The requirement that the term $t$ be free for $x$ in $A$ rules out the case of a substitution resulting in new variable binding. The sort of case we want to rule out is substituting $x$ for $y$ in $\forall x Rxy$, to give us $\forall x Rxx$. Such a substitution does not do what we would expect and results in unwanted variable capture. Before substitution, there is a free variable, namely the $y$, but after substitution, all variables are bound by the one quantifier. When we say $\forall x Rxy$, we are saying that $R$ relates everything to $y$. Intuitively speaking, if we apply this fact (that everything is related to $y$) instead to the object $x$, we would like to say that $R$ relates everything to $x$. But that is not what $\forall x Rxx$ says. This says that everything is related by $R$ to *itself*.

Limiting substitution to cases where there is no variable capture will not restrict us in any way because, as we will see, we can re-letter bound variables in a way that results in a logically equivalent formula, which does not capture the substituted variable. If we wanted to apply the fact that $\forall x Rxy$ to the object we had been calling $x$, then we could restate the fact as $\forall z Rzy$ instead and then substitute $x$ for $y$ in this formula, to conclude $\forall z Rzx$, which says what we were wanting to say. In general, if we have a quantified formula, $\forall x A(x)$ or $\exists x A(x)$, that we want to re-letter, we can pick a variable, say $z$, not occurring in $A$, and use $\forall z((A)[x/z])$ or $\exists z((A)[x/z])$, which will be equivalent to the respective formula that we started with. Changing bound variables uniformly in this manner results in *alphabetic variants*, formulas that differ only in choice of bound variables and are equivalent.

**Definition 69 (Alphabetic variant)**   *Let $A$ be a formula. The formula $A'$ is an* ALPHABETIC VARIANT *of $A$ iff there is a sequence of formulas $\langle B_1, \ldots, B_n \rangle$ such that $A = B_1$, $A' = B_n$, and if $n \geq 2$, for $i$ such that $1 \leq i < n$, $A_{i+1}$ is obtained from $A_i$ by replacing an occurrence of a subformula of the form $\forall x C(x)$ with $\forall y C(y)$, where $y$ is a variable not occurring in $C(x)$, or by replacing an occurrence of a subformula of the form $\exists x C(x)$ with $\exists y C(y)$, where $y$ is a variable not occurring in $C(x)$.*

Using this definition, we can see that $\exists y Fy \wedge \forall z(Rzu \to Fz)$ is an alphabetic variant of $\exists z Fz \wedge \forall x(Rxu \to Fx)$. We can show this by constructing a sequence of formulas beginning with the latter and ending with the former. In particular, the sequence

$$\langle \exists z Fz \wedge \forall x(Rxu \to Fx),\ \exists y Fy \wedge \forall x(Rxu \to Fx),\ \exists y Fy \wedge \forall z(Rzu \to Fz) \rangle$$

will work. Alphabetic variants are used to avoid problems with variable capture in substitutions. As we will see below, alphabetic variants are equivalent. As formulas, they differ only in the labeling of their bound variables. Their free variables are the same.

* * *

We will end this section with some final definitions and conventions. First, a helpful convention will be a shorthand for describing substitution when we are thinking of inference rules in general, without talking about individual specific formulas. We will use the notation $A(x)$ to indicate that the formula $A$ may contain the variable $x$ free. If $A(x)$ is such a formula, then we will write $A(b)$ for the formula $(A(x))[x/b]$. Substitution is a key concept for understanding the rules for quantifiers, and we will use substitution to define the key idea of an instance of a quantified expression.

**Definition 70 (Instance)**   *If $\forall x A(x)$ is a formula and $b$ is a term free for $x$ in $A$, then $A(b)$ is an* INSTANCE *of the universal quantifier. Similarly, if $\exists x A(x)$ is a formula and $b$ is a term free for $x$ in $A$, then $A(b)$ is an* INSTANCE *of the existential quantifier.*

So, for example, $\exists yLya$ is an instance of the formula $\forall x \exists yLyx$ (where we have substituted the name $a$ for the variable $x$) and $Lza$ is an instance of $\exists yLya$ (where we have substituted the variable $z$ for the variable $y$).

* * *

With all of those definitions out of the way, we have everything we need in order to consider how we might extend our proof system to reason with predicates, names, and quantifiers.

## 10.2 Natural Deduction for CQ

We called the natural deduction system for classical propositional logic **C**. This proof system has introduction and elimination rules for each of the connectives $\wedge$, $\rightarrow$, $\vee$, $\neg$, and an elimination rule for $\bot$, and the special rule *DNE*, to make the logic complete for classical valuations. Our task in the rest of this chapter is to explore how to extend this system to help us define proofs for predicate logic.

The addition of predicates and names to our language does not require us to expand our repertoire of rules. There are no specific inference principles governing names or predicates as such.[109]

The distinctive *logic* of predicates, names, and quantifiers results from the inference principles involving the quantifiers. Just like the connectives and modal operators, we have introduction and elimination rules for $\forall$ and for $\exists$, and it is the special properties of these rules that allow us to do so much more with predicate logic than we can do with propositional logic alone. Since the quantifiers do all the extra work in our logic, we will label the proof system **CQ**, for classical logic with (first-order) quantifiers.

*Terminological aside*: We have not yet explained what makes first-order predicate logic *first-order*. The idea is that objects are the first level of complexity in a hierarchy. In atomic statements like *Fa* and *Rxy*, we have objects at the first level and properties of those objects at the second level. The names pick out the objects and the predicates pick out the properties. We have quantifiers that range over the objects, so these are called *first-order* quantifiers. If we were to have quantifiers that ranged over *predicates*, these would be *second-order* quantifiers. In a second-order logic, we could say $\exists X(Xa \wedge Xb)$ to say that there was some property shared by $a$ and $b$ or $\exists X(Xa \wedge \neg Xb)$ to say that $a$ and $b$ differed in some way.

We do not have to stop at the second level. *Higher-order logics* allow for properties of properties (level 3), properties of those properties (level 4), and so on, *ad infinitum*, as well as quantifiers that range over the higher-order properties at each level. Higher-order logics are sometimes called *type theories*, because the predicates in the language are stratified into levels according to their *types*. For an exploration and defense of second-order logic, you

---

109. That is because the predicates and names have no special logical properties governed by inference rules. If we were to add to our language a special *logical* predicate, like the identity predicate "=," then we would have to consider inference rules governing this predicate. But we have no space in this chapter to consider proofs involving the special properties of the identity predicate.

cannot miss Stewart Shapiro's book *Foundations without Foundationalism* (1991). *End of terminological aside*

So, our proof system for CQ will use all the rules for the propositional connectives and *DNE*. We add to these rules new rules for the quantifiers. As with our modal proof systems, two of the rules are straightforward. The idea behind the $\forall E$ rule is that if everything is an $A$, then $t$ is $A$ (where $t$ is a term). This works for *any* term. The idea behind $\exists I$ goes in reverse. If $t$ is an $A$ (whatever the term $t$ might be), then something is an $A$. So, we have rules of these shapes:

$$\frac{\forall x A(x)}{A(t)} \ \forall E \qquad \frac{A(t)}{\exists x A(x)} \ \exists I$$

The rule $\forall E$ says that a universal quantifier implies any of its instances and the rule $\exists I$ says that every instance implies an existential quantifier.

The two remaining rules require some explanation. One way to understand how the rules work is to think that whatever it takes to introduce $\forall x A(x)$ has to be powerful enough to allow us to infer anything that follows from $\forall x A(x)$. That is, it must be powerful enough to allow us to infer the instances $A(a)$, $A(b)$, and so on. Now, a naive way of thinking about how to do this would be to take the introduction rule for $\forall$ to have this shape:

$$\frac{A(t_1) \quad A(t_2) \quad A(t_3) \quad \vdots}{\forall x A(x)} \ \forall I?$$

To prove that everything has the property $A$, for each thing, you prove that *it* has the property $A$. Unfortunately, this rule is both too *long* (how *many* premises is that?—no one proves general claims in this way; it takes too much work!) and too *short* (there is no guarantee that we have a different term for each different object—in fact, it is pretty clear that we *don't* have terms that pick out each individual object in the universe). So, this rule is not a good candidate.

No, the correct rule to introduce a universal quantifier is rather like the rule from the previous chapter to introduce a necessity statement. We do not need to prove many instances of the quantified expression. We need to prove just one—but we need to prove it in a special way. We need to prove $A(b)$ in such a way that this proof would *also* work were it applied to $c$ or to $t$ for any term $t$ we can substitute into the quantifier. In other words, if we have proved $A(b)$ in such a way that is totally general, and applies not only to $a$, but for *everything*. The rule will have this shape:

$$\frac{\overset{\displaystyle X}{\underset{\displaystyle \overset{\Pi}{\nabla}}{\phantom{X}}}}{\dfrac{A(b)}{\forall x A(x)}} \ \forall I$$

The conclusion of this rule depends on the same assumptions that the premise of the rule does. There is a *side condition* on the rule, just like the side condition on $\Box I$. Before explaining the side condition, let's see an example of the rule at work, which will show us how the side condition does its job.

$$\frac{\dfrac{\forall x(Fx \rightarrow Gx)}{Fa \rightarrow Ga} \; \forall E \qquad \dfrac{\forall xFx}{Fa} \; \forall E}{\dfrac{Ga}{\forall xGx} \; \forall I} \; \rightarrow E$$

This is a proof from $\forall x(Fx \rightarrow Gx)$ and $\forall xFx$ to $\forall xGx$. If we stop at the penultimate inference, we have a proof from $\forall x(Fx \rightarrow Gx)$ and $\forall xFx$ to $Ga$. It is clear that we could have rewritten this proof to be a proof to $Gb$ instead, or to $Gc$ or to $Gy$ or $Gz$ or $Gt$ for any term $t$ at all. All we would have had to do is to apply the $\forall E$ inferences to the new term instead of to $a$. The proofs would have had the same premises $\forall x(Fx \rightarrow Gx)$ and $\forall xFx$, and we would have got to the different conclusions. The premises here in our proof do not involve the term $a$ at all, so since none of the rules of our proof system treat $a$ as special, the proof could be transformed from a proof of $Ga$ into a proof for $Gt$ for any other term $t$, without altering the premises. So, the proof of $Ga$ is *general*, and on this basis, we infer $\forall xGx$. That is how our proof works. The proof to the conclusion $Ga$ satisfies our side condition, which we can now state in full:

[$\forall$-EIGENVARIABLE CONDITION] In an application of $\forall I$, the displayed term $b$ in the premise $A(b)$ cannot occur either in the conclusion of $\forall I$ or in any assumption in $X$ upon which the premise of the application of $\forall I$ depends.

The term $b$ in the condition will be called the *eigenvariable* for the $\forall I$ step.

\* \* \*

The key idea with the eigenvariable condition is that the object we pick out with our term is, in some sense, *arbitrary*, since we have made no assumption about it in our proof. So, it is not the kind of name we use for objects about which we have special information. In this sense, it acts rather like a variable. We could have used variables for this role, but it is traditional in many proof systems to use names or to introduce a special class of variables, distinct from the variables bound by quantifiers.

\* \* \*

The rule $\exists E$ will obey a restriction similar to that of $\forall I$. To work out what follows *from* an existential claim, we work out what follows from an instance of the claim. But as with the $\lozenge E$ rule, we have a side condition. The condition appropriate to $\exists$ is that we cannot rely on any particular features of the instance in question apart from the fact that it is an instance of the existential itself. So, the rule $\exists E$ has a shape very similar both to that of $\lozenge E$ and to that of $\vee E$:

$$\frac{\exists xA(x) \qquad \begin{array}{c} [A(b)]^1 \quad X \\ \diagdown \; \Pi \; \diagup \\ B \end{array}}{B} \; \exists E^1$$

To infer something from $\exists xA(x)$, we make the assumption that $A(b)$, where we have no other information about $b$. So, in effect, we are naming some arbitrary object with property

*A*, "*b*" and seeing what we can conclude. To enforce the arbitrariness, we require that the conclusion, *B*, does not contain the term *b*, and no open assumption of Π contains *b* either, apart from the assumption *A*(*b*), discharged by the rule.

[∃-EIGENVARIABLE CONDITION] In an application of ∃*E*, the term *b* in the discharged assumption cannot occur in ∃*xA*(*x*), *B* or in *X*.

We will refer to the conjunction of the ∀-eigenvariable condition and the ∃-eigenvariable condition together as the EIGENVARIABLE CONDITION. We will say that a proof satisfies the eigenvariable condition just in case each ∃*E* and ∀*I* step satisfies the eigenvariable condition.

$$* * *$$

We can put these rules together to show how we can infer ∃*yGy* from ∀*x*(*Fx* → *Gx*) and ∃*xFx*.

$$\cfrac{\exists xFx \qquad \cfrac{\cfrac{\cfrac{\forall x(Fx \to Gx)}{Fa \to Ga}\ \forall E \qquad [Fa]^1}{\cfrac{Ga}{\exists yGy}\ \exists I}\ \to E}{\exists yGy}\ \exists E^1}{\exists yGy}$$

Here, the application of ∃*E* in the last step of the proof satisfies the eigenvariable condition, because the formulas ∃*yGy*, ∀*x*(*Fx* → *Gx*), and ∃*xFx* do not contain any occurrence of the term *a*. So, since we can prove ∃*yGy* from ∀*x*(*Fx* → *Gx*) and *Fa*, it also follows from ∀*x*(*Fx* → *Gx*) and ∃*xFx*, since we appealed to no special features of *a* other than the stipulated fact that *Fa*.

$$* * *$$

The eigenvariable condition is important for ensuring that we can derive only valid conclusions, but it might be difficult to tell what sorts of arguments are ruled out by the condition. It will be helpful to provide some illustrative examples of violations of the eigenvariable condition. The first example is a violation of the ∀-eigenvariable condition.

$$\cfrac{\cfrac{Fa \land Ga}{Fa}\ \land E}{\forall xFx}\ \forall I^?$$

This argument concludes that everything is *F* on the basis of the assumption that a particular thing is both *F* and *G*. The mere fact that some object is *F* is not going to be sufficient to conclude that everything is *F*.

The ∃-eigenvariable condition is a bit more involved, so we will illustrate violations of its parts. Consider the following example:

$$\cfrac{\exists xRbx \qquad \cfrac{[Rbb]^1}{\exists xRxx}\ \exists I}{\exists xRxx}\ \exists E^{?1}$$

In this example, the term substituted for the bound variable of the existential quantifier is the same as the term in the quantified major premise. This argument takes us from the assumption that $b$ bears $R$ to something to the conclusion that something bears $R$ to itself. Not every nonempty binary relation, however, is reflexive. Consider the two-place relation "$x$ is younger than $y$ by at least a day." While many people bear this relation to other people, no one bears this relation to themselves.

To illustrate the need for the next part of the $\exists$-eigenvariable condition, consider the following example:

$$\dfrac{\dfrac{(Rab \to Rba) \land \exists x Rax}{\exists x Rax} \land E \qquad \dfrac{\dfrac{\dfrac{(Rab \to Rba) \land \exists x Rax}{Rab \to Rba} \land E \qquad [Rab]^1}{Rba} \to E}{\exists x Rxa} \exists I}{\exists x Rxa} \exists E^{?1}$$

This proof violates the $\exists$-eigenvariable condition in that the term substituted for the quantified variable appears in an assumption of the proof ending in the minor premise. The problem is that while $a$ may bear $R$ to something, that thing may not be $b$. It can turn out that nothing bears $R$ to $a$, even though $a$ bears $R$ to something.

The need for the final portion of the $\exists$-eigenvariable condition is easiest to illustrate when paired with the universal quantifier rule.

$$\dfrac{\dfrac{\exists x Fx \qquad [Fa]^1}{Fa} \exists E^{?1}}{\forall x Fx} \forall I$$

This proof violates the condition at the $\exists E$ step in that the term substituted for the variable bound by the existential quantifier appears in the minor premise. The assumption $Fa$ is discharged by the $\exists E$ rule, setting aside the eigenvariable condition for the example, so the only assumption is $\exists x Fx$. This means that the $\forall$-eigenvariable condition is satisfied at the final step. But, this would let us show, from the assumption that something or other is $F$, that everything is $F$. This would be a disastrous conclusion. Problems like this can arise even without the universal quantifier around, as the following proof illustrates:

$$\dfrac{\exists x Gx \qquad \dfrac{\dfrac{\dfrac{\exists x Fx \qquad [Fa]^1}{Fa} \exists E^{?1} \qquad [Ga]^2}{Fa \land Ga} \land I}{\exists x(Fx \land Gx)} \exists I}{\exists x(Fx \land Gx)} \exists E^2$$

The $\exists$-eigenvariable condition is violated at the upper $\exists E$ step. To see why the conclusion does not follow, suppose that $Fx$ means $x$ is an even number and $Gx$ means $x$ is an odd number. From the assumption that there is an even number and an odd number, it does not follow that some number is both even and odd.

$$* * *$$

We will close this section by showing that these rules are well balanced, to the extent that they allow us to eliminate detours. First, let's consider a detour through an introduction and elimination of a universal quantifier.

$$
\begin{array}{c}
X \\
\nabla\!\!\!\!\!\diagdown \Pi \diagup \\
\hline
\underline{A(a)} \quad \forall I \\
\underline{\forall x A(x)} \\
A(t) \quad \forall E
\end{array}
$$

Here, the detour formula $\forall x A(x)$ was introduced by a $\forall I$ step from the premise $A(a)$, which was proved from $X$ by a proof $\Pi$ satisfying the eigenvariable condition. This means that the premises $X$ do not contain the term $a$, and neither does the conclusion $\forall x A(x)$. This means that if we replace the term $a$ throughout the proof $\Pi$ by $t$, the result, $\Pi[a/t]$, is a proof from $X$ (the assumptions are unchanged as they do not contain $a$) to the conclusion $A(t)$. So, we can eliminate our detour by using this shorter proof, from $X$ to $A(t)$.

$$
\begin{array}{c}
\Pi[a/t] \\
A(t)
\end{array}
$$

The same holds for a proof with an $\exists I/\exists E$ detour.

$$
\begin{array}{cc}
\Pi' & \\
\underline{A(t)} \quad \exists I & [A(b)]^1 \\
\underline{\exists x A(x)} & \begin{array}{c}\Pi \\ B\end{array} \\
\hline
B & \exists E^1
\end{array}
$$

Since the eigenvariable condition is satisfied by the $\exists E$ step, we know that the other premises of the proof $\Pi$ (other than $A(b)$) and the conclusion $B$ do not contain the eigenvariable $b$. So, if we globally replace the $b$ in $\Pi$ by the term $t$, the result is still a proof from the same premises (except for $A(b)$, which now becomes $A(t)$) to the conclusion $B$. So, we can smooth out the detour with this simpler proof:

$$
\begin{array}{c}
\Pi' \\
A(t) \\
\Pi[b/t] \\
B
\end{array}
$$

with the same premises and conclusion, and which does not go through the detour formula $\exists x A(x)$.

There is one subtlety to the two preceding arguments. If the term $t$ is a variable, say $z$, that is bound in $\Pi$, then the substitutions $\Pi[a/z]$ and $\Pi[b/z]$ may not be defined, if $z$ is not free for $a$ or $b$ in some formula in $\Pi$. All is not lost. We can show that all the bound occurrences of $z$ can be shifted to a new variable that doesn't otherwise occur in $\Pi$. The substitutions $\Pi[a/z]$ and $\Pi[b/z]$ can then be carried out as above. The assumptions of the resulting proofs may differ from those of the original proofs, but the difference will only be in relabeling some bound occurrences of $z$ by a fresh variable. That difference is not a significant one.

$$* * *$$

To summarize, the quantifier rules are the following:

$$\frac{\forall x A(x)}{A(t)} \ \forall E \qquad \frac{\begin{array}{c} X \\ \diagdown\!\!\diagup \\ \Pi \\ A(b) \end{array}}{\forall x A(x)} \ \forall I \qquad \frac{A(t)}{\exists x A(x)} \ \exists I \qquad \frac{\exists x A(x) \qquad \begin{array}{c} [A(b)]^1 \quad X \\ \diagdown\!\!\diagup \\ \Pi \\ B \end{array}}{B} \ \exists E^1$$

The $\forall I$ rule has the side condition that the $\forall$-eigenvariable condition is satisfied:

[$\forall$-EIGENVARIABLE CONDITION] In an application of $\forall I$, the displayed term $b$ in the premise $A(b)$ cannot occur either in the conclusion of $\forall I$ or in any assumption in $X$ upon which the premise of the application of $\forall I$ depends.

The $\exists E$ rule has the side condition that the $\exists$-eigenvariable condition is satisfied:

[$\exists$-EIGENVARIABLE CONDITION] In an application of $\exists E$, the displayed term $b$ in the discharged assumption cannot occur in $\exists x A(x)$, $B$ or in $X$.

A CQ proof is a tree built from the rules for C and the rules $\forall I, \forall E, \exists I$, and $\exists E$, where the eigenvariable conditions are satisfied. Now that we have introduced our proof system and verified that detours can be simplified, let's explore its properties, to see what we can prove, using these rules for quantifiers.

## 10.3 Features of Proofs with Quantifiers

As with our other proof systems, we use a notation to represent provability. Here, we will use $X \vdash_{CQ} A$ to say that there is a CQ proof with $A$ as its conclusion and with premises chosen from the set $X$.

For the first provability facts, let's verify how the quantifiers interact with negation. The quantifiers obey *De Morgan* principles, much like those involving conjunction and disjunction and those involving $\Box$ and $\Diamond$.

**Theorem 35 (De Morgan principles for quantifiers)** *The following four provability claims are true:*

1. $\forall x \neg A(x) \vdash_{CQ} \neg \exists x A(x)$
2. $\neg \exists x A(x) \vdash_{CQ} \forall x \neg A(x)$
3. $\exists x \neg A(x) \vdash_{CQ} \neg \forall x A(x)$
4. $\neg \forall x A(x) \vdash_{CQ} \exists x \neg A(x)$

*Proof.*  For (1):

$$\cfrac{\cfrac{\cfrac{\forall x \neg A(x)}{\neg A(b)}\ \forall E \qquad [A(b)]^1}{\bot}\ \neg E}{\cfrac{[\exists x A(x)]^2 \qquad\qquad\qquad\qquad \bot}{\cfrac{\bot}{\neg \exists x A(x)}\ \neg I^1}\ \exists E^1}$$

For (2):

$$\cfrac{\cfrac{\neg \exists x A(x) \qquad \cfrac{[A(b)]^1}{\exists x A(x)}\ \exists I}{\cfrac{\bot}{\neg A(b)}\ \neg I^2}\ \neg E}{\forall x \neg A(x)}\ \forall I$$

For (3):

$$\cfrac{\cfrac{\exists x \neg A(x) \qquad \cfrac{[\neg A(b)]^1 \qquad \cfrac{[\forall x A(x)]^2}{A(b)}\ \forall E}{\bot}\ \neg E}{\bot}\ \exists E^1}{\cfrac{\bot}{\neg \forall x A(x)}\ \neg I^2}$$

Finally, for (4), we have the one principle that requires an appeal to *DNE* to complete.

$$\cfrac{\neg \forall x A(x) \qquad \cfrac{\cfrac{\cfrac{[\neg \exists x \neg A(x)]^2 \qquad \cfrac{[\neg A(b)]^1}{\exists x \neg A(x)}\ \exists I}{\bot}\ \neg E}{\cfrac{\neg \neg A(b)}{A(b)}\ DNE}\ \neg I^1}{\forall x A(x)}\ \forall I}{\cfrac{\cfrac{\bot}{\neg \neg \exists x \neg A(x)}\ \neg I^2}{\exists x \neg A(x)}\ DNE}\ \neg E$$

$\square$

The quantifiers have distinctive interactions with the other connectives, in addition to negation.

**Theorem 36**  *The following are provable:*

1. $\forall x A(x) \wedge \forall x B(x) \vdash_{CQ} \forall x (A(x) \wedge B(x))$
2. $\exists x (A(x) \vee B(x)) \vdash_{CQ} \exists x A(x) \vee \exists x B(x)$
3. $C \to \forall x B(x) \vdash_{CQ} \forall x (C \to B(x))$, *provided x is not free in C*
4. $\forall x (A \to C) \vdash_{CQ} \exists x A(x) \to C$, *provided x is not free in C*

*Proof.* For (1), the universal quantifier distributes over *conjunction*.

$$\cfrac{\cfrac{\cfrac{\forall xA(x) \land \forall xB(x)}{\forall xA(x)} \land E}{A(b)} \forall E \quad \cfrac{\cfrac{\forall xA(x) \land \forall xB(x)}{\forall xB(x)} \land E}{B(b)} \forall E}{\cfrac{A(b) \land B(b)}{\forall x(A(x) \land B(x))} \forall I} \land I$$

It is a useful exercise to convince yourself why the existential quantifier does not distribute over conjunction in the way that the universal quantifier does in the preceding proof, and to compare this intuition with what goes on when you attempt to construct a proof from $\exists xA(x) \land \exists xB(x)$ to $\exists x(A(x) \land B(x))$. For (2), we have this proof:

$$\cfrac{\exists x(A(x) \lor B(x)) \quad \cfrac{[A(c) \lor B(c)]^3 \quad \cfrac{\cfrac{[A(c)]^1}{\exists xA(x)} \exists I}{\exists xA(x) \lor \exists xB(x)} \lor I \quad \cfrac{\cfrac{[B(c)]^2}{\exists xB(x)} \exists I}{\exists xA(x) \lor \exists xB(x)} \lor I}{\exists xA(x) \lor \exists xB(x)} \lor E^{1,2}}{\exists xA(x) \lor \exists xB(x)} \exists E^3$$

It is also a useful exercise to see why there is no way to construct a proof from $\forall x(A(x) \lor B(x))$ to $\forall xA(x) \lor \forall xBx$.

For (3), we will choose a name $b$ that does not occur in the formula $C$, to satisfy the eigenvariable condition.

$$\cfrac{\cfrac{\cfrac{\cfrac{C \to \forall xB(x) \quad [C]^1}{\forall xB(x)} \to E}{B(b)} \forall E}{C \to B(b)} \to I^1}{\forall x(C \to B(x))} \forall I$$

For (4), we will assume that the name $b$ does not occur in $C$.

$$\cfrac{\cfrac{[\exists xA(x)]^2 \quad \cfrac{\cfrac{\forall x(A(x) \to C)}{A(b) \to C} \forall E \quad [A(b)]^1}{C} \to E}{C} \exists E^1}{\exists xA(x) \to B} \to I^2$$

$\square$

In the proof of (4), the assumption that $x$ is not bound in $C$ by the universal quantifier is important. If $x$ is bound in $C$, then the $\forall E$ step will result in $A(b) \to C(b)$, which means that the application of $\exists E$ will violate the $\exists$-eigenvariable condition. The assumption noted in the proof of (3) permits the $\forall I$ step to satisfy the condition.

$$* * *$$

If a variable is already bound, then binding it again has no effect. We have the following equivalences:

- $\vdash_{CQ} \forall x \forall x A(x) \leftrightarrow \forall x A(x)$
- $\vdash_{CQ} \exists x \exists x A(x) \leftrightarrow \exists x A(x)$

Here is a proof for the first equivalence (the second is left as an exercise).[110]

$$\cfrac{\cfrac{[\forall x \forall x A(x)]^1}{\forall x A(x)}\ \forall E \qquad \cfrac{[\forall x A(x)]^2}{\forall x \forall x A(x)}\ \forall I}{\forall x \forall x A(x) \leftrightarrow \forall x A(x)}\ {\leftrightarrow}I^{1,2}$$

In this proof, the inference steps $\forall E$ and $\forall I$ look strange, but they make some kind of sense. In the $\forall E$ step, we have substituted a term (it doesn't matter which) to construct an instance of $\forall x \forall x A(x)$. There are no instances of $x$ free in $\forall x A(x)$, so the substitution has no effect. The resulting instance is just $\forall x A(x)$. Similarly, in the $\forall I$ inference, we can consider $\forall x A(x)$ as an instance of $\forall x \forall x A(x)$, and we can choose the term substituted (which does not actually *appear*) to be one that does not occur in $\forall x \forall x A(x)$, so this step satisfies the eigenvariable condition. So, this counts as a proof.

What this highlights is the following fact: the quantified variables are keeping track of positions in a formula. Any variable not otherwise occurring in a formula can be used to do the job, as captured in the following fact:

**Lemma 23** *In the following facts, $y$ is not bound by any other quantifier in $A(x)$, and, as usual, $A(y)$ is the result of substituting $y$ for $x$ in $A(x)$.*

1. $\forall x A(x) \vdash_{CQ} \forall y A(y)$
2. $\exists x A(x) \vdash_{CQ} \exists y A(y)$

*Proof.* For (1):

$$\cfrac{\cfrac{\forall x A(x)}{A(b)}\ \forall E}{\forall y A(y)}\ \forall I$$

For (2):

$$\cfrac{\exists x A(x) \qquad \cfrac{[A(b)]^1}{\exists y A(y)}\ \exists I}{\exists y A(y)}\ \exists E^1$$

$\square$

This theorem doesn't mean that you can freely switch out variables that are bound by another quantifier in a formula. Doing so can result in a variable being bound by another quantifier. For example, take $\forall x \forall y Rxy$. This says that everything bears $R$ to everything. If we were to switch the $x$ to a $y$, the result would be $\forall y \forall y Ryy$, which is equivalent to $\forall y Ryy$, which says that everything bears $R$ to itself, a weaker claim than the claim that everything is related to everything. In terms of our definition of substitution, $x$ is not free for $y$ in $\forall y Rxy$. One can prove a more general claim about alphabetic variants.

---

110. We don't have the biconditional in the language, but we can use the definitions from section 2.3 to expand the proof to not use a biconditional.

**Theorem 37 (Equivalence of alphabetic variants)**  *Let A be a formula and A′ an alphabetic variant of it. Then, $A \vdash_{CQ} A'$ and $A' \vdash_{CQ} A$.*

*Proof.*  The proof is left as an exercise.                                                           □

**Theorem 38**  *The following are provable:*

1. *If $A(t) \vdash_{CQ} B$, then $\forall x A(x) \vdash_{CQ} B$.*
2. *If $A \vdash_{CQ} B(t)$, then $A \vdash_{CQ} \exists x B(x)$.*
3. *If $X \vdash_{CQ} B(c)$, then $X \vdash_{CQ} \forall x B(x)$, provided c does not occur in X.*
4. *If $X, A(c) \vdash_{CQ} B$, then $X, \exists x A(x) \vdash_{CQ} B$, provided c does not occur in B or in X.*

*Proof.*  For (1), suppose $\Pi$ is a proof of $A(t) \vdash_{CQ} B$. Then,

$$\frac{\forall x A(x)}{A(t)} \, \forall E$$
$$\Pi$$
$$B$$

is a proof of the consequent.

For (2), we can extend the proof of the conclusion with a $\exists I$ step.

For (3), suppose $\Pi$ is a proof of $X \vdash_{CQ} B(c)$. Since $c$ does not occur in $X$, $c$ does not occur in any assumptions of $\Pi$. So, we can extend $\Pi$ with a $\forall I$ step ending in $\forall x B$ that satisfies the eigenvariable condition.

Finally, for (4), suppose $\Pi$ is a proof of $X, A(c) \vdash_{CQ} B$. Since $c$ does not occur in $X$, $c$ does not occur in any assumptions of $\Pi$. So, we can extend the proof in the following way, which satisfies the eigenvariable condition:

$$\frac{\exists x A(x) \qquad\qquad B}{B} \, \exists E^1$$

with $[A(c)]^1 \quad X$ over $\Pi$ yielding $B$.

□

To end this section, let's consider a proof in which quantifiers are nested. Let's see how we can construct a proof from $\exists y \forall x Lxy$ to the conclusion $\forall x \exists y Lxy$. Since the premise is an existentially quantified formula, we can try proving the conclusion from the assumption $\forall x Lxa$ and then eliminate the existential quantifier. The proof should have this outline:

$$[\forall x Lxa]^1$$
$$\vdots$$
$$\frac{\exists y \forall x Lxy \qquad \forall x \exists y Lxy}{\forall x \exists y Lxy} \, \exists E^1$$

Since $a$ occurs neither in the premise $\exists y \forall x Lxy$ nor in the conclusion $\forall x \exists y Lxy$, this satisfies the eigenvariable condition. So, to complete the proof, we need to figure out how to get from $\forall x Lxa$ to $\forall x \exists y Lxy$. Notice that we can easily get from $Lxa$ to $\exists y Lxy$ (that's an $\exists I$ step), so all we need to do is to unwrap the universal quantifier from $\forall x Lxa$ and then place it back on, once we've put in the existential quantifier in between. But that's easy. It is a

$\forall E$ step to peel off the universal quantifier, a quick $\exists I$ to put on an existential, and then a $\forall I$ to put the universal quantifier on top.

$$
\cfrac{
  \cfrac{
    \cfrac{
      \cfrac{
        \cfrac{[\forall xLxa]^1}{Lba}\;\forall E
      }{\exists yLby}\;\exists I
    }{\forall x\exists yLxy}\;\forall I
  \qquad \exists y\forall xLxy \quad }{\forall x\exists yLxy}\;\exists E^1
$$

The $\forall I$ step satisfies its eigenvariable condition because *its* eigenvariable (the $b$) does not occur in the assumption on which the subproof rests, the $\forall xLxa$.

* * *

Finally, let's consider a more complex proof, illustrating many of the principles we have seen in a more complex piece of reasoning. The proof below leads from the three premises $\forall x\forall y\forall z((Rxy \wedge Ryz) \to Rxz)$, $\forall x\forall y(Rxy \to Ryx)$, and $\forall x\exists yRxy$ to the conclusion $\forall xRxx$.

$$
\cfrac{
  \cfrac{\forall x\exists yRxy}{\exists yRay}\;\forall E \qquad
  \cfrac{
    \cfrac{
      \cfrac{
        \cfrac{
          \cfrac{\forall x\forall y\forall z((Rxy \wedge Ryz) \to Rxz)}{\forall y\forall z((Ray \wedge Ryz) \to Raz)}\;\forall E
        }{\forall z((Rab \wedge Rbz) \to Raz)}\;\forall E
      }{(Rab \wedge Rba) \to Raa}\;\forall E \qquad
      \cfrac{[Rab]^1 \qquad
        \cfrac{
          \cfrac{
            \cfrac{\forall x\forall y(Rxy \to Ryx)}{\forall y(Ray \to Rya)}\;\forall E}{Rab \to Rba}\;\forall E \qquad [Rab]^1
          }{Rba}\;{\to}E}{Rab \wedge Rba}\;\wedge E
    }{Raa}\;{\to}E
  }{\cfrac{\cfrac{Raa}{\forall xRxx}\;\forall I}{}}
}{}\;\exists E^1
$$

The proof ends with an application of $\exists E$ and an application of $\forall I$. Before the $\exists E$ step, we have a rather straightforward proof from the universal premises $\forall x\forall y\forall z((Rxy \wedge Ryz) \to Rxz)$ and $\forall x\forall y(Rxy \to Ryx)$, as well as the premise $Rab$, to the conclusion $Raa$. This proof involves choosing the appropriate instances of the universally quantified premises, in order to derive $Raa$ from $Rab$. Once we've done that, we appeal to the final universal premise, $\forall x\exists yRxy$ to justify $\exists yRay$, and since the conclusion $Raa$ of our other proof does not involve the term $b$, and $b$ appears nowhere other than the premise $Rab$, which is an instance of the existential $\exists yRay$, we can complete the $\exists E$ step, discharging the appeals to $Rab$. The result is now a proof from the universal premises (none of which include the name $a$) to the conclusion $Raa$. So, $a$ satisfied the eigenvariable condition in this proof, and we can use $\forall I$ to deduce $\forall xRxx$.

This proof shows us that if the relation $R$ is *transitive* (in other words, if $\forall x\forall y\forall z((Rxy \wedge Ryz) \to Rxz)$) and *symmetric* (if $\forall x\forall y(Rxy \to Ryx)$) and *serial* (that is, $\forall x\exists yRxy$, which states that each thing is related to *something* by $R$), then $R$ must be reflexive. This proof is a good test that you are on your way to mastery of the concepts introduced in this chapter. If you can read and understand this proof, and if you can see how the eigenvariable conditions are satisfied by the $\exists E$ and $\forall I$ steps, then you are well on the way to understanding some of the power of proofs in first-order predicate logic.

## 10.4 Discussion

First-order predicate logic is more powerful than propositional logic. The kind of generality expressed by quantified expressions allows for a piece of information to be applied in many different ways. A universally quantified claim $\forall x A(x)$ is a gift that can keep giving. There is no finite bound on the number of different things that we can infer *from* $\forall x A(x)$, in one step. The more we learn about what things there are, the claim that $\forall x A(x)$ keeps applying, again and again, to the new objects as we find them. Whenever we conclude another existentially quantified statement, we have more information about what objects there are—and more things to which we can apply the universally quantified statements. When we put these things together, especially in nested statements like $\forall x \exists y R x y$, we have *especially* powerful claims from which we can prove many different things. There is an important sense in which the power of first-order predicate logic involves *infinity*, in that we now have the tools at hand to express with a *finite* expression something that not only has infinitely many consequences but that has infinitely many consequences *about infinitely many things*.

To make this claim precise, we will need to turn to models for first-order predicate logic, and that will be our topic in the next chapter. But before we get there, we have the space to say a little more about what we have done so far and what is special about the language of predicate logic and the proof system we have explored.

* * *

There are two common types of formulas used with the quantifiers to represent claims as they naturally appear in English. Claims of the form "Every $F$ is $G$" are represented as $\forall x(Fx \to Gx)$ and claims of the form "Some $F$ is $G$" are represented as $\exists x(Fx \wedge Gx)$. It is much harder, in English, to say things of the form "$\forall x Fx$" and "$\exists x Ax$." When we say "Every*thing* is $F$" or "Some*thing* is $F$," this seems to involve a predicate "*thing*" that applies to absolutely everything. The expressions we naturally say in our everyday language seem to have the form "Every $F$ is $G$" and "Some $F$ is $G$," and it is not so clear that these expressions are *fundamentally* formed from a *unary* quantifier and a connective ($\to$, in the case of the universal quantifier, and $\wedge$, in the case of the existential quantifier). There is some reason to explore a language like ours, but in which the fundamental quantifiers have the form of *binary* quantifiers "Every $F$ is $G$" and "Some $F$ is $G$." See the challenge questions at the end of the chapter to explore this issue further.

Not all expressions like "Every $F$ is $G$" and "Some $F$ is $G$" are easily modeled in our formal language. Claims like "Kittens are cute," which have no explicit quantifier, can be understood in at least two different ways. One is the *universal reading*, which we can paraphrase as "*All* kittens are cute" or "*Every* kitten is cute." This would be formalized as $\forall x(Kx \to Cx)$. The second reading is the *generic reading*, which says something about kittens generically. The logic of generics is much more complicated, as they can admit exceptions, which the universal quantifier cannot. For example, as a generic statement, "Birds fly," is true, even though penguins and cassowaries are birds that do not fly, whereas "All birds fly" is false. We won't say any more on generics in this text, although there is much more to say on the topic, but if you'd like to know more, it is an area of active research that you can explore: see, for example, Pelletier and Asher (1997), Leslie (2007), and Sterken (2017).

* * *

It is very tempting to think that the correct form of an $\exists E$ rule would be to infer $A(b)$ from $\exists x A(x)$, provided that the term $b$ used in the conclusion $A(b)$ is *new*. In other words, to make an inference rather like this:[111]

$$\frac{\exists x A(x)}{A(b)} \; \exists ?$$

As a matter of fact, there are some proof systems where this kind of rule is used. *Tableaux* proofs are one prominent example (see Smullyan (1968) or Restall (2006) for introductions to tableaux proofs). There is an obvious appeal to such a rule. If we have come to $\exists x A(x)$ in our reasoning, it is very tempting to add a name to our vocabulary to name such an object and proceed to conclude $A(b)$, using that new name $b$. This is, of course, appealing. However, if we think of this principle as an inference rule, we could compose it with other rules, like this:

$$\frac{\dfrac{\exists x A(x)}{A(b)} \; \exists ?}{\forall x A(x)} \; \forall I$$

which would, of course, be a catastrophe. We shouldn't be able to conclude that *everything* is $F$ from the assumption that something is. No, the step from $\exists x A(x)$ to an instance $A(b)$ (where $b$ is new) is certainly a step in reasoning, but it isn't a valid inference. After all, it is quite possible that $\exists x A(x)$ be true without $A(b)$ being true—it could have been something *else*, other than $b$, that satisfies the condition $A(x)$. In our proof system, the step from $\exists x A(x)$ to $A(b)$ is not an inference step but the link between the major premise and the discharged assumption in the application of the $\exists E$ rule:

$$\frac{\exists x A(x) \qquad \begin{array}{c} [A(b)]^1 \\ \Pi \\ B \end{array}}{B} \; \exists E^1$$

Here, we make the *assumption* that $A(b)$ holds—and deduce something from that assumption—and then this assumption is discharged on the basis of the existential claim $\exists x A(x)$. There is a logical link between $\exists x A(x)$ and $A(b)$, but that link is not one of there being a valid inference from $\exists x A(x)$ to $A(b)$.[112]

* * *

We will close this chapter by picking up an idea from part I of the book. Recall there we started with natural deduction for intuitionistic logic ($\vdash_I$) and obtained natural deduction for classical logic ($\vdash_C$) by adding *DNE*. We can obtain a natural deduction system for quantified intuitionistic logic, which will be called IQ, by adding the quantifier rules to the

---

111. Getting these rules in a natural deduction proof system is not easy. Pelletier (1999) contains a fascinating discussion of the ways different logic texts have treated the existential quantifier, including cases where the rules get things wrong.

112. For an approach to quantifiers different from the one developed here, see Fine (1985).

system for I. Alternatively, we can obtain it by dropping the *DNE* rule from CQ. That system is nice in many ways, including being able to prove a normalization theorem. There is a quantified analog to the disjunction property, the *existence property*: if $\vdash_L \exists x A(x)$, then for some term $t$, $\vdash_L A(t)$. If IQ proves an existential statement, then it also proves a particular instance as a witness. This is an important part of what is meant when intuitionistic logic is called *constructive*.[113] Just as C lacks the disjunction property, CQ lacks the existence property. There are many points of difference between IQ and CQ, but getting into them in detail will take us too far afield. When you are constructing proofs in CQ, keep in mind when you use *DNE* and when it seems that that rule is necessary for the proof.

## 10.5  Key Concepts and Skills

☐ You need to understand the difference between predicates, names, variables, and quantifiers.

☐ You should be able to read and understand formulas in the language of predicate logic.

☐ You should understand the definition of *substitution*, when a term is free for a variable in a formula, and why the definition of substitution is restricted to the cases where the terms substituted in the formula are free for the variable substituted *for*.

☐ You should be able to *read* tree proofs using any or all of the quantifier rules, as well as be able to check that the eigenvariable condition is satisfied when the $\forall I$ and $\exists E$ rules are used.

☐ You should be able to *construct* simple tree proofs using all the inference rules for CQ.

The $\forall$-eigenvariable condition says that in an application of $\forall I$, the displayed term $b$ in the premise $A(b)$ cannot occur either in the conclusion of $\forall I$ or in any assumption in $X$ upon which the premise of the application of $\forall I$ depends.

The $\exists$-eigenvariable condition says that in an application of $\exists E$, the term $b$ in the discharged assumption cannot occur in $\exists x A(x)$, $B$ or in $X$.

## 10.6  Questions for You

### Basic Questions

1. For each of the following sentences, find a formula in the language of propositional logic that best represents its structure. Use the following dictionary: $Fx$ is *x is a footballer*, $Bx$ is *x is a biped*, $a$ is *Sócrates*, $Txy$ is *x is taller than y*, $Oxy$ is *x is older than y*, and $Hx$ is *x is happy*.

---

113. Dummett (2000, 6–8) highlights the disjunction and existence properties as key to constructive proofs.

   i. Some footballer isn't a biped.

   ii. No footballer is a biped.

   iii. Some footballer is taller than Sócrates.

   iv. No footballer is taller than $y$.

   v. $z$ is taller than any biped older than Sócrates.

   vi. No footballer is taller than any biped older than Sócrates.

   vii. Every footballer taller than Sócrates is older than $y$.

  viii. $z$ is older than some footballer who isn't a biped but is happy.

   ix. Every footballer taller than Sócrates is older than some footballer who isn't a biped but is happy.

2. Check that you understand how substitution works by performing these substitutions:

   i. $(Fx)[x/a]$

   ii. $(Fx)[y/a]$

   iii. $(\forall x(Fx \to Lxy))[y/b]$

   iv. $(\exists xFx \land Gx)[x/y]$

   v. $((Fx \to \exists y(Lxy \land (Gz \lor \forall xHx)))[x/z])[z/a]$

   What problem is there with performing the substitution $(\forall x(Fx \to Lxy))[y/x]$?

3. Are these trees *proofs*?

   Check whether the rules are correctly applied and, particularly, whether the eigenvariable conditions are satisfied by each $\forall I$ and $\exists E$ step.

$$\frac{\dfrac{\dfrac{\forall x(Fx \land \exists y\neg Fy)}{Fa \land \exists y\neg Fy}\forall E}{\exists y\neg Fy}\land E \quad \dfrac{[\neg Fa]^1 \quad \dfrac{\dfrac{\forall x(Fx \land \exists y\neg Fy)}{Fa \land \exists y\neg Fy}\forall E}{Fa}\land E}{\bot}\neg E}{\bot}\exists E^1$$

$$\frac{\dfrac{\forall x(Fx \lor Gx)}{Fa \lor Ga}\forall E \quad \dfrac{\dfrac{[Fa]^1}{\forall xFx}\forall I}{\forall xFx \lor \forall xGx}\lor I \quad \dfrac{\dfrac{[Ga]^2}{\forall xGx}\forall I}{\forall xFx \lor \forall xGx}\lor I}{\forall xFx \lor \forall xGx}\lor E^{1,2}$$

$$\frac{\forall x(Fx \lor Gx)}{Fa \lor Ga}\forall E \quad \dfrac{[Fa]^2}{Fa \lor Ha}\lor I \quad \dfrac{\dfrac{\dfrac{\dfrac{\dfrac{[Ga]^3 \ [\neg Ha]^1}{Ga \land \neg Ha}\land I}{\exists x(Gx \land \neg Hx)}\exists I \quad \neg\exists x(Gx \land \neg Hx)}{\bot}\neg E}{\neg\neg Ha}\neg I^1}{\dfrac{Ha}{Fa \lor Ha}\lor I}DNE}{\dfrac{Fa \lor Ha}{\forall x(Fx \lor Hx)}\forall I}\lor E^{2,3}$$

4. Construct proofs for the following arguments:
   i. $\forall x(Fx \rightarrow Gx), \forall x(Gx \rightarrow Hx) \succ \forall x(Fx \rightarrow Hx)$
   ii. $\exists xFx \succ \forall x(Fx \rightarrow Gx) \rightarrow \exists xGx$
   iii. $\exists xFx, \forall xGx \succ \exists x(Fx \wedge Gx)$
   iv. $p \wedge \exists xFx \succ \exists x(p \wedge Fx)$
   v. $\forall x\forall y\forall z((Rxy \wedge Ryz) \rightarrow Rxz), \neg\exists xRxx \succ \forall x\forall y(Rxy \rightarrow \neg Ryx)$
   Warning: Proofs for the final one will be significantly longer than proofs for the rest.
5. Construct proofs for the following arguments. (Beware, for these you may need to employ *DNE*, and they might be quite complicated.)
   i. $\forall x(p \vee Fx) \succ p \vee \forall xFx$
   ii. $\neg\exists x(\neg Fx \wedge \neg Gx), \neg\exists x(Gx \wedge \neg Hx) \succ \forall x(Fx \vee Hx)$
   iii. $\succ \exists x(Fx \rightarrow \forall yFy)$

**Challenge Questions**
1. In this chapter, we have introduced quantifiers in the traditional way. This is the dominant tradition in philosophy, mathematics, and computer science. If we were starting from scratch, we might have done things differently, to stay closer to the way quantifiers are used in our own languages, by introducing *binary* quantifiers

$$All\, x(A(x):B(x)) \qquad Some\, x(A(x):B(x))$$

to stand for "All *A*s are *B*s" and "Some *A*s are *B*s."

   Of course, we could define $All\, x(A(x):B(x))$ using the unary quantifier, as $\forall x(A(x) \rightarrow B(x))$ and we could define $Some\, x(A(x):B(x))$ as $\exists x(A(x) \wedge B(x))$, but let's not do that. Instead, define them directly, using their own introduction and elimination rules, without appealing to $\forall$ and $\exists$, or to the connectives $\rightarrow$ and $\wedge$.
2. Once you have answered the previous question, show that your rules for *All* and *Some* allow for the elimination of detours.
3. Given your answer to question 1, show how you could also define the traditional quantifiers $\forall$ and $\exists$ in terms of the binary quantifiers *All* and *Some*. Could you also define $\wedge$ and $\rightarrow$ using the binary quantifiers?
4. Given your answers to questions 1–3, consider this: what reasons might we have for preferring the old quantifiers over the new quantifiers, or vice versa?
5. Provide a detailed proof of theorem 37.
6. You might have noticed the parallel between the rules for $\forall$ and $\exists$, on the one hand, and the rules for $\square$ and $\lozenge$, on the other. One way to make the similarity between modal rules and quantifier rules more precise is define a translation of the modal language into the language of predicate logic in such a way that S5 *proofs* can be translated into CQ proofs. In this exercise, you will work through the details of this: first, define the *world translation* of a modal formula *A* into a predicate logic formula $w(A)$ in this way.

   We pick a variable (say $x$) from the language of predicate logic. The idea behind the translation is that for each formula *A* in the modal language, $w(A)$ will be a formula in which the variable $x$ may be free, and we could think of the value of $x$ as the world at which the formula is evaluated.

For each atom $p$ in the modal language, select a corresponding one-place predicate letter $P$ to represent it in the first-order language. Then we define the world translation $w(A)$ for each formula $A$ inductively, like this:

- $w(p) = Px$
- $w(\bot) = \bot$
- $w(A \wedge B) = w(A) \wedge w(B)$
- $w(A \vee B) = w(A) \vee w(B)$
- $w(A \rightarrow B) = w(A) \rightarrow w(B)$
- $w(\neg A) = \neg w(A)$
- $w(\Box A) = \forall x\, w(A)$
- $w(\Diamond A) = \exists x\, w(A)$
- If $X$ is a set of formulas, $w(X)$ is the set of world translations of those formulas.

So, for example, $w(\Diamond p \wedge \Diamond \Box q)$ is $\exists x Px \wedge \exists x \forall x Qx$.

Show how to transform any **S5** proof for $X \succ A$ into a **CQ** proof for $w(X) \succ w(A)$.

HINT: Do this by induction on the structure of the proof for $X \succ A$, by showing how an inference rule in the proof system for **S5**, when the premises and conclusions are translated by $w$, corresponds to a rule (or a few rules chained together) in the proof system for **CQ**. The cases for the propositional connectives should be straightforward. The cases for $\Box$ and $\Diamond$ need careful attention.

# 11 Models for Predicate Logic

The language of first-order predicate logic is much richer than the language of propositional logic. We have seen how to add to our inference rules in order to give an account of *proof* in this language. Our task in this chapter is to expand our account of *models*.

Recall that a Boolean valuation for the language of propositional logic is an assignment of truth values to the atoms, which can then be extended to assign truth values to each formula in the language. We use models to define counterexamples for our arguments, and we take *valid* arguments to be those without counterexamples. A system of proofs and a collection of models are connected by the *soundness* theorem if no argument with a proof also has a counterexample in that collection. The systems satisfy the *completeness* theorem if every argument either has a proof *or* a counterexample.

In this chapter, we will extend our account of models to model the language of predicate logic, with an aim to defining counterexamples to any arguments that have no proofs. This task is significantly harder than the case with propositional logic, for the language is richer. To interpret the language, we need to interpret *predicates*, *names*, *variables*, and *quantifiers*, in addition to the connectives of the language of propositional logic.

## 11.1 Defining Models

Models for first-order logic are more complicated than truth tables. Since the language of predicate logic has more structure, we need to interpret more *things* in order to assign a truth value to each formula in the language. As with propositional logic, we need to assign a truth value to every atomic formula. Now, an atomic formula can have the shape

$$Ft_1 \cdots t_n$$

where $F$ is an $n$-ary predicate, and $t_1$ to $t_n$ are terms. We would like the result of interpreting such a formula to be a truth value, and we would like the interpretation of the formula $Ft_1 \cdots t_n$ to depend in some systematic way on the interpretation of the predicate $F$ and the terms $t_1$ to $t_n$. If we focus for a moment on *names*, it seems fair to start with taking the interpretation of a name to be some *object*. (Whatever interpreting a name involves, it seems reasonable to take it to involve naming some*thing*.) So, if the interpretation supplies some object for each of the terms $t_1$ to $t_n$, we are left with interpreting the $n$-place predicate $F$. Now an answer for this can become clear. However we interpret the predicate $F$, it needs to give us a truth value for each list of $n$ objects. That way, we can take the list of objects $o_1, \ldots, o_n$ interpreting the terms $t_1, \ldots t_n$ and use them as inputs for the interpretation of

$F$, which will deliver a truth value. Now, providing a truth value for *every* list of $n$ objects is a tall order. (How many objects are there to choose from?) We can exploit the idea that a model is a *model* of how things are (or how things could have been) by allowing the objects to be taken from among a smaller selection of objects—a *domain* of possible values for our terms to take. So, that is how we will proceed. A model for a first-order language will have two parts, a domain of objects and an interpretation function assigning values to our names and to our predicates.

**Definition 71 (Model)**   *A* MODEL *M for a first-order language, or a first-order model, is a pair* $\langle D, I \rangle$ *where $D$ is a nonempty set, called the* DOMAIN, *and $I$ is a function on the language such that for each name $a$ in the language, $I(a) \in D$, and if $F$ is an n-ary predicate, then $I(F): D^n \mapsto \{0, 1\}$, where $D^n$ is the set of sequences of length $n$ of objects from $D$, where $I(\bot) = 0$.*[114]

The interpretation function tells us what the names and predicates pick out in the model. This is enough to interpret closed atomic formulas (those involving no free variables), and we can interpret the propositional connectives in the familiar way. Let's look at an example, interpreting a language with the one-place predicates $F$ and $G$, the two-place predicate $R$, the zero-place predicate (proposition) $p$, and the names $a$ and $b$.

Let's select, as an example, our domain to be the set $D = \{m, n, o\}$. Our language, for the sake of this example, has two names, so we can interpret each name by one object in the domain. Let's say that $a$ names the object $n$ and $b$ names the object $o$. That means that $I(a) = n$ and $I(b) = o$. (That leaves $m$ an object in our domain without a name in the language we are interpreting. Our language has fewer names than there are things to name in this domain.)[115] To interpret $p$, we need a truth value (let's say that it's false in this model). To interpret the predicates $F$ and $G$, we need an assignment of a truth value for each object. Let's say that $F$ is true of $m$ and $n$ and false of $o$, while $G$ is false of $m$ and true of $n$ and $o$. That is, $I(F)(m) = 1$, $I(F)(n) = 1$, and $I(F)(o) = 0$.[116] Similarly, $I(G)(m) = 0$, $I(G)(n) = 1$, and $I(G)(o) = 1$.

Finally, to interpret $R$, we assign truth values to each sequence of objects of length 2. We might say that $R$ is *true* of $(m, m)$ ($R$ relates the object $m$ to itself) and false of $(m, n)$ ($R$ doesn't relate $m$ to $n$) and so on, for each of the *nine* choices for sequences of length 2. Instead of listing these things one by one in a paragraph of text, it is simplest to represent all of this information in a series of tables, like this:

| | $I$ |
|---|---|
| $a$ | $n$ |
| $b$ | $o$ |

| | $I$ |
|---|---|
| $p$ | $0$ |

| | $F$ | $G$ |
|---|---|---|
| $m$ | $1$ | $0$ |
| $n$ | $1$ | $1$ |
| $o$ | $0$ | $1$ |

| $R$ | $m$ | $n$ | $o$ |
|---|---|---|---|
| $m$ | $1$ | $0$ | $1$ |
| $n$ | $1$ | $0$ | $0$ |
| $o$ | $1$ | $0$ | $1$ |

---

114. We will follow the convention that $D^1$ is just $D$, so we interpret a one-place predicate by a function $D \mapsto \{0, 1\}$, which assigns a truth value to each object. In addition, $D^0$ is a set containing just one object (the empty sequence $\langle \rangle$), so a 0-ary, read "nullary," predicate—an atomic formula with no terms—can be interpreted by the choice of a truth value for that formula.

115. In *this* model, the names name different objects. There is no requirement that this be the case in every model. Each name must name an object in the domain. This means that we *could* have multiple names for one thing.

116. Notice, $I(F)$ is itself a *function*. So, we represent the fact that we apply $I(F)$ to the object $m$ to get the value 1 as "$I(F)(m) = 1$."

To read the table for $R$, take the row header to be the first argument and the column header to be the second argument. The 1 in the $n$-row and the $m$-column says that $I(R)(n, m) = 1$. The 0 in the $m$-row and the $n$-column says that $I(R)(m, n) = 0$.

With all of this, we can interpret closed quantifier-free formulas in our language. To interpret $Fa$, find $I(a)$ (that's $n$), and feed this value into $I(F)$. That is, we check $I(F)(I(a))$, which is $I(F)(n)$, which is 1. So, $Fa$ is *true* according to this model. Similarly, to interpret $Rab$, we check $I(R)(I(a), I(b))$, which is $I(R)(n, o)$, which is 0. $Rab$ is false. So, $\neg Rab$ is true, $Fa \rightarrow Rab$ is false, and so on.

* * *

Nothing so far tells us how to interpret our variables or our quantifiers. There is a reason for that. Variables *vary*. They have no fixed interpretation in a model. If I know what the names and predicates mean, I can figure out what sentences involving those names and predicates mean. If I want to know whether $Fx$ is true, nothing in the interpretation of the *language* will tell me that. To interpret $Fx$, we need to know what $x$ is—and whatever $x$ *is* will vary from context to context. That is the job of a variable. To interpret $Fx$, the interpretation of the *language* will tell me how to interpret $F$, but something else needs to tell me how to interpret the variables. This job will be done by an *assignment of values*—or, an *assignment* for short.

**Definition 72 (Assignments)**   *Let M be a model with domain D. An* ASSIGNMENT *v is a function from the variables in the language to D.*
   *We will write $I(\cdot, v)$ for the interpretation function I together with assignment v.*

Given an assignment, we can interpret all other atomic sentences in our language. The closed sentences are interpreted in the language, and the variables are handled by the assignment.

So, if $v_1(x) = m$, $v_2(x) = n$, and $v_3(x) = o$, then for our interpretation $I$ given in the tables above, we have
$$I(Fx, v_1) = 1, \quad I(Fx, v_2) = 1, \quad I(Fx, v_3) = 0.$$
As $x$ varies from value to value in the different assignments, the value of the statement $Fx$ varies along with it. $Fx$ is true, given the assignment that sends $x$ to $m$ and the assignment that sends $x$ to $n$, but it is false if $x$ is assigned the object $o$. We can represent how the predicate $F$ classifies different objects in the domain, even when the language being interpreted has no names for those objects. One way to understand variables and assignments is by way of the analogy with pronouns. If the valuation $v_1$ assigns the object $m$ to the variable $x$ (where our language has no name for the object in question), we can think of $I(Fx, v_1)$ as saying "$F$ is true of *that*," where "*that*" is a pronoun picking out the object $m$. We can represent this classification in a diagram like this:

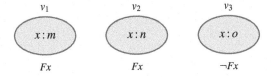

using the same notation as we used for possible worlds models.[117] Here, though, the *assignments* do the job that the worlds do in modal logic. The formula $Fx$ varies in truth depending on the value taken by the variable $x$. According to $v_1$ or to $v_2$, $Fx$ is true. According to $v_3$, $Fx$ is false.

Now we can see how we might interpret a statement like $\exists x Fx$, or $\forall x Fx$. We check the value of $Fx$ as we allow the values of the bound variable $x$ to vary across the domain, by varying the assignment function to send $x$ to each different object in turn. So, $\exists x Fx$ is true (according to our model $M$) if there is *something* in the domain we can assign as the value of $x$ to make $Fx$ true, and $\forall x Fx$ is true (according to $M$) if for *whatever* in the domain we assign as the value of $x$, $Fx$ is true. In our example, $\exists x Fx$ is true, while $\forall x Fx$ is false.

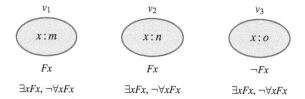

If we are interpreting a sentence with other variables too (say, $\exists x Fxy$), we need to vary the values of the variable we have quantified (here $x$) while keeping the values of other variables fixed (here $y$). To describe exactly what is going on when we interpret quantified expressions, it is helpful to use the notion of a *variant* assignment.

**Definition 73 (Variants)**   *An assignment $u$ is an $x$-VARIANT of an assignment $v$ iff $v(y) = u(y)$, for all variables $y$ distinct from $x$. We will write $u \sim_x v$ when $u$ is an $x$-variant of $v$. We will write $v_d^x$ for the function that assigns to $x$ the value $d$ and assigns to all other variables $y$ the same value assigned by $v$. That is:*

$$v_d^x(x) = d$$

$$v_d^x(y) = v(y), \quad \text{if the variable } y \text{ is not } x.$$

*So, $v_d^x$ is the unique $x$-variant $v$, which assigns $x$ the value $d$.*

With variants, we can describe the shifts in assignments appropriate to evaluating each different quantifier. Consider the variables $x$ and $y$ and the way the formula $Rxy$ is interpreted as $x$ and $y$ vary. We can represent these assignments in a grid, and we will indicate whether $Rxy$ holds in that assignment by writing $Rxy$ or $\neg Rxy$ underneath the oval.[118]

---

117. The similarity between the diagrams for the double indexed models and assignments for first-order models is suggestive, and we encourage you to think about it. Van Benthem (1996, chap. 9; 2010, chap. 27) develops the analogy between assignment shifts and modal operators in detail.

118. Compare this to the table for $R$ on page 218.

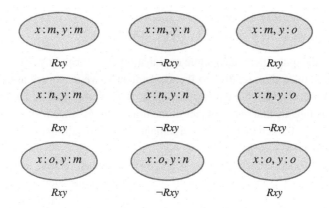

Now, if we quantify over the variable $x$, we are evaluating up and down a *column*, as we allow $x$ to vary while keeping $y$ fixed. If we quantify over the variable $y$, we are evaluating left and right across a *row*, as we allow $y$ to vary while keeping $x$ fixed. In this diagram, the $x$-variants of an assignment are the assignments in its *column*. The $y$-variants of an assignment are the assignments in its *row*.

$$* * *$$

Now we have everything we need to describe exactly what it is for a formula to be true in a model $M = \langle D, I \rangle$, *given an assignment $v$.*

First, we interpret *terms*:

- $I(t, v) = I(t)$, if $t$ is a name, and
- $I(t, v) = v(t)$, if $t$ is a variable.

In other words, to interpret a name, we consult the interpretation function $I$ from the model $M = \langle D, I \rangle$, to see what object the model assigns to the name. To interpret a variable, we consult the local assignment of values to variables.

This, then, gives us enough information to interpret all of our atomic formulas.

- $I(Ft_1 \cdots t_n, v) = I(F)(I(t_1, v), \ldots, I(t_n, v))$.

We look up the value assigned to the predicate $F$ and apply this function to the values assigned to the terms, whether they are names or variables (or a combination of the two).

Interpretations extend to truth-functional formulas in a straightforward way.

- $I(\neg A, v) = 1$ iff $I(A, v) = 0$
- $I(A \land B, v) = 1$ iff $I(A, v) = 1$ and $I(B, v) = 1$
- $I(A \lor B, v) = 1$ iff $I(A, v) = 1$ or $I(B, v) = 1$
- $I(A \rightarrow B, v) = 1$ iff $I(A, v) = 0$ or $I(B, v) = 1$
- $I(\bot, v) = 1$ never

The quantifiers are where the action is, at least for assignments. Evaluating quantified formulas will require the use of variant assignments.

- $I(\forall x A, v) = 1$ iff for every $u$ where $u \sim_x v$, $I(A, u) = 1$
- $I(\exists x A, v) = 1$ iff for some $u$ where $u \sim_x v$, $I(A, u) = 1$

The clauses for the quantifiers need some explanation. The formula $\forall x F x$ says that everything is $F$. To evaluate that, relative to the assignment $v$, we use every assignment $u$ that agrees with $v$ on every variable except perhaps $x$. If $Fx$ comes out as 1 relative to every such $u$, then this tells us that every object in the domain is $F$, since if there is some object $o$ such that $I(F)(o) = 0$, then there is an assignment $w(x) = o$ resulting in $I(Fx, w) = 0$, which is sufficient for $I(\forall x F x, v) = 0$.

The formula $\exists x F x$ says that something is $F$. To evaluate that, relative to an assignment $v$, we need to check every $u$ such that $u \sim_x v$. If there is at least one such $u$ where $I(Fx, u) = 1$, then that shows that something is $F$, namely $u(x)$, which is sufficient for $I(\exists x F x, v) = 1$.

\* \* \*

Let's return to our model that we introduced in the tables on page 218. To see how quantifiers and assignments work, we'll interpret $Gy$, $\forall x F x$, $\exists y R x y$, and $\forall x \exists y R x y$ in this model, relative to an assignment $v$, which sends $v(x) = m$ and $v(y) = n$.

- $Gy$: To evaluate $I(Gy, v)$, we note that $v(y) = n$, and $I(G)(n) = 1$. So $Gy$ is true (relative to $v$). In the model, $G$ is true *of n*.
- $\forall x F x$: $I(\forall x F x, v) = 1$ iff for all $u \sim_x v$, $I(Fx, u) = 1$. The three variant assignments are $v_m^x$, $v_n^x$, and $v_o^x$, which send $x$ to $m$, $n$, and $o$, respectively, but keep $y$ evaluated as $n$. The assignment $v_m^x$ yields

$$I(Fx, v_m^x) = I(F)(v_m^x(x)) = I(F)(m) = 1,$$

the assignment $v_n^x$ yields

$$I(Fx, v_n^x) = I(F)(v_n^x(x)) = I(F)(n) = 1,$$

and the assignment $v_o^x$ yields

$$I(Fx, v_o^x) = I(F)(v_o^x(x)) = I(F)(o) = 0.$$

Since $v_o^x \sim_x v$ and $I(Fx, v_o^x) = 0$, $I(\forall x F x, v) = 0$. This formula is *false*.

The preceding explanation involved a lot of *words*. Here is a way to represent this reasoning in a diagram, in a way that might be familiar to you. We have one ellipse here for each different $x$-variant of $v$:

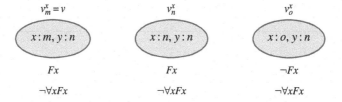

The universally quantified statement is false (at each assignment) because we have *at least one* assignment (here, the one assigning $x$ the value $o$) that makes $Fx$ false.

- $\exists yRxy$: This formula has the variable $x$ free, while the variable $y$ is bound. $I(\exists yRxy, v) = 1$ iff $I(Rxy, u) = 1$ for some $u \sim_y v$. We see that $v(x) = m$, and the value of $x$ constant between $v$ and $u$ ($u$ allows the value of $y$ to vary, but the value of $x$ is kept stable), so $u(x) = m$ too. We see from the table that $I(R)(m, u(y)) = 1$ when $u(y)$ is either $m$ or $o$. For *these* $y$-variant assignments, $I(Rxy, u) = 1$. So, $I(\exists yRxy, v) = 1$. Our formula is *true*. Here is this reasoning represented in a diagram. We use the same assignments, but now we actually *use* the value of $x$ as well as the value of $y$.

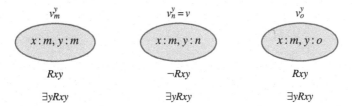

- $\forall x\exists yRxy$: This formula is true in our model too. We need to repeat the reasoning from the previous case while varying the value of $x$ across our assignments as well. Checking the diagram below, we see that indeed, $\exists yRxy$ is true in the model for each assignment of values to $x$, so $\forall x\exists yRxy$ is also true.

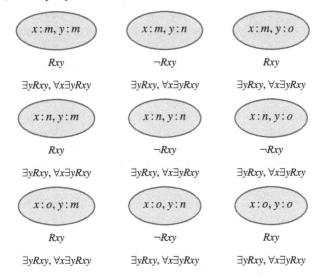

To round out the picture of how quantifiers are interpreted, look at the contrast between evaluating $\exists yRxy$ (which checks across each row of the diagram) and $\exists xRxy$ (which checks up and down each column).

## 11.2   Features of Models

In this section, we will explore some of the features of models and assignments of values to variables. The main upshot of this section is that if two assignments $v$ and $u$ agree on what values they assign to the free variables of $A$, then $I(A, v) = I(A, u)$. In other words, the interpretation of a formula $A$ in a model and assignment does not depend on any of the variables not free in $A$. So, in particular, if a formula is *closed*, its interpretation depends only on the model and not on any assignment of values to the variables. This fact might seem *obvious*. (If it does seem obvious to you, that's good—it means you understand the definitions.) But regardless, it is a good check on our statement of the definitions that what should be obvious is actually true. So, let's work our way to verifying this result.

We will begin by noting some features of assignments.

**Lemma 24**   *Let $a$ be a name, $x$ a variable, and $M$ a model. Then for any assignments $v$ and $u$, $I(a, v) = I(a, u)$ and $I(x, v_o^x) = I(x, u_o^x)$ for any object $o \in D$.*

*Proof.*   Let $a$ be a name, $x$ a variable, and $M$ a model. Then, $I(a, v) = I(a) = I(a, u)$. For the second part, let $o \in D$ be any object in the domain. Then,

$$I(x, v_o^x) = v_o^x(x) = o = u_o^x(x) = I(x, u_o^x),$$

as desired.                                                                                                 □

**Lemma 25**   *Let $M$ be a model and $v$ an assignment. For all assignments $u$ such that $u \sim_x v$, there is an object $o \in D$ such that $u = v_o^x$.*
   *For all objects $o$, there is an assignment $w$ such that $w \sim_x v$ and $w(x) = o$.*

*Proof.*   Let $M$ be a model and $v$ an assignment. For the first part, let $u$ be an assignment such that $u \sim_x v$. Then, $u = v_{u(x)}^x$, where $u(x)$ is the desired object. For the second part, let $o \in D$ be an object. Let $w$ be $v_o^x$. By definition, $w \sim_x v$ and $w(x) = o$.                                 □

This lets us give an equivalent clause for the universal and existential quantifiers.

**Lemma 26** *Let M be a model. $I(\forall xA, v) = 1$ iff for all $d \in D$, $I(A, v_d^x) = 1$. Similarly, $I(\exists xA, v) = 1$ iff for some $d \in D$, $I(A, v_d^x) = 1$.*

*Proof.* For the left-to-right direction of the universal quantifier case, assume $I(\forall xA, v) = 1$. Then, for all $w \sim_x v$, $I(A, w) = 1$. Let $d \in D$ be arbitrary. As $v_d^x \sim_x v$, $I(A, v_d^x) = 1$. Therefore, for all $d \in D$, $I(A, v_d^x) = 1$.

For the right-to-left direction, assume for all $d \in D$, $I(A, v_d^x) = 1$. Let $w$ be an arbitrary assignment such that $w \sim_x v$. By lemma 25, there is a $d \in D$, such that $w = v_d^x$. Then, $I(A, w) = 1$. Since $w$ was arbitrary, we can conclude that $I(\forall xA, v) = 1$.

The reasoning for the existential quantifier is similar. □

**Lemma 27** *Let M be a model. For all assignments v and u such that $v(x) = u(x)$, for all variables x occurring freely in A, $I(A, v) = I(A, u)$.*

*Proof.* Let $M$ be a model and $v$ and $u$ assignments, and suppose $v(x) = u(x)$, for all variables $x$ occurring freely in $A$. The proof is by induction on the construction of the formula $A$.

If $A$ is atomic, then it is of the form $Ft_1, \ldots, t_n$. $I(F, v) = I(F, u)$ and $I(t_i, v) = I(t_i, u)$, for each $i \leq n$, since if $t_i$ is a name, by lemma 24, the interpretations are identical, and if $t_i$ is a variable, by assumption, $v(t_i) = u(t_i)$. Therefore, $I(Ft_1 \cdots t_n, v) = I(Ft_1 \cdots t_n, u)$.

If $A$ is $\bot$, then $I(\bot, v) = 0 = I(\bot, u)$, and the result is immediate.

For the inductive hypothesis, we will assume that for all subformulas $B$ of $A$, for all variables $x$ occurring freely in $A$, then $I(B, v) = I(B, u)$.

Suppose $A$ is of the form $\neg B$. By the inductive hypothesis, $I(B, v) = I(B, u)$. Suppose $I(B, v) = I(B, u) = 1$. Then, $I(\neg B, v) = 0$ and $I(\neg B, u) = 0$, so $I(\neg B, v) = I(\neg B, u)$. If $I(B, v) = I(B, u) = 0$, the argument is similar.

Suppose $A$ is of the form $B \wedge C$. By the inductive hypothesis, $I(B, v) = I(B, u)$ and $I(C, v) = I(C, u)$. Then, we argue as follows:

$$I(B \wedge C, v) = 1 \quad \text{iff} \quad I(B, v) = 1 \text{ and } I(C, v) = 1$$
$$\text{iff} \quad I(B, u) = 1 \text{ and } I(C, u) = 1$$
$$\text{iff} \quad I(B \wedge C, u) = 1$$

The second line is justified by the inductive hypothesis.

The cases where $A$ is of the form $B \vee C$ or $B \rightarrow C$ are similar to the conjunction case.

Suppose $A$ is of the form $\forall xB$. Note that we cannot suppose that $v(x) = u(x)$, since $x$ is not free in $A$. However, we do know that, for all variables $y$ free in $B$, $v_d^x(y) = u_d^x(y)$, for any $d \in D$, so by the inductive hypothesis, $I(B, v_d^x) = I(B, u_d^x)$.

$$I(\forall xB, v) = 1 \quad \text{iff} \quad \text{for all } d \in D, I(B, v_d^x) = 1$$
$$\text{iff} \quad \text{for all } d \in D, I(B, u_d^x) = 1$$
$$\text{iff} \quad I(\forall xB, u) = 1$$

The case where $A$ is of the form $\exists xB$ is similar to the universal quantifier case, so with those done, we will declare our lemma proved. □

Lemma 27 says that if an interpretation of a formula $A$ differs on assignments $v$ and $u$, then $v$ and $u$ must differ on some variable occurring freely in $A$.

Finally, we will show that terms contribute to the value of formulas only by way of their value. We will show that one can switch out a variable for another term with the same interpretation as that variable, without changing the interpretation of the formula.

**Lemma 28 (Semantic substitution)**   *Given any model M, assignment v, formula A(y), and term t that is free for y in A(y), such that $I(t, v) = d$, we have $I(A(t), v) = I(A(y), v_d^y)$.*

*Proof.*   Let $M$ be a model and $v$ an assignment. Let $A(y)$ be a formula, $t$ a term free for $y$ in $A(y)$, where $I(t, v) = d$. The proof will be by induction on the structure of $A(y)$.

Suppose $A(y)$ is an atomic formula, $Fs_1 \cdots s_n$, where $y$ is one of the terms $s_i$. So, $A(y)$ is $Fs_1 \cdots s_{i-1} y s_{i+1} \cdots s_n$, and replacing $y$ by $t$, we see that $A(t)$ is $Fs_1 \cdots s_{i-1} t s_{i+1} \cdots s_n$. It will get quite tedious to write $Fs_1 \cdots s_{i-1} y s_{i+1} \cdots s_n$ through the rest of the proof, so let's consider from now the case where $y$ only occurs as the first term $s_1$, and the reasoning will apply equally to the cases where $y$ occurs elsewhere in the formula.

Now, $I(Ft\, s_2 \cdots s_n, v) = I(F)(I(t, v), I(s_2, v), \ldots, I(s_n, v))$. Since $I(t, v) = I(y, v_d^y)$, we have

$$I(F)(I(t, v), I(s_2, v), \ldots, I(s_n, v))$$
$$= \quad I(F)(I(y, v_d^y), I(s_2, v), \ldots, I(s_n, v)).$$

The right-hand side is identical to $I(Fy\, s_2 \cdots s_n, v_d^y)$—since $I(s_i, v) = I(s_i, v_d^y)$, since $y$ was assumed not to occur in $s_i$—as desired.

The cases for $A$ of the forms $\neg B$, $B \wedge C$, $B \vee C$, or $B \to C$ are handled by the inductive hypothesis, given the fact that the connectives do not shift the assignment of values to variables.

For the existential quantifier case, let $A(y)$ be the quantified formula $\exists x B(y)$. If $y$ is not free in $\exists x B$, then $v$ and $v_d^y$ agree on all variables free in $\exists x B$ and $(\exists x B)[y/t] = \exists x B$. Therefore, $I(\exists x B(y), v) = I(\exists x B(y), v_d^y) = I(\exists x B(t), v)$.

If $y$ is free in $\exists x B$, then $I(\exists x B(y), v) = 1$ iff $I(B(y), v_e^x) = 1$ for some $e \in D$. Since $t$ is free for $y$ in $\exists x B(y)$, the inductive hypothesis applies, so $I(B(y), (v_e^x)_d^y) = 1$ iff $I(B(t), v_e^x) = 1$. Since $x$ and $y$ are different variables, $(v_e^x)_d^y = (v_d^y)_e^x$, and so $I(B(y), (v_e^x)_d^y) = 1$ iff $I(B(y), (v_d^y)_e^x) = 1$. It follows that $I(\exists x B(t), v) = 1$ iff $I(B(t), v_e^x) = 1$ for some $e \in D$, and this holds iff $I(\exists x B(y), v_d^y) = 1$, as desired.

The reasoning for the universal quantifier is similar, except we replace "for some $e \in D$" by "for *all* $e \in D$." So, the proof of our lemma is complete.                     □

The results of this section show that terms and assignments of values to variables interact in straightforward ways with the evaluations of formulas as true or false in models. All of these results will be useful in the next section, when we turn our attention toward validity and counterexamples.

## 11.3  Counterexamples and Validity

As in the previous chapters dealing with models, we can define a notion of counterexample and validity appropriate to models for first-order predicate logic.

**Definition 74 (Counterexample, validity)**   *A pair of a model M and an assignment v is a* COUNTEREXAMPLE *to the argument $X \succ A$ iff $I(B, v) = 1$, for each $B \in X$, and $I(A, v) = 0$.*

*An argument $X \succ A$ is* VALID *iff it has no counterexamples. When the argument $X \succ A$ is valid, we write* $X \models_{FO} A$.

An equivalent definition of validity is possible: $X \models_{FO} A$ if and only if for every model $M$ and assignment $v$, if $I(X, v) = 1$, then $I(A, v) = 1$. We will not prove this equivalence here, as it is straightforward.

$$* * *$$

In the previous chapter, we demonstrated some facts about the provability relation $\vdash_{CQ}$. We will show that the same facts hold concerning validity.

**Theorem 39** *The following are true:*

1. $\forall x A \land \forall x B \models_{FO} \forall x(A \land B)$
2. $\exists x(A \lor B) \models_{FO} \exists x A \lor \exists x B$
3. $A \rightarrow \forall x B \models_{FO} \forall x(A \rightarrow B)$, *where $x$ is not free in $A$*
4. $\forall x(A \rightarrow B) \models_{FO} \exists x A \rightarrow B$, *provided $x$ is not free in $B$*

As with propositional logic, the facts about provability were demonstrated by extending and manipulating proofs. For validity, the reasoning involves paying attention to the different things that a model (and assignment) *satisfies*.

*Proof.* For (1), let $M$ be a model and $v$ an assignment such that $I(\forall x A \land \forall x B, v) = 1$. It follows that $I(\forall x A, v) = I(\forall x B, v) = 1$. So, for all $w \sim_x v$, $I(A, w) = I(B, w) = 1$. Therefore, $I(A \land B, w) = 1$, so $I(\forall x(A \land B), w) = 1$.

The argument for (2) is similar to the preceding argument, swapping $\forall$ and $\exists$, $\land$ and $\lor$, and premise and conclusion.

For (3), let $M$ be a model and $v$ an assignment such that $I(A \rightarrow \forall x B, v) = 1$, where $x$ is not free in $A$. Suppose, for reductio, that $I(\forall x(A \rightarrow B), v) = 0$, so there is a $w \sim_x v$ such that $I(A \rightarrow B, w) = 0$. Therefore, $I(A, w) = 1$ and $I(B, w) = 0$. Since $x$ is not free in $A$, $I(A \rightarrow \forall x B, v) = I(A \rightarrow \forall x B, w)$. As $I(A, w) = 1$, it follows that $I(\forall x B, w) = 1$. But this is a contradiction, as it implies that $I(B, w) = 1$.

For (4), let $M$ be a model and $v$ an assignment such that $I(\forall x(A \rightarrow B), v) = 1$, where $x$ is not free in $B$. Suppose, for reductio, that $I(\exists x A \rightarrow B, v) = 0$, so $I(\exists x A, v) = 1$ while $I(B, v) = 0$. Since $I(\exists x A, v) = 1$, there is a $w \sim_x v$ such that $I(A, w) = 1$. By assumption, $I(\forall x(A \rightarrow B), v) = 1$, so $I(A \rightarrow B, w) = 1$. Therefore, $I(B, w) = 1$. Since $w$ agrees with $v$ on the free variables of $B$, $I(B, w) = I(B, v)$, which is a contradiction. $\square$

We will highlight some more features of validity that will be useful in the proof of soundness for the quantifier rules.

**Lemma 29** *The following validity claims are true:*

1. $\forall x A(x) \models_{FO} A(t)$, *where $t$ is free for $x$ in $A$*
2. $A(t) \models_{FO} \exists x A(x)$, *where $t$ is free for $x$ in $A$*

*Proof.* For (1), suppose that $M$ is a model and $v$ an assignment such that $I(\forall x A(x), v) = 1$. Let $I(t, v) = d$. It follows that $I(A, v_d^x) = I(A(t), v)$. As $v_d^x \sim_x v$, $I(A(t), v) = 1$. The argument for (2) has the same shape as the argument for (1), swapping $\exists$ for $\forall$ and premise for conclusion. $\square$

**Lemma 30**  *If* $X \models_{FO} B(a)$, *then* $X \models_{FO} \forall x B(x)$, *provided the term a does not occur in X and a is free for x in B(x).*

*Proof.*  Suppose $X \models_{FO} B(a)$, $a$ does not occur in $X$, and $a$ is free for $x$ in $B(x)$. For reductio, suppose there is a model $M$ and assignment $v$ such that $I(X, v) = 1$ and $I(\forall x B(x), v) = 0$. This means that there is some $x$-variant $u$ of $v$ such that $I(B(x), u) = 0$. Notice that the term $a$ is not present in $X$ or in $B(x)$. So, consider the variant interpretation $I'$, which agrees with $I$, except for setting $I'(a) = u(x)$. In other words, we let $a$ name whatever the variable $x$ picks out on the assignment where $I(B(x), u) = 0$. By lemma 28, $I'(B(a), u) = 0$ too. Since $x$ does not occur free in $B(a)$, $I'(B(a), u) = I'(B(a), v) = 0$. Since $a$ is not present in any of the formulas in $X$, $I(X, v) = I'(X, v) = 1$, so this model and assignment is a counterexample to $X \succ B(a)$. But this contradicts the assumption that $X \models_{FO} B(a)$. Therefore, $X \models_{FO} \forall x B(x)$.  □

**Lemma 31**  *If* $X \models_{FO} \exists x B(x)$ *and* $Y, B(a) \models_{FO} C$, *then* $X, Y \models_{FO} C$, *provided a does not occur in Y,* $\exists x B(x)$, *or C, and a is free for x in B(x).*

*Proof.*  Suppose $X \models_{FO} \exists x B(x)$ and $Y, B(a) \models_{FO} C$, $a$ does not occur in $Y$, $\exists x B(x)$, or $C$, and $a$ is free for $x$ in $B(x)$. For reductio, suppose that there is a model $M$ and assignment $v$ such that $I(X, Y, v) = 1$ but $I(C, v) = 0$. Since $I(X, v) = 1$, $I(\exists x B(x), v) = 1$. Therefore, there is an assignment $u \sim_x v$ such that $I(B(x), u) = 1$. Let $I'$ be an interpretation just like $I$ except that $I'(a) = u(x)$, so $I'(B(a), u) = I'(B(x), u) = 1$. Since $x$ is not free in $\exists x B(x)$, $x$ is not free in $B(a)$. Therefore, $I'(B(a), u) = I'(B(a), v)$. As $a$ does not occur in $Y$, $I'(Y, v) = I(Y, v) = 1$. Since $I'(Y, v) = 1$ and $I'(B(a), v) = 1$, $I'(C, v) = 1$. Since $a$ does not occur in $C$, $I'(C, v) = I(C, v)$. But by assumption, $I(C, v) = 0$, which is a contradiction. Therefore, $X, Y \models_{FO} C$.  □

Notice that the eigenvariable condition played an important role in lemmas 30 and 31. In both cases, we began with an assumed counterexample, $I$ and $v$, and shifted to a new assignment $u$ and interpretation $I'$. Lemma 28 lets us go back and forth between evaluations relative to $v$ and $u$, and the eigenvariable condition lets us shift between $I$ and $I'$. Without the assumptions that the relevant arguments obey the eigenvariable condition, we would be unable to obtain the contradictions needed for the reductio arguments.

* * *

In the previous chapter, we supplied some examples of trees that violated the $\forall$- and $\exists$-eigenvariable conditions and so failed to be CQ proofs. Now that models are available, it will be useful to present some counterexamples. The first argument was $Fa \wedge Ga \succ \forall x Fx$. Let $M_1$ be the model such that $D = \{c, d\}$ and $I$ is defined as

|   | $I(F)$ | $I(G)$ |
|---|--------|--------|
| $c$ | 0 | 0 |
| $d$ | 1 | 1 |

with $I(a) = d$. Then, $I(Fa \wedge Ga, v) = 1$ but $I(\forall x Fx, v) = 0$, for any assignment $v$. This serves as a counterexample to the argument $\exists x Fx \succ \forall x Fx$ as well.

The model $M_2$ will be the model such that $D = \{c, d\}$ and $I$ is defined as follows.

| | $I(F)$ | $I(G)$ |
|---|---|---|
| $c$ | 0 | 1 |
| $d$ | 1 | 0 |

This model provides a counterexample to $\exists x Fx, \exists x Gx \succ \exists x(Fx \wedge Gx)$. Let $v$ be an assignment. Then, $I(\exists x Fx, v) = 1$, since $I(Fx, v_d^x) = 1$, and $v(=, \exists x Gx)1$, since $I(Gx, v_c^x) = 1$. For every assignment $u \sim_x v$, $I(Fx \wedge Gx, u) = 0$, so $I(\exists x(Fx \wedge Gx), v) = 0$.

For the next model, $M_3$, let $D = \{c, d, e\}$, with $I(R)$ as

| $I(R)$ | $c$ | $d$ | $e$ |
|---|---|---|---|
| $c$ | 0 | 0 | 1 |
| $d$ | 0 | 0 | 1 |
| $e$ | 0 | 0 | 0 |

$I(a) = c$, and $I(b) = d$. This is a counterexample to $\exists x Rbx \succ \exists x Rxx$, since $I(\exists x Rbx, v) = 1$ but $I(\exists x Rxx, v) = 0$, for any assignment $v$.

$M_3$ is also a counterexample to $(Rab \rightarrow Rba) \wedge \exists x Rax \succ \exists x Rxa$. Let $v$ be any assignment. Since $I(Rab, v) = 0$, $I(Rab \rightarrow Rba, v) = 1$. It is the case that $I(\exists x Rax, v) = 1$ as well, but $I(\exists x Rxa, v) = 0$. In this model, $a$ bears $R$ to something apart from $b$ while not bearing $R$ to $b$, so the conditional does not force $b$ to bear $R$ to $a$, and so nothing bears $R$ to $a$.

$$* * *$$

We can piece these results together to conclude that provability, $\vdash_{CQ}$, is sound for validity, $\vDash_{FO}$.

**Theorem 40 (Soundness)** *For any set $X$ of formulas, and for any formula $A$ in the language of first-order predicate logic, if $X \vdash_{CQ} A$, then $X \vDash_{FO} A$.*

*Proof.* This is a straightforward induction on the construction of the proof $\Pi$ for $X \succ A$. Extending proofs with the propositional rules (including *DNE*) preserves validity, as we showed when we proved soundness for propositional logic. The remaining rules to consider are the quantifier rules, so let's take them in turn.

For $\forall E$, if we have a proof for $X \succ \forall x A(x)$, which we have already shown is valid, we have $X \vDash_{FO} \forall x A(x)$, by the inductive hypothesis. If we extend that proof using $\forall E$, we have a proof from $X$ to $A(t)$ for some term $t$ (provided that $t$ is free for $x$ in $A(x)$). By lemma 29, we have $\forall x A(x) \vDash_{FO} A(t)$, and so, $X \vDash_{FO} A(t)$, as desired.

Similarly, for $\exists I$, if we have a proof for argument from $X$ to $A(t)$, which we have already shown is valid, we have $X \vDash_{FO} A(t)$. If we extend that proof using $\exists I$, we have a proof from $X$ to $\exists x A(x)$ (provided that $t$ is free for $x$ in $A(x)$). By lemma 29, we have $A(t) \vDash_{FO} \exists x A(x)$, and so, $X \vDash_{FO} \exists x A(x)$, as desired.

For $\forall I$, if we have a proof for $X \succ A(a)$, where $a$ is a term not present in the premises $X$, and where $a$ is free for $x$ in $A(x)$, then $X \vDash_{FO} A(a)$. If we extend that proof using $\forall I$, we have a proof from $X$ to $\forall x A(x)$. To show that $X \vDash_{FO} \forall x A(x)$, we use lemma 30, completing this case.

For $\exists E$, suppose we have a proof for $X \succ \exists x A(x)$ and a proof for $Y, A(a) \succ C$, where $a$ does not occur in $Y$, $\exists x A(x)$, or $C$, and $a$ is free for $x$ in $A(x)$. Extending these proofs with $\exists E$ results in a proof of $X, Y \succ C$. From the inductive hypothesis, $X \models_{\mathsf{FO}} \exists x A(x)$ and $Y, A(a) \models_{\mathsf{FO}} C$. We can then use lemma 31 to conclude $X, Y \models_{\mathsf{FO}} C$, which is what we needed to prove.

We conclude that the soundness theorem is proved.                                    □

That suffices for soundness. Soundness has some quick corollaries. One, which will be used in the proof of the completeness theorem, is that alphabetic variants are equivalent according to models. In particular, if $A$ and $A'$ are alphabetic variants, then for any model $M = \langle D, I \rangle$ and assignment $v$, $I(v, A) = I(v, A')$. This follows from soundness together with theorem 37.

Let us turn to the completeness theorem.

**Theorem 41 (Completeness)**   *For any set $X$ of formulas, and any formula $A$ in the language of first-order predicate logic, if $X \models_{\mathsf{FO}} A$, then $X \vdash_{\mathsf{CQ}} A$.*

As with the other proofs of completeness earlier in the text, this proof is rather advanced. So, we attach a   WARNING LABEL   to the rest of this section. If you have not mastered the completeness theorems of the previous chapters, you would do well to skip ahead to section 11.4 (page 234). If you are prepared for one last completeness proof, read on. We will start with an outline of the proof method. Like the completeness proof for classical propositional logic, the key idea is that for any $X$ and $A$ where $X \nvdash_{\mathsf{CQ}} A$, we can "fill out" our set $X$ into a $\mathsf{CQ}$-maximal $A$-avoiding set $X'$. By lemmas 9 and 10, the formulas in $X'$ act just like the formulas that are assigned *true* by a Boolean valuation function. This time, we will use our set $X'$ to define a *model* for first-order predicate logic. For that, we need to pick out a domain, and we will use the language to define the domain. Since each term in the language has to pick out an object in a domain (given an assignment of values to the variables), we take the domain of our model to *be* all the terms in the language. Each name $c$ will name itself. There is also a *trivial* assignment of values to the variables where each variable is assigned itself as a value. We will interpret the predicates so that $F t_1 \cdots t_n$ is assigned as true (on the trivial assignment) if and only if $F t_1 \cdots t_n$ is in our set $X'$. The constraints of being a $\mathsf{CQ}$-maximal $A$-avoiding set will *almost* be enough to enable us to show that this is, indeed, a model that makes a formula true (relative to that trivial assignment) if and only if it is in the set $X'$, but we will need to tweak our construction just a little to make things work. The wrinkle is with the behavior of the quantifiers. The $\forall E$ and $\exists I$ rules behave well. For any $\mathsf{CQ}$-maximal $A$-avoiding set $X'$, if $\forall x B(x)$ is in $X'$, then $B(t)$ is in $X'$ for any term $t$, since $\forall x B(x) \vdash_{\mathsf{CQ}} B(t)$. Similarly, if $B(t)$ is in $X'$, then so is $\exists x B(x)$. The set $X'$ respects *half* of the truth conditions for the quantifiers. It is the other half that proves more difficult.

There is, in general, no bar to having $B(t) \in X'$ for absolutely every term $t$ (including every variable), without having $\forall x B(x) \in X'$. Let $B(x)$ be a particularly simple formula $Fx$, for a one-place predicate $F$. The set containing each $Ft$ for *every* term $t$, whether a name or a variable, is not strong enough to prove $\forall x Fx$. Consider the following model:

$$D = \{c, d\} \qquad \begin{array}{c|c} & I(F) \\ \hline c & 1 \\ d & 0 \end{array}$$

interpreting a language in which *every* name is assigned the value $c$ and under the assignment $v$ that assigns every variable the value $c$. Relative to that assignment, each formula of the form $Ft$ is true, and the conclusion $\forall x Fx$ is *false*. Given the soundness theorem, there is no way to prove $\forall x Fx$ from all those premises. So, there is a CQ-maximal $\forall x Fx$-avoiding set, including $Ft$ for *every* term $t$. So, CQ-maximal sets like this are not quite enough to act as models for predicate logic. To fill the gap, we need for our set to not only be *CQ-maximal* but to be *witnessed*.

**Definition 75 (Witnessed sets)** *A set $X$ of formulas in the language of first-order predicate logic is said to be* WITNESSED *iff whenever $\forall x B(x) \notin X$, then there is some term $t$ that is free for $x$ in $B$ such that $B(t) \notin X$, and whenever $\exists x B(x) \in X$, then there is some term $t$ that is free for $x$ in $B$ such that $B(t) \in X$.*

The crucial step in our proof, then, is to show that whenever $X \nvdash_{\mathsf{CQ}} A$, then there is some maximal $A$-avoiding *witnessed* set $X'$ where $X' \nvdash_{\mathsf{CQ}} A$ and $X \subseteq X'$. This is not *quite* true as it stands, for we have already seen a counterexample. If $X$ is large enough to contain all the terms in the language (like our set consisting of every statement of the form $Ft$, for each term $t$), then maybe it has so many opinions that we cannot construct a witnessed maximal $\forall x Fx$-avoiding set extending it. So, we limit our claim to a slightly more modest one, which we can prove. The technique will be a slight modification of the method used to demonstrate lemma 12 (see page 100).

**Lemma 32** *If $X \nvdash_{\mathsf{CQ}} A$, and there is an unending supply of terms $t_1, t_2, \ldots$, not present in $X$ or in $A$, then there is some* witnessed *maximal $A$-avoiding set $X'$ where $X \subseteq X'$.*

The proviso that we have a supply of unused terms is trivially satisfied if the starting set $X$ is finite, for our language does have an infinite supply of variables, and any finite set $X$ of formulas can use only finitely many of them.

*Proof.* As before, we select an *enumeration* of the language, like this:

$$B_1, B_2, B_3, B_4 \ldots$$

We define $X'$ step-by-step in a sequence $X_0, X_1, X_2, \ldots$. We start with $X_0 = X$. To define $X_n$ for each $n > 0$, we consider the formula $B_n$. If $X_{n-1}, B_n \nvdash_{\mathsf{CQ}} A$, then $X_n = X_{n-1} \cup \{B_n\}$, unless $B_n$ starts with an existential quantifier: if $B_n = \exists x C(x)$ for some variable $x$ and formula $C(x)$, then we set $X_n = X_{n-1} \cup \{B_n, C(t_m)\}$ where $t_m$ is a term (from our list $t_1, t_2, \ldots$) not occurring in $X_{n-1} \cup \{B_n\}$ or in $A$. So, we add to our set not only the existentially quantified statement but a *witness* to it, where the witness involves a fresh term, one not used in our set so far. Otherwise, $X_n = X_{n-1}$. So, we have this increasing sequence:

$$X = X_0 \subseteq X_1 \subseteq X_2 \subseteq X_3 \subseteq \cdots$$

of sets of formulas. The reasoning from lemma 12 applies almost unchanged. At each step of the way, the set $X_n$ is $A$-avoiding, since $X_0$ starts off avoiding $A$, and each $X_n$ is defined to be the same as $X_{n-1}$ unless the new formula added (that is, $B_n$) can be added without including $A$. Well, with one exception. We must check that if $B_n$ has the form $\exists x C(x)$, then adding the witness doesn't make $A$ provable. We need to verify that if $X_{n-1}, \exists x C(x) \nvdash_{CQ} A$, then $X_{n-1}, \exists x C(x), C(t_m) \nvdash_{CQ} A$, provided that $t_m$ is not present in $X_{n-1}, \exists x C(x)$ or in $A$. But suppose we had a proof $\Pi$ from $X_{n-1}, \exists x C(x), C(t_m)$ to $A$. We could extend it with one step of $\exists E$, like this:

$$
\dfrac{\exists x C(x) \qquad \dfrac{X_{n-1} \quad \exists x C(x) \quad [C(t_m)]^1}{\displaystyle \Pi} \atop A}{A}\; \exists E^1
$$

since the term $t_m$ satisfies the eigenvariable condition at this step (it is not present in the premises or the conclusion of the proof $\Pi$), so the whole proof shows that $X_{n-1}, \exists x C(x) \vdash_{CQ} A$, contradicting our assumption. So, there is no proof $\Pi$ from $X_{n-1}, \exists x C(x), C(t_m)$ to $A$ either.

So, even with this extra step whenever we add an existentially quantified formula to our pile of formulas, each $X_n$ is $A$-avoiding. The set $X'$ is defined as $\bigcup_{n=0}^{\infty} X_n$, the union of all of the sets in the series.

It remains to show that $X'$, defined like this, is a witnessed maximal $A$-avoiding set. First, $X'$ is $A$-avoiding, since if we had $X' \vdash_{CQ} A$, then there is some proof from $X'$ to $A$. But a proof is finite, so the assumptions used in that proof, since they are all in $X'$, are all in some set $X_n$ constructed along the way. But each $X_n$ is $A$-avoiding, so there is no proof from $X'$ to $A$ either.

The argument for maximality is similar to the maximality argument of chapter 6. To show that $X'$ is a *maximal* $A$-avoiding set, we want to show that if $B \notin X'$, then $X' \vdash_{CQ} A$. As before, this is easy to show since $B$ appears on our list somewhere, let's say it is $B_n$. Since $B_n \notin X'$, it follows that $B_n \notin X_n$, which means that when $B_n$ came up for consideration, it was not added to $X_n$. That is, we saw that $X_{n-1}, B_n \vdash_{CQ} A$. So, since $X_{n-1} \subseteq X'$, then $X', B_n \vdash_{CQ} A$ too. So, $X'$ is a maximal $A$-avoiding set, as we wanted to show.

Finally, to show that $X'$ is *witnessed*, we need to show that if $\exists x C(x) \in X'$, then $C(t) \in X'$ for some term $t$, and whenever $\forall x C(x) \notin X'$, then $C(t) \notin X'$ for some term $t$. The first part is immediately satisfied by design. If $\exists x C(x) \in X'$, then it was added at some stage of the process, and so, a witness $C(t_m)$ was added at the same time. For the second part, if $\forall x C(x) \notin X'$, then by the negation-completeness of $X'$ (lemma 10), $\neg \forall x C(x) \in X'$, and hence, $\exists x \neg C(x) \in X'$, so, by the same reasoning as before, $\neg C(t) \in X'$ for some term $t$, and by the consistency of $X'$ (again, lemma 10), $C(t) \notin X'$, as desired.                   $\square$

The final step of our completeness proof, then, will be to show that if $X'$ is a witnessed maximal $A$-avoiding set, then there is some model that makes *exactly* the formulas in $X'$ true. We start with the definition:

**Definition 76 (The term model for a set)**   *The* TERM MODEL *for a set $X'$ of formulas is defined to be $\langle D, I \rangle$ where we set $D$ to be the set of all terms occurring in $X'$, and for each $n$-place predicate $F$ and for all terms $t_1, \ldots, t_n \in D$, we set $I(F)(t_1, \ldots, t_n) = 1$ iff $Ft_1 \cdots t_n \in X'$.*

In the term model, since the terms *are* the domain, there is a natural assignment of values to variables: let each variable have *itself* as a value. This is the IDENTITY ASSIGNMENT of values to variables, which we call *id*.

We want to prove the following lemma concerning term models: if $X'$ is a witnessed maximal $A$-avoiding set of formulas, then in the term model for $X'$, for each formula $B$, $I(B, id) = 1$ iff $B \in X'$.

**Lemma 33** *If $X'$ is a witnessed maximal $A$-avoiding set of formulas, then in the term model for $X'$, for each formula $B$, $I(B, id) = 1$ iff $B \in X'$.*

*Proof.* As usual with lemmas like this, we prove it by induction on the construction of the formula. If $B$ is atomic, of the form $Ft_1 \cdots t_n$, then we have, by the definition of satisfaction in models $I(Ft_1 \cdots t_n, id) = 1$ iff $I(F)(I(t_1, id), \ldots, I(t_n, id)) = 1$, and in our particular model, this holds iff $I(F)(t_1, \ldots, t_n) = 1$ iff $F(t_1 \cdots t_n) \in X'$, so the base case is proved.

The cases for the propositional connectives (and $\bot$) are exactly the same as in the propositional completeness proof (see lemma 11 and its proof, on page 99), so we will not repeat them here.

The cases for the quantifiers are where the extra work lies. Let's tackle the universal quantifier first. By the definition of models, $I(\forall xB, id) = 1$ iff $I(B, v) = 1$ for every $x$-variant $v$ of $id$. By lemma 26, for each $x$-variant $v$, $v = id_t^x$, for some $t \in D$. We will use this fact to show that $I(\forall xB, id) = 1$ iff $\forall xB \in X'$.

For the right-to-left direction, suppose $\forall xB \in X'$ but $I(\forall xB, id) \neq 1$. Then there is some $x$-variant $id_t^x$ of $id$ such that $I(B, id_t^x) \neq 1$. If $t$ is free for $x$ in $B$, then from lemma 28, $I(B, id_t^x) = I((B)[x/t], id)$. We can then use the inductive hypothesis to conclude that $(B)[x/t] \notin X'$, which contradicts the facts that $\forall xB(x) \vdash_{CQ} B(t)$ and $\forall xB(x) \in X'$. If $t$ is not free for $x$ in $B$, then we can take an alphabetic variant $C$ such that $t$ is free for $x$ in $C$. Since alphabetic variants are satisfied by the same assignments, $I(B, id_t^x) = I(C, id_t^x)$. From lemma 28, $I(C, id_t^x) = I((C)[x/t], id)$, so by the inductive hypothesis, $(C)[x/t] \notin X'$. It then follows that $\forall xC \notin X'$; otherwise, $(C)[x/t] \in X'$. Since $\forall xC$ is an alphabetic variant of $\forall xB$, by theorem 37, $\forall xB \notin X'$, which contradicts an assumption. Therefore, $I(\forall xB, id) = 1$.

Conversely, since $X'$ is witnessed if $\forall xB \notin X'$, then there is some term $t$ that is free for $x$ in $B$ such that $(B)[x/t] \notin X'$, so by the inductive hypothesis, $I((B)[x/t], id) \neq 1$. By lemma 28, $I(B, id_t^x) \neq 1$, so $I(\forall xB, id) \neq 1$, as desired.

The case for the existential quantifier has the same form as the case for the universal quantifier, so we leave the details as an exercise for the reader. $\qquad\square$

Now we can put all these pieces together, and we have the completeness theorem for proofs for first-order predicate logic (theorem 41).

*Proof.* Suppose $X \nvdash_{CQ} A$. Check the terms in our language featured in $X \cup \{A\}$. If all but finitely many terms occur in $X \cup \{A\}$, then add a fresh sequence of names $c_1, c_2, \ldots$ to form a larger language to build our model. If there are infinitely many terms not occurring in the set $X \cup \{A\}$, there is no need to expand the language for the next step. The condition for lemma 32 (page 231) is thus satisfied (at least in the new language), so it follows that there is a witnessed maximal $A$-avoiding set $X'$ where $X \subseteq X'$. By lemma 33 (page 233), the term model for $X'$ is a model in which, under the identity assignment *id*, every member of $X$ is true and $A$ is false. So, it follows that $X \not\models_{FO} A$. In other words, if $X \models_{FO} A$, then $X \vdash_{CQ} A$.                                                                                     □

This completeness proof is more involved than either the completeness proof for classical logic or the completeness proof for the propositional modal logics S4 or S5. However, the basic details are the same. In each case, the model is constructed out of the raw materials provided by the language. As in the case of propositional logic, we construct a single set of formulas to provide a counterexample to the argument from $X$ to $A$—a maximal $A$-avoiding set of formulas is almost always enough. The extra details involve the domain of quantification. The domain of the term model consists of all the terms of the language, but in some rare cases (where almost all[119] the terms in the language feature in the set $X$), then we need to supplement our language with a supply of new terms to be used as witnesses to the quantified expressions in our set.

This technique is used again and again in completeness proofs of different logical systems. If, for example, we extended our language to involve the *identity* predicate (see page 237), then we would need to modify this proof, since our set $X'$ might contain statements of the form $t_1 = t_2$, which would mean that the distinct terms $t_1$ and $t_2$ name the same object in the domain. So, we cannot take the domain of our model to be the set of all the terms in the language. In that case, we need to show that the terms in the language form *equivalence classes*, $[t]$ of all terms $s$ where $s = t \in X'$, and the domain of the terms model is the set of all equivalence classes. The details need to be changed, but the general structure is the same. We use properties of the *proof systems* to show that we can construct sets of formulas with the features corresponding to the set of truths in our model (or *sets* of truths, if our model contains more than one world), and then we show that we can construct a model that satisfies just that set of formulas.

### 11.4   The Power and Limits of Predicate Logic

We will end this chapter with a few observations about the significance of these results, on what we can do with proofs and models for first-order predicate logic—and what we can't do with them. There is much more that could be said about the philosophical significance of formal logic and of first-order predicate logic in particular. These comments will be a brief guide to some of the issues and connections that are worth exploring.

\* \* \*

The first point to make concerns proofs and, in particular, what proofs in CQ can tell us about proofs of *general* statements. You may have heard people say that you can never prove a *universal* claim, such as that all wombats are cute. Or that you cannot refute an existence claim, such as that there are living thylacenes. We have seen that statements like this are not, in general, true. There are plenty of universal claims that we can prove—not only can we prove universal claims from *premises* (if we assume $\forall xFx$ and $\forall x(Fx \to Gx)$,

---

119. "Almost all" is a technical term here. "Almost all" means "all but finitely many." So, if we had a set of terms of our language that left out only 7 of them, or 3088, or one million of them, that set would contain almost all of the terms. But a set that contained all of the names and all of the variables except $x_2, x_4, x_6, \ldots$, and so on, would *not* contain almost all of the terms.

we can prove—from them—that $\forall x Gx$, say), but we can even prove universal claims from *no* premises ($\forall x(Fx \lor \neg Fx)$, for example). Similarly, we can refute existential claims by proving statements of the form $\neg \exists x A(x)$, by proving $\perp$ after having assumed $\exists x A(x)$, and discharging that assumption with a $\neg I$ inference.

The crucial feature about these proofs that slogans like "you can't prove a universal statement" and "you can't refute an existence claim" ignore is the special form of rules like $\forall I$ and $\exists E$. These rules explain what is involved in concluding a universal claim and what is involved in making use of an existential claim in a proof. *That* is how you can come to conclude a universal generalization, not by proving some of its instances and making a leap from there (that may be a good way to reason to what is *reasonable* to believe, but that isn't a deductively valid inference, with no possible counterexamples), but if we have proved $A(t)$ where the term $t$ is *arbitrary*, then it is fair to conclude $\forall x A(x)$, since our proof concerning $t$ could apply to *anything*, given that we have assumed nothing about what $t$ is.

Those who find those slogans appealing are not thinking of inferences like that, but of the reasonableness—or unreasonableness—of inferring $\forall x Fx$ from instances, like $Fa$, $Fb$, $Fc$, ... and there, we can agree that no inference like that is free of counterexamples. No matter how many objects we assume have the property $F$, there is some model we can construct that makes the premises $Fa$, $Fb$, $Fc$, ... true and that has some *other* object in the domain that does not satisfy the predicate $F$. This model will make the premise statements $Ft$ (for all those terms $t$) true but will make the putative conclusion $\forall x Fx$ *false*. Any virtue had by the inference of *inductive generalization* (the inference from some number of premises $A(a)$, $A(b)$, $A(c)$, ... to the conclusion $\forall x A(x)$) is neither the virtue of provability nor validity. There might be a kind of inductive reasonableness to such an inference, but it is not, in general, valid. Living without this inference of generalization does not inhibit our ability to prove things or to construct counterexamples, even when we reason about general claims. Perhaps inferences like inductive generalization can be a guide to what models we might prefer or which theories are more "reasonable," but those distinctions are getting at different issues than those marked out by concerns of provability, consistency, or deductive validity.

$$* * *$$

In chapter 7, we introduced the concept of the proposition, $\|A\|$, expressed by the formula $A$ in a model. There is an analog in the first-order case, namely, the EXTENSION of a formula. The extension of a formula $A(x)$, whose only free variable is $x$, in a model $\langle D, I \rangle$ is

$$\|A(x)\| = \{d \in D : I(A(x), v_d^x) = 1\}.$$

The notion of the extension of a formula in a model can be extended to formulas with more than one free variable, provided that we impose some ordering on the variables. For this definition, we will extend the notion of an $x$-variant of an assignment to multiple distinct variables, so that the assignment $v_{d_1, \ldots, d_n}^{x_1, \cdots, x_n}(y) = v(y)$ when $y$ is not one of the $x_1, \ldots, x_n$ and otherwise $v_{d_1, \ldots, d_n}^{x_1, \cdots, x_n}(x_i) = d_i$. Given that we have settled on some ordering on variables, the extension of a formula $A(x_1, \ldots, x_n)$ whose free variables are among $x_1, \ldots, x_n$, where $n \geq 2$, is

$$\|A(x_1, \ldots, x_n)\| = \{ \langle d_1, \ldots, d_n \rangle \in D^n : I(A(x_1, \ldots, x_n), v_{d_1, \ldots, d_n}^{x_1, \cdots, x_n}) = 1\}.$$

There are a few things to note about extensions of formulas.

First, just as distinct formulas can express the same proposition, distinct formulas can have the same extension. For example, in any model, $\|Fx \wedge Gx\| = \|Gx \wedge Fx\|$. In a particular model, distinct formulas can have the same extension, as in the model with $D = \{m, n, o\}$ and the following interpretations of $F$ and $G$.

|   | F | G | H |
|---|---|---|---|
| m | 1 | 1 | 0 |
| n | 1 | 1 | 0 |
| o | 0 | 0 | 1 |

In this model, $Fx$ and $Gx$ have the same extension. They are true of the same things. Similarly, $\neg Gx$ and $Hx$ have the same extensions, as do $Hx \wedge Fx$ and $Gx \wedge \neg Gx$.

Second, when two formulas have the same extension, we say that they are EXTENSIONALLY EQUIVALENT. When two (unary) formulas $A(x)$ and $B(x)$ are extensionally equivalent in a model, $\forall x(A(x) \leftrightarrow B(x))$ will be true in the model. When two (unary) formulas $A(x)$ and $B(x)$ are extensionally equivalent in all models, $\forall x(A(x) \leftrightarrow B(x))$ will be valid. The extensional equivalence of two formulas is an important aspect of meaning, but it falls short of stronger semantic connections, such as synonymy. For example, the expressions "$x$ is an Indigenous prime minister of Australia before 2019" and "$x$ is a female president of the United States before 2019" have the same extension, taking the set of people as the domain and interpreting the predicates as in English. They both have the empty extension, namely, $\emptyset$. It isn't *necessary* that those two expressions have the same extension, though. It is possible that politics in Australia could have gone differently, resulting in at least one Indigenous prime minister before 2019. Similarly, it is possible that elections in the United States could have gone differently, resulting in at least one female president before 2019. We can say that these two expressions have something in common, namely their extensions, but they also differ in at least one aspect of meaning, an intensional aspect, which might be put as saying they express different *properties*. The property of being an Indigenous prime minister of Australia before 2019 is distinct from the property of being a female president of the United States before 2019.

Finally, in a given model, when a subset $X$ of the domain is the extension of a formula $A(x)$, we say that the set $X$ is DEFINABLE and that it is DEFINED BY THE FORMULA $A(x)$. A natural question for a logician is, in a given model, what subsets of the domain are definable in a given language. We know that the empty set and the whole domain are both definable in any language with a unary predicate. In some models, whether a particular set is definable is a difficult question. One model where questions of definability are important is the standard model of arithmetic. The language for this model has three-place predicates $S$ and $P$ for being the sum of two numbers and being the product of two numbers, as well as a two-place relation, $<$, of being less than and names for every natural number.[120] The domain is the set of natural numbers, $\{0, 1, 2, \ldots\}$, and the language is interpreted in the standard way, for example, $S(a, b, c)$ is true iff the numbers picked out by $a$ and $b$ add up

---

120. It is more common to use function symbols for sum and product, rather than relations, and to include a function symbol for the successor function. Using relations here will not affect the points we are making.

to the number picked out by $c$. Many infinite sets of numbers are definable in this model. It was an important discovery of the first half of the twentieth century that there are infinite sets of natural numbers that are not definable in the language of arithmetic. There are many such sets, and it turns out that some undefinable sets are more complex than others.

\* \* \*

We have mentioned that one distinctive feature of first-order predicate logic is the expressive power made available by nested quantifiers. This expressive power played a very important role in making mathematics rigorous in the eighteenth and nineteenth centuries. The work of pioneering mathematicians such as Bernard Bolzano (1781–1848) and Karl Weierstrass (1815–1897) involved giving formal *definitions* of informal notions such as the convergence of a series of numbers, as well as the continuity of a function of real numbers, in terms of more basic mathematical relations such as order and distance, using the quantifiers $\forall$ and $\exists$. For example, a series of numbers

$$a_1, a_2, a_3, \ldots$$

*converges* to a limiting value $a$ if and only if *for any* ($\forall$) distance $\epsilon > 0$, there is ($\exists$) some number $n$ along the sequence such that for any ($\forall$) further item $a_m$ in the sequence where $m \geq n$, the distance between $a_m$ and the limiting value $a$ is less than $\epsilon$.

This definition has alternating (that is, $\forall \exists \forall$) quantifiers, and the development of these analyses of limits, continuity, and the rigorization of the calculus arose alongside the logical tools necessary to understand the logical relationships involving claims of these forms. It is not going too far to say that the development of logic, motivated by the concerns to appropriately analyze such mathematical definitions, directly paved the way to the development of *analytic philosophy* in the early twentieth century. (J. Alberto Coffa's *The Semantic Tradition from Kant to Carnap* [1993] is a good place to read some of the history and the intellectual trajectory of this development, from Kant's attempts to chart the synthetic a priori, and the development of possibilities for logical analysis, from Bolzano through Frege, Russell and Wittgenstein, and on to Carnap.)

\* \* \*

One topic we have not had space to explore is the logic of the *identity* predicate. It is natural to reserve a special two-place predicate letter in our language, to be interpreted like this: $I(s=t, v) = 1$ if and only if $I(s, v) = I(t, v)$. That is, the formula $s = t$ is true in a model (with an assignment) if and only if the terms $s$ and $t$ denote the same object in the domain. With the identity predicate, it is possible to express many things. Bertrand Russell is famous for using first-order predicate logic's quantifiers and the identity predicate to analyze *definite descriptions* (B. Russell 1905), sentences of the form *the F is a G*, as saying

$$\exists x((Fx \wedge \forall y(Fy \rightarrow y = x)) \wedge Gx)$$

On this way of thinking, to say that *the F is a G* is to say that there is something that is $F$, and it is the *only F* (anything that is $F$ is identical to that thing) and it is also $G$.

With identity in the language, we have yet greater expressive possibilities. Not only can we say things like "$x$ is the one and only $F$" by saying $Fx \wedge (\forall y)(Fy \rightarrow y = x)$, but we can express *cardinality* quantifiers for all finite numbers. For example, we can say that there are exactly three objects with property $F$, like this:[121]

$$\exists x_1 \exists x_2 \exists x_3 \big((Fx_1 \wedge Fx_2 \wedge Fx_3) \wedge (x_1 \neq x_2 \wedge x_1 \neq x_3 \wedge x_2 \neq x_3)$$
$$\wedge \forall y(Fy \rightarrow (y = x_1 \vee y = x_2 \vee y = x_3))\big)$$

This says that there are three things with property $F$ (they are all distinct from each other), and anything with property $F$ is one of those three things. What goes for "exactly three" could go for "exactly $n$" for any finite number $n$. But there is no way to generalize this to say that there are infinitely many things with property $F$ because the sentences in our language are finite.

* * *

This limitation, that there is no way in our language to say that there are infinitely many things with property $F$, is not a contingent feature of our language. It is a consequence of its fundamental design constraints. This follows from a significant consequence of the soundness and completeness theorems: the compactness theorem.

**Theorem 42 (Compactness)**　*If $X \models_{FO} A$, then there has to be some* finite *subset $X'$ of $X$ such that $X' \models_{FO} A$.*

This would not be obvious at all, if all we had to go on were the definition of validity in terms of the absence of counterexamples. Some models have finite domains. Some models have infinitely large domains. There is no telling whether an infinite number of statements is required to narrow down the field of our models to just those in which the formula $A$ is true. Regardless, if we state this in terms of provability, the result is obvious.

*Proof.* If $X \vdash_{CQ} A$, then there is some CQ proof whose assumptions are in $X$ and whose conclusion is $A$. Since the proof is a finite tree, it appeals to only finitely many assumptions from $X$. Call the set of those assumptions $X'$. This is a finite subset of $X$, and we have $X' \vdash_{CQ} A$ as desired.

Now, since $\vdash_{CQ}$ is sound and complete for $\models_{FO}$, we see the same result for FO validity. The compactness theorem is proved.　　　　　　　　　　　　　　　　　　　□

With the compactness theorem in hand, we can show that, while we can define *finite* cardinality quantifiers in our language, we cannot define *infinite* cardinality quantifiers.

**Theorem 43 (Inexpressibility of infinity)**　*There is no sentence $A$ in the language of predicate logic (with identity) that is true in all and only models where there are* infinitely *many objects with property $F$.*

*Proof.* Suppose that there were a sentence $A$ that is true in every model in which there are infinitely many objects with property $F$. Consider the set $X$, consisting of the following sentences:

1. $A_1$: $\exists x_1 Fx_1$ (there is at least one object with property $F$)

---

121. To save space, we have left out some parentheses in extended conjunctions or disjunctions. You can fill them in however you like.

2. $A_2$: $\exists x_1 \exists x_2 (Fx_1 \land Fx_2 \land x_1 \neq x_2)$ (there are at least two objects with property $F$)
3. $A_3$: $\exists x_1 \exists x_2 \exists x_3 \big((Fx_1 \land Fx_2 \land Fx_3) \land (x_1 \neq x_2 \land x_1 \neq x_3 \land x_2 \neq x_3)\big)$ (there are at least three objects with property $F$)
4. $A_4, A_5, A_6, \ldots$ and so on.

Given that the only models in which *all* of the sentences $A_1, A_2, \ldots A_n, A_{n+1}, \ldots$ are true are models in which there are infinitely many objects with property $F$, if the sentence $A$ is true in all models with infinitely many objects with property $F$, we have to have

$$A_1, A_2, A_3, \ldots \models_{\text{FO}} A.$$

So, by the compactness theorem, there is some finite subset $X'$ of the set $\{A_1, A_2, A_3, \ldots \}$ such that $X' \models_{\text{FO}} A$. But take the set $X'$. From its members, take the *longest* sentence $A_n$, which we can do as $X'$ is finite, and so there is such a longest sentence. It follows that in any model in which there are $n$ objects with property $F$, all of the sentences $A_i$ in $X'$ (which have the form $A_i$ where $i \leq n$) are *true* in any such model. Take the model with *exactly n* objects, each of which has property $F$. This model makes every formula in $X'$ true, and so, since $X' \models_{\text{FO}} A$, we have $A$ true there too. In this model, $F$ is true of only finitely many objects (exactly $n$). So, since $A$ was a sentence true in every model where there are *infinitely* many $F$s, we have shown that for any sentence true in any model with infinitely many $F$s, it must be true in some models with finitely many $F$s too. That is, we have shown that there is no sentence true in *all* and *only* the models with infinitely many $F$s. □

This result is the first of a number of *limitative* results concerning what can be expressed in the language of first-order predicate logic. For more results in this vein, and much more *powerful* and *significant* limitations on what we can do with first-order predicate logic, we need to ascend the mountain first scaled by Kurt Gödel, when he proved his *incompleteness* theorems. To cover those, we would need another book, and we will leave them for another time. In the next chapter, we provide some references for readers interested in learning more about the limitative results.

### 11.5 Key Concepts and Skills

☐ You should understand the definition of a *model* for a first-order language (with a domain and an interpretation of each predicate and each name in the language) and of an *assignment* of values to the variables. You should also understand why we need both a model and an assignment to assign truth values to every formula.

☐ You should be able to check the truth or falsity of a formula of first-order logic in a model.

☐ You should understand the concept of validity in first-order predicate logic and be able to construct counterexamples to simple invalid arguments and, for simple valid arguments, be able to demonstrate that they are indeed valid.

☐ You should understand the soundness and completeness theorems and have a grasp of how the soundness theorem is proved.

A MODEL $M$ for a first-order language is a pair $\langle D, I \rangle$ where $D$ is a nonempty set, called the DOMAIN, and $I$ is a function on the language such that for each name $a$ in the language, $I(a) \in D$, and if $F$ is an $n$-ary predicate, then $I(F): D^n \mapsto \{0, 1\}$, where $D^n$ is the set of sequences of length $n$ of objects from $D$.

- $I(\neg A, v) = 1$ iff $I(A, v) = 0$
- $I(A \wedge B, v) = 1$ iff $I(A, v) = 1$ and $I(B, v) = 1$
- $I(A \vee B, v) = 1$ iff $I(A, v) = 1$ or $I(B, v) = 1$
- $I(A \rightarrow B, v) = 1$ iff $I(A, v) = 0$ or $I(B, v) = 1$
- $I(\bot, v) = 1$ never
- $I(\forall x A, v) = 1$ iff for every $u$ where $u \sim_x v$, $I(A, u) = 1$
- $I(\exists x A, v) = 1$ iff for some $u$ where $u \sim_x v$, $I(A, u) = 1$

A pair of a model $M$ and an assignment $v$ is a COUNTEREXAMPLE to the argument $X \succ A$ iff $I(B, v) = 1$, for each $B \in X$, and $I(A, v) = 0$.

## 11.6   Questions for You

### Basic Questions

1. Consider a first-order language with names $a$ and $b$ and with predicates $M$ (one-place), $B$ (two-place). Take a model with domain $D = \{rock, paper, scissors\}$, where $I$ interprets the names and predicates like this:

| | $I$ | | $M$ | | $B$ | rock | paper | scissors |
|---|---|---|---|---|---|---|---|---|
| | | rock | 1 | | rock | 0 | 0 | 1 |
| $a$ | rock | paper | 0 | | paper | 1 | 0 | 0 |
| $b$ | paper | scissors | 1 | | scissors | 0 | 1 | 0 |

   i. First, check the following closed formulas for truth or falsity in the model:

$$Ma \quad Mb \quad Bab \quad Bab \vee Bba$$

   ii. Now, list out the nine different assignments of values to the variables $x$ and $y$ in a three-by-three grid, and assess the following open formulas for truth or falsity at each assignment of values to the variables:

$$Mx \quad My \quad Bxy \quad My \wedge Bxy \quad \exists y Bxy \quad \exists y (My \wedge Bxy)$$

   iii. Finally, assess the following closed formulas for truth or falsity in the model:

$$\forall x \exists y Bxy \quad \exists x \forall y Bxy \quad \forall x (Mx \vee \exists y (My \wedge Bxy))$$

2. Consider the language with a two-place predicate $R$. Consider the domain $D = \{k, l, m\}$ and the three different models for our language given by these three different ways to

interpret $R$ on that domain:

| $I_1(R)$ | $k$ | $l$ | $m$ |
|---|---|---|---|
| $k$ | 0 | 0 | 1 |
| $l$ | 1 | 0 | 0 |
| $m$ | 0 | 1 | 0 |

| $I_2(R)$ | $k$ | $l$ | $m$ |
|---|---|---|---|
| $k$ | 0 | 1 | 1 |
| $l$ | 1 | 0 | 1 |
| $m$ | 1 | 1 | 0 |

| $I_3(R)$ | $k$ | $l$ | $m$ |
|---|---|---|---|
| $k$ | 0 | 1 | 1 |
| $l$ | 0 | 0 | 1 |
| $m$ | 0 | 0 | 0 |

So, we have three models: $M_1 = \langle D, I_1 \rangle$, $M_2 = \langle D, I_2 \rangle$, and $M_3 = \langle D, I_3 \rangle$. For each of the following arguments, establish which (if any) of $M_1$, $M_2$, and $M_3$ are *counterexamples* to those arguments. (If the argument involves formulas with free variables, specify the assignment of values to those variables used in your counterexample.)

  i. $Rxy \succ Ryx$

  ii. $Rxy, Ryz \succ Rxz$

  iii. $\succ \forall x \forall y (Rxy \rightarrow Ryx)$

  iv. $\forall x \exists y Rxy \succ \forall x Rxx$

  v. $\forall x \exists y Rxy \succ \exists x \forall y Rxy$

3. Construct models (on a domain of whatever size you like) to give counterexamples to the following arguments—if they have counterexamples—and if they do not, explain why they do not, either by showing directly that there is no such counterexample or by providing a **CQ** *proof* for the argument.

  i. $\exists x Fx, \exists x Gx \succ \exists x (Fx \wedge Gx)$

  ii. $\exists x Fx, \forall x Gx \succ \exists x (Fx \wedge Gx)$

  iii. $\succ \exists x (Fx \rightarrow \forall y Fy)$

  iv. $\forall x \exists y Rxy \succ \forall y \exists x Rxy$

  v. $\forall x \forall y (Rxy \rightarrow \exists z (Rxz \wedge Rzy)), \forall x \exists y Rxy, \forall x \neg Rxx \succ \bot$

4. Consider a model with the domain $D = \{0, 1, 2, 3, \dots\}$ of all the natural numbers. Consider the language with the one-place predicate $O$ (for *is odd*) true of all the odd numbers and the two-place predicate $S$ (for *smaller than*) such that $Sxy$ is true of $x$ and $y$ just when the number assigned to $x$ is smaller than the number assigned to $y$. And let's add a lot of names $a_0$, $a_1$, $a_2$, and so on, naming 0, 1, 2, and so forth. Evaluate the following formulas in this model:

  i. $Oa_1, Oa_2, Sa_1a_1, Sa_1a_2, Sa_2a_1$

  ii. $\forall y Sa_0 y$

  iii. $\exists x \forall y Sxy$

  iv. $\exists x Sxa_1$

  v. $\exists y Sa_{10\,000} y$

  vi. $\forall x \exists y Sxy$

  vii. $\forall x (Ox \rightarrow \exists y Syx)$

  viii. $\exists y \forall x (Ox \rightarrow Syx)$

  ix. $\forall x \forall y (Sxy \rightarrow \exists z (Sxz \wedge Szy))$

## Challenge Questions

1. This question continues on from challenge question 6 in the previous chapter (see page 215). In that question, we showed that S5 proofs can be systematically translated into CQ proofs. In this, show that for any possible worlds model $\langle W, V \rangle$ that is a counterexample to an argument $X \succ A$, it is possible to define a first-order model $M$ that is a counterexample to the world translation $w(X) \succ w(A)$ of that argument.

   HINT: To do this, take the domain of the model $M$ to be the set $W$ of worlds, and interpret the predicate $P$ corresponding to the atom $p$ as true *of* the worlds *at* which the atom $p$ is true. Show by induction on the structure of a formula $A$ that $A$ is true at $w$ (in the possible worlds model) if and only if $A$ is true in the first-order model $M$ under any assignment that assigns the variable $x$ the world $w$.

2. This question continues on from challenge question 1 in the previous chapter (see page 215). In that question, we considered the binary quantifiers *All* and *Some* and proof rules for them. Give *truth* conditions, showing when formulas of the form

$$All\, x(A(x) : B(x)) \qquad Some\, x(A(x) : B(x))$$

   are true in a model $M$ (given an assignment $v$). These truth conditions should follow very closely the truth conditions for the unary quantifiers $\forall$ and $\exists$.

   Once you have given those truth conditions, consider another binary quantifier:

$$Most\, x(A(x) : B(x))$$

   which means that most (let's say that means "more than half") of the things that are $A$ are also $B$.

   Specify truth conditions for the *Most* quantifier. (Feel free to just consider models with a finite domain. We won't get into questions of what it means for most *numbers* to have some property.)

   Once you have your truth conditions specified, consider the argument from the premise

$$Most\, x(Px : Most\, y(Wy : Lxy))$$

   (most people love most wombats)[122] to the conclusion

$$Most\, y(Wy : Most\, x(Px : Lxy))$$

   (most wombats are such that most people love them). Can you find a counterexample?

3. Following the proof of the soundness theorem, we noted that a consequence of the theorem is that alphabetic variants are equivalent according to models. Provide a direct proof of this fact, without appealing to the soundness theorem. In other words, prove that if $A$ and $A'$ are alphabetic variants, then for any model $M = \langle D, I \rangle$ and any assignment $v$, $I(A, v) = I(A', v)$.

4. Some formulas are only true in infinite models. Consider the following three formulas in a language with a single binary predicate, $R$.

   [TRANSITIVITY] $\forall x \forall y \forall z((Rxy \wedge Ryz) \to Rxz)$

   [ANTISYMMETRY] $\forall x \forall y(Rxy \to \neg Ryx)$

---

122. We owe this example to John Slaney.

[SERIALITY] $\forall x \exists y Rxy$

Let the conjunction of these three formulas be $C$. First, specify a model with an infinite domain in which $C$ is true. Next, prove that any model in which $C$ is true has an infinite domain. Finally, show that $C$ is not true in any *finite model*, that is, a model where the domain contains a finite number of objects.

5. A formula is in PRENEX FORM iff it is of the form $Q_1 \nu_1 \cdots Q_n \nu_n A$, where $A$ contains no quantifiers, and each $Q_i$ is either $\forall$ or $\exists$, and each $\nu_i$ is a variable. In this problem, you will show that every formula is logically equivalent to a formula in prenex form.

   Say that $A$ and $B$ are LOGICALLY EQUIVALENT in FO iff $A \models_{FO} B$ and $B \models_{FO} A$. First, show the following pairs of formulas are logically equivalent in FO.

   - $\forall x(A \to B)$ and $\exists x A \to B$, where $x$ is not free in $B$.
   - $\forall x(A \to B)$ and $A \to \forall x B$, where $x$ is not free in $A$.
   - $\forall x(A \vee B)$ and $\forall x A \vee B$, where $x$ is not free in $B$.
   - $\exists x(A \wedge B)$ and $\exists x A \wedge B$, where $x$ is not free in $B$.
   - $\exists x(A \to B)$ and $A \to \exists x B$, where $x$ is not free in $A$.
   - $\forall x(A \wedge B)$ and $\forall x A \wedge \forall x B$.
   - $\exists x(A \vee B)$ and $\exists x A \vee \exists x B$.
   - $\forall x \neg A$ and $\neg \exists x A$.
   - $\exists x \neg A$ and $\neg \forall x A$.

   Next, use the above equivalences, along with the equivalence of alphabetic variants, to show that every formula is logically equivalent to a formula in prenex form.

   Finally, given a formula $Q_1 \nu_1 \cdots Q_n \nu_n A$ in prenex form, show that it is logically equivalent to a formula $Q_1 \nu_1 \cdots Q_n \nu_n B$ in prenex form where $B$ is in disjunctive normal form.

6. For this question, you will consider two ways of establishing the validity of one formula, $\exists x(\exists y Fy \to Fx)$.[123] First, show that $\models_{FO} \exists x(\exists y Fy \to Fx)$. Next, provide a formal proof for $\vdash_{CQ} \exists x(\exists y Fy \to Fx)$. (You cannot appeal to the completeness theorem for this.) Was one way of establishing the validity of this formula easier than the other? What do you think accounts for this disparity?

---

123. This formula is important in Allen Hazen's (1987) discussion of alternative rules for the existential quantifier.

# 12 Coda

In this final chapter, we will do two things. One is bring together threads from the final two parts of the book, combining modal logic and first-order logic. The other is to provide some suggestions to the reader interested in learning some more logic.

## 12.1 Quantified Modal Logic

In part II, we covered modal logic with singular operators for necessity ($\Box$) and possibility ($\Diamond$). In part III, we covered predicate logic and the universal ($\forall$) and existential ($\exists$) quantifiers. These two topics can be brought together, as in *quantified modal logic*. Indeed, this topic arose in an example at the start of chapter 8.

- It is possible for everything that is *actually* red to be shiny.

There we focused on the *propositional* logic involved. In this section, we will start to investigate the interaction between quantifiers and modalities by presenting simple models for quantified modal logic.[124]

Models for first-order logic have the form $M = \langle D, I \rangle$ where $D$ is a domain of objects, and $I$ interprets the predicates and names. The simplest way to be able to interpret the modal operators of $\Box$ and $\Diamond$ with these models is to add a set $W$ of worlds and to allow the predicates to vary in interpretation from world to world. Putting these together will give us our definition.

**Definition 77 (CDQS5 model)** *A* CDQS5 MODEL *is a triple* $M = \langle D, W, I \rangle$ *where* $D$ *is a nonempty set of objects (the domain),* $W$ *is a nonempty set of worlds, and* $I$ *is an interpretation function such that for each name* $a$, *and for all worlds* $w, w' \in W$, $I(a, w) \in D$, *and* $I(a, w) = I(a, w')$, *and for each n-ary predicate* $F$, $I(F): D^n \times W \mapsto \{0, 1\}$.

"CDQS5" stands for "constant domain quantified S5." It is so called because for each model, the domain of objects does not vary from world to world. Constant domain quantified S5 is one of the simplest ways to combine possible worlds and a domain of quantification.

In the CDQS5 models, $I$ interprets *names* as denoting *objects*, and each name denotes the same object in each world. Predicates, however, can vary in interpretation from world

---

124. We are going to leave the addition of the actuality operator to a challenge question.

to world. An $n$-ary predicate is interpreted by assigning a truth value for every $n$-tuple of objects and a choice of a world. That is:

- For each $n$-place predicate $F$ and for each $d_1, \ldots, d_n$ from $D$ and for each $w$ in $W$, $I(F)(d_1, \ldots, d_n, w)$ is a truth value, either 0 or 1.

As with models of first-order logic, an assignment $v$ is a function that assigns each variable a value, which is to say that for each variable $x$, $v(x)$ is in $D$. And as before, we can interpret any term, whether a variable or a name, by assigning

- $I(t, w, v) = I(t, w)$, if $t$ is a name.
- $I(t, w, v) = v(t)$, if $t$ is a variable.

Since worlds do not affect the interpretation of names and variables, we can, without ambiguity, drop the world parameter from the interpretation of terms, writing $I(t, v)$, rather than $I(t, w, v)$. We can extend our interpretation $I$ to assign truth values to each formula, given a world $w$ and an assignment $v$, like this:

- $I(Ft_1 \cdots t_n, w, v) = 1$ iff $I(F)(I(t_1, v), \ldots, I(t_n, v), w) = 1$.
- $I(A \land B, w, v) = 1$ iff $I(A, w, v) = 1$ and $I(B, w, v) = 1$.
- $I(A \lor B, w, v) = 1$ iff $I(A, w, v) = 1$ or $I(B, w, v) = 1$.
- $I(A \to B, w, v) = 1$ iff $I(A, w, v) = 0$ or $I(B, w, v) = 1$.
- $I(\neg A, w, v) = 1$ iff $I(A, w, v) = 0$.
- $I(\bot, w, v) = 1$ never.
- $I(\Box A, w, v) = 1$ iff for every world $w' \in W$, $I(A, w', v) = 1$.
- $I(\Diamond A, w, v) = 1$ iff for some world $w' \in W$, $I(A, w', v) = 1$.
- $I(\forall x A, w, v) = 1$ iff for every $v'$ such that $v' \sim_x v$, $I(A, w, v') = 1$.
- $I(\exists x A, w, v) = 1$ iff for some $v'$ such that $v' \sim_x v$, $I(A, w, v') = 1$.

As with modal logic and first-order logic, we use models, worlds, and assignments to define counterexamples to arguments.

**Definition 78 (Counterexample, validity)** *A* CDQS5 COUNTEREXAMPLE *to* $X \succ A$ *is a* CDQS5 *model* $\langle D, W, I \rangle$, *an assignment* $v$, *and a world* $w \in W$ *such that* $I(B, w, v) = 1$, *for all* $B \in X$, *but* $I(A, w, v) = 0$.
    *An argument is* CDQS5 VALID *iff there are no* CDQS5 *counterexamples to it.*
    *When the argument* $X \succ A$ *is valid, we write* $X \models_{\mathsf{CDQS5}} A$.

The logic CDQS5 is the result of a fairly natural way of combining validity for S5 and validity for FO.

$$* \, * \, *$$

Let us look at an example of a CDQS5 model. Let the set of worlds $W = \{w_1, w_2\}$ and let the domain $D = \{a, b, c\}$. Interpret the predicates $F$ and $G$ as in the following tables.

| $I(F)$ | $a$ | $b$ | $c$ |
|--------|-----|-----|-----|
| $w_1$  | 1   | 0   | 1   |
| $w_2$  | 0   | 1   | 0   |

| $I(G)$ | $a$ | $b$ | $c$ |
|--------|-----|-----|-----|
| $w_1$  | 1   | 1   | 0   |
| $w_2$  | 0   | 1   | 0   |

We will evaluate some formulas in this model, relative to an assignment $v$ that has $v(x) = a$ and $v(y) = b$.

- $Fy$: If we evaluate this at $w_1$, we get $I(Fy, w_1, v) = I(F)(v(y), w_1) = 0$. If we evaluate it at $w_2$, we get $I(Fy, w_2, v) = I(F)(v(y), w_2) = 1$.

- $\forall y \Diamond Fy$: If we evaluate this at $w_1$, we have $I(\forall y \Diamond Fy, w_1, v) = 1$ iff for all $v' \sim_y v$, $I(\Diamond Fy, w_1, v') = 1$. Where $v'(y) = a$ or $v'(y) = c$, $I(Fy, w_1, v') = 1$, so $I(\Diamond Fy, w_1, v') = 1$. If $v'(y) = b$, then $I(Fy, w_1, v') = 0$, but $I(Fy, w_2, v') = 1$, so $I(\Diamond Fy, w_1, v') = 1$. Therefore, $I(\forall y \Diamond Fy, w_1, v) = 1$.

- $\Diamond \forall y Fy$: Unlike the previous formula, in this formula, $\Diamond$ has scope over the universal quantifier. In this model, $I(\Diamond \forall y Fy, w_1, v) = 1$ iff there is some world $w' \in W$ such that $I(\forall y Fy, w', v) = 1$. An inspection of the table for $F$ in this model reveals that $I(\forall y Fy, w_1, v) = 0$ and $I(\forall y Fy, w_2, v) = 0$. Therefore, $I(\Diamond \forall y Fy, w_1, v) = 0$.

- $\forall x(Gx \lor \Diamond Gx)$: In this model, $I(\forall x(Gx \lor \Diamond Gx), w_1, v) = 1$ iff for all $v' \sim_x v$, $I(Gx \lor \Diamond Gx, w_1, v') = 1$. As we can see in the tables above, $I(Gx, w_1, v') = 1$ where $v'(x) = a$ or $v'(x) = b$. If $v'(x) = c$, however, $I(Gx, w_1, v') = 0$, and we also have $I(\Diamond Gx, w_1, v') = 0$, as $I(Gx, w_2, v') = 0$. Therefore, $I(\forall x(Gx \lor \Diamond Gx), w_1, v) = 0$.

The second and third examples can be read as saying that while in this model, everything is possibly $F$, it is not the case that possibly everything is $F$. Much as $\forall x(Fx \lor Gx)$ says something different than $\forall x Fx \lor \forall x Gx$, $\forall x \Diamond Fx$ says something different than $\Diamond \forall x Fx$.

The interactions between the quantifiers and the modal operators are subtle and important in quantified modal logic. For **CDQS5** models, we have the following facts.

**Theorem 44 (Validity facts)** *The following are true:*

1. $\Box \forall x A \models_{CDQS5} \forall x \Box A$
2. $\forall x \Box A \models_{CDQS5} \Box \forall x A$
3. $\Diamond \exists x A \models_{CDQS5} \exists x \Diamond A$
4. $\exists x \Diamond A \models_{CDQS5} \Diamond \exists x A$

*Proof.* For (1), suppose there is a counterexample, so that for some model $\langle D, W, I \rangle$, some world $w \in W$, and some assignment $v$, $I(\Box \forall x A, w, v) = 1$ and $I(\forall x \Box A, w, v) = 0$. $I(\forall x \Box A, w, v) = 0$ implies that for some $v' \sim_x v$, $I(\Box A, w, v') = 0$. This implies that for some $u \in W$, $I(A, u, v') = 0$. On the other hand, $I(\Box \forall x A, w, v) = 1$ implies that for all worlds $w' \in W$, $I(\forall x A, w', v) = 1$, so in particular, $I(\forall x A, u, v) = 1$. This, in turn, implies that for all assignments $v'' \sim_x v$, $I(A, u, v'') = 1$, so in particular, $I(A, u, v') = 1$. This, however, is a contradiction as $I(A, u, v') = 0$. Therefore, there is no such counterexample.

The remainder are left as an exercise.

$\Box$

The second and third arguments (or rather, the *formulas* corresponding to the arguments, replacing the turnstile with a conditional: $\forall x \Box Fx \to \Box \forall x Fx$ and $\Diamond \exists x Fx \to \exists x \Diamond Fx$) are famous in modal logic. They were formulated by Ruth Barcan Marcus in her pioneering work on quantified modal logic in the 1940s.[125] Their converses, $\Box \forall x Fx \to \forall x \Box Fx$ and

---

125. Ruth Barcan Marcus published her first work under the name "Ruth C. Barcan" (1946).

$\exists x \Diamond Fx \to \Diamond \exists x Fx$, are known as the *converse Barcan formulas*. In the rest of this section, we will explore the status of these *Barcan formulas/arguments*.[126]

The validity of the Barcan and converse Barcan formulas is connected to the fact that the **CDQS5** models have a single domain that remains constant from world to world. The converse Barcan formulas are also straightforward to prove, given the standard rules for the quantifiers and the modal operators, as the following example indicates.

$$\frac{\dfrac{\dfrac{\dfrac{\dfrac{[\Box \forall x Fx]^1}{\forall x Fx}\, \Box E}{Fa}\, \forall E}{\Box Fa}\, \Box I}{\forall x \Box Fx}\, \forall I}{\Box \forall x Fx \to \forall x \Box Fx}\, \to I^1$$

Here, since the assumption $\Box \forall x Fx$ is a necessitive, the $\Box I$ side condition is satisfied, and since the name $a$ does not appear in the assumption, the eigenvariable condition for $\forall I$ is satisfied. This is a legitimate proof, provided the proposed modification is adopted. What about the Barcan formula? In challenge question 1, we ask you to modify the definition of "modal formula" to permit proofs of the Barcan and converse Barcan formulas using a combination of rules from **S5** and **CQ**.

If we think of our **CDQS5** models as representing how things might be, constant domain models seem to say that the same things exist in all worlds. Yet, this is not *obviously* true: it seems conceivable that there are possibilities that contain things that genuinely do not exist in our world. For example, there may be a different possible world where the thylacines did not die out and hundreds of thylacines—creatures that do not exist in our world—wander Australia. It also seems conceivable that there are some things in our world that just do not exist in another world, such as a world in which none of the things that are pukekos *in our world* exist.

Changing the models to make the Barcan formulas *invalid*, however, is not particularly straightforward, and neither is it simple to modify our proof rules so as to rule out derivations for both the Barcan formulas and their converses. In the next section, we will provide some references for the reader interested in pursuing *varying domain* quantified modal logics.

## 12.2 Suggestions for Further Reading

We will conclude with a few suggestions for the reader who wants to continue studying logic. There are many topics that we covered in this book: proof theory, intuitionistic logic, nonclassical logics, modal logic, philosophy of logic, and model theory.[127] We will briefly

---

126. Ruth Barcan Marcus (1921–2012) was an American philosopher whose work ranged over logic and the philosophy of language, including work on quantified modal logic and on moral dilemmas. A collection of her essays was published in 1993 (Barcan Marcus 1993).

127. As the reader might have noticed, these areas overlap significantly, and attempts to separate them are, to some degree, artificial.

describe some of the areas, as well as one important one we did not cover, and make some suggestions for things to read.

If you want to keep learning about logic but you are not sure what you are interested in, perhaps the best thing to do is look through some of the general introductions and handbooks out there. Read (1994) is a classic, and Goble (2001) has good coverage of topics. The *Handbook of Philosophical Logic* (2001), edited by Gabbay and Guenthner, is a multivolume handbook containing survey articles on a wide variety of topics. There are many great philosophical logic textbooks; some recent ones include Sider (2010) and MacFarlane (2020). Humberstone (2011) covers a wide variety of topics in great detail and rewards revisiting. Let us now turn to the more specific topics.

If you enjoyed learning about the proof systems and things like normalization, then you might enjoy *proof theory*. Proof theory is the branch of logic that studies properties of proof systems. Natural deduction systems came up in several chapters, but chapter 4 was the one most focused on proof theory. Apart from the kind of natural deduction systems we have presented, there are many kinds of proof systems that are useful and interesting. If you want to read more about proof theory, have a look at Prawitz (1965), Negri and von Plato (2001), Restall (2000), Paoli (2002), von Plato (2013), Bimbó (2014), or Mancosu, Galvan, and Zach (2021).

There are many philosophical issues arising in proof theory. One is the interpretation of sequents. Restall (2005) gives a normative reading of sequents according to which they tell us about coherent combinations of assertions and denials.

If you enjoyed the chapters on necessity, possibility, and actuality, then you might enjoy learning more *modal logic*. There are a range of topics one can study in modal logic, including applications of modal logic to different domains. Garson (2018) presents a nice overview of the basics of modal logic and some of the applications, including temporal logic and provability logic.

There are a lot of great books on modal logic, from more philosophical and more formal perspectives. Some good introductory books are Chellas (1980), Hughes and Cresswell (1996), van Benthem (2010), Garson (2013), and Zach (2019). Humberstone (2016) is a good intermediate text, as is Blackburn, de Rijke, and Venema (2002). For more on epistemic logic, Hintikka (1962) is a classic and van Ditmarsch, van der Hoek, and Kooi (2007) is another good text, with a more recent overview by Pacuit (2013a, 2013b). For further reading on temporal, or tense, logic, see Prior (1967) and van Benthem (1991). Both Garson (2013) and Sider (2010) contain more material on quantified modal logic, and Williamson (2013) develops the topic in great philosophical detail.

If you would like to explore the interaction of proof theory and modal logic, there are many resources available on the proof theory of modal logics. Negri (2011) and Poggiolesi and Restall (2012) have overviews of different types of proof systems out there for modal logics. Wansing (1998), Poggiolesi (2010), and Indrzejczak (2010) develop proof systems for a range of modal logics. Bednarska and Indrzejczak (2015) examine different approaches to proof systems for S5 in hypersequent systems.

In the initial chapters, we focused on *intuitionistic logic*, which came up again in later chapters. There is a great deal of work on intuitionistic logic, both philosophical and technical. Good overviews of technical issues in the area are offered by van Dalen (2002) and

Moschovakis (2018). Gabbay (1981) studies intuitionistic logic via models, and Dummett (2000) provides a detailed development of intuitionistic logic and mathematics.

Michael Dummett was a major proponent of intuitionistic logic. Dummett (1975) develops the philosophical basis of intuitionistic logic, and Dummett (1991) articulates a philosophical criticism of classical logic on the basis of features of intuitionistic logic.

While much of the book was focused on classical logic, *nonclassical logics* arose in a few places. Of the nonclassical logics, intuitionistic logic received the most attention, but we also discussed K3 and LP, in chapter 6. If you enjoyed some of the material on nonclassical logics, there is a wealth of further topics to explore. For many areas of nonclassical logic, a lot of philosophical and technical work has been done, so it is good to identify an area that seems interesting and follow up references in those areas.

Priest (2008) is a good, user-friendly textbook that covers a wide range of nonclassical logics, including intuitionistic logic, K3, and LP. It goes on to cover topics that we did not, such as *fuzzy logics*, many-valued logics where truth comes in degrees, and *relevant logics*, logics with conditional connectives that express entailment. For more on fuzzy logics, see Hájek (1998) and Cintula, Fermüller, and Noguera (2017). For more on relevant logics, see Dunn and Restall (2002) and Bimbó (2007), which offer great overviews of the area. Anderson and Belnap (1975), especially the first few sections, has many philosophical gems, although it is not a book to read straight through.

Restall (2000) develops the proof theory and a range of models for nonclassical logics, known as *substructural logics*. In terms of our natural deduction systems, these logics are ones that restrict the discharge policy in different ways, such as banning vacuous discharge or banning multiple discharge.

Three-valued logics have been used for many philosophical purposes. Kripke (1975) uses K3 to address the semantic paradoxes, and Priest (1979) uses LP to provide an alternative approach to the paradoxes. Anderson, Belnap, and Dunn (1992, §81) motivates a four-valued logic, FDE, in terms of reasoning about contradictory and partial information. FDE has important connections to relevant logics, and it is discussed by Anderson and Belnap (1975). It has features of K3 and LP, and one can actually obtain those logics through restrictions on FDE models.

Cobreros et al. (2012) and Cobreros et al. (2013) and Ripley (2013a, 2013b) develop *nontransitive logics*, logics where one can have both $A \succ B$ and $B \succ C$ valid without $A \succ C$ being valid. These logics are used to address issues arising from vagueness and from the semantic paradoxes. French (2016) defends a logic in which reflexivity, $A \succ A$, is invalid, motivated by the semantic paradoxes. There are other nonclassical approaches to the semantic paradoxes, such as the noncontractive approach of Zardini (2011), which uses substructural logics. Gupta and Belnap (1993); Field (2008); Halbach (2011); Beall, Glanzberg, and Ripley (2018); and Halbach and Leigh (2020) are good references for more on the semantic paradoxes.

The *philosophy of logic* covers a variety of topics that deal with philosophical issues arising from logic, proofs, and models. We will mention three topics that have come up in this book: consequence, pluralism, and meaning.

Throughout the book, we have been concerned with consequence relations, such as $\vdash_I$ and $\models_{CQ}$. A major topic in the philosophy of logic addresses the nature of logical consequence, namely, what is it for a conclusion to follow from some premises. Beall, Restall, and Sagi (2019) and Dutilh Novaes (2020) are introductions to some contemporary and historical discussions concerning logical consequence.

We have presented multiple logics in this book. Which logic is the correct or true one? *Monists* respond to this question by arguing for their chosen logic. On the other hand, *pluralists* reject the presupposition of uniqueness, thinking that there are multiple logics that are correct. Beall and Restall (2005) is a great place to start reading about the debate about logical pluralism, and Cook (2010) and G. Russell (2019) are good overviews of the debate.

One of the issues that we raised in chapter 6 was what determined the meaning of the logical connectives. Inferentialism provides one answer to this question, that rules of inference determine meaning. An early form of inferentialism said that *any* collection of rules determines the meaning of a connective. This idea was criticized by Prior (1961) using his connective tonk, which was introduced on page 112. Belnap (1962) gave an influential response, saying that the rules need to satisfy some additional conditions to define a connective. Taking proof rules to determine meanings has developed into the area of proof-theoretic semantics, for which see Schroeder-Heister (2018).

If you found the models particularly interesting, you might want to look into *model theory*. Model theory studies properties of models and classes of models, as well as models for specific *theories*, such as the theory of the natural numbers, the theory of the real numbers, or the theory of sets. Much work on model theory focuses on models for classical logic, particularly in connection to different areas of mathematics. Doets (1996), Hodges (1997), Enderton (2001), and Tent and Ziegler (2012) provide good introductions to the area. Not all work in model theory is concerned with classical logic and theories. For example, Gabbay (1981) explores model theory for intuitionistic logic.

Once you have a bit of model theory under your belt, you can engage with the many philosophical issues that arise from models. Walsh and Button (2018) provide excellent coverage of many issues. Some background in set theory is useful for engaging with formal and philosophical work on model theory, especially work on models for set theories. Hrbacek and Jech (1999) provide a good introduction to formal aspects of set theory, and Potter (2004) develops philosophical issues along with an alternative set theory.

The last topic we will highlight, *computability and incompleteness*, is one that we did not cover in this book. Anyone that has made it through this book should, however, have enough logical background to dive into this material.

The study of formal theories of arithmetic is important in logic. There are some simple axioms that one can give for the concepts of successor, addition, and multiplication that capture all the arithmetic you learned in primary school and much more. These axioms let one define a notion of *computation* that agrees with other conceptions of computability, such as Turing machines. In a surprising discovery, Kurt Gödel showed that the axioms of arithmetic are insufficient for deriving certain arithmetical truths. The axioms of arithmetic are *incomplete*. Gödel's incompleteness results have far-reaching consequences, and they are one bit of logic that pops up from time to time in popular culture.

Smullyan (1992), Boolos, Burgess, and Jeffrey (2007), P. Smith (2013), and Zach (2020) provide excellent overviews of incompleteness phenomena and the necessary background on computability and arithmetic. There are numerous philosophical issues surrounding incompleteness, many of which are discussed by Franzén (2005).

## 12.3  Key Concepts and Skills

☐ You should be able to determine the truth or falsity of formulas of quantified modal logic when provided with a **CDQS5** model.

☐ You should be able to evaluate simple arguments stated in quantified modal logic.

☐ You should be able to understand ways in which scope distinctions matter when both quantifiers and modalities are in the language.

---

A **CDQS5** MODEL is a triple $M = \langle D, W, I \rangle$ where $D$ is a nonempty set of objects (the domain), $W$ is a nonempty set of worlds, and $I$ is an interpretation function such that for each name $a$, and for all worlds $w, w' \in W$, $I(a, w) \in D$, and $I(a, w) = I(a, w')$, and for each $n$-ary predicate $F$, $I(F): D^n \times W \mapsto \{0, 1\}$.

- $I(Ft_1 \cdots t_n, w, v) = 1$ iff $I(F)(I(t_1, v), \ldots, I(t_n, v), w) = 1$.
- $I(A \wedge B, w, v) = 1$ iff $I(A, w, v) = 1$ and $I(B, w, v) = 1$.
- $I(A \vee B, w, v) = 1$ iff $I(A, w, v) = 1$ or $I(B, w, v) = 1$.
- $I(A \rightarrow B, w, v) = 1$ iff $I(A, w, v) = 0$ or $I(B, w, v) = 1$.
- $I(\neg A, w, v) = 1$ iff $I(A, w, v) = 0$.
- $I(\bot, w, v) = 1$ never.
- $I(\Box A, w, v) = 1$ iff for every world $w' \in W$, $I(A, w', v) = 1$.
- $I(\Diamond A, w, v) = 1$ iff for some world $w' \in W$, $I(A, w', v) = 1$.
- $I(\forall x A, w, v) = 1$ iff for every $v'$ such that $v' \sim_x v$, $I(A, w, v') = 1$.
- $I(\exists x A, w, v) = 1$ iff for some $v'$ such that $v' \sim_x v$, $I(A, w, v') = 1$.

---

## 12.4  Questions for You

### Basic Questions

1. Consider the model with worlds $W = \{w_1, w_2\}$ and domain $D = \{a, b, c\}$. Interpret the predicates $F$ and $G$ like this:

| $I(F)$ | $a$ | $b$ | $c$ |
|---|---|---|---|
| $w_1$ | 1 | 0 | 1 |
| $w_2$ | 0 | 1 | 1 |

| $I(G)$ | $a$ | $b$ | $c$ |
|---|---|---|---|
| $w_1$ | 1 | 0 | 0 |
| $w_2$ | 0 | 1 | 0 |

For the first part of this question, explain why, under any assignments, $\Box \exists x Gx$ is *true* at worlds $w_1$ and $w_2$, and why $\exists x \Box Gx$ is *false* at worlds $w_1$ and $w_2$. Make your explanations as clear and explicit as possible, showing how the status $\Box \exists x Gx$ at each world depends on the value of $\exists x Gx$ at each world and how this depends on the value of $Gx$ for each possible assignment of values to the variable $x$ and, similarly, how $\exists x \Box Gx$ at each world depends on the value of $\Box Gx$ at each world for each assignment of values to the variable $x$ and how this depends on the values of $Gx$ at each world for each assignment of values.

When you have completed the explanations for $\Box \exists x Gx$ and $\exists x \Box Gx$, check the status of each of the formulas

- $\Diamond \forall x Fx$, and
- $\forall x \Diamond Fx$

at each world in our model too.

2. Prove that the following hold, to complete the proof of theorem 44:

    i. $\forall x \Box A \vDash_{\text{CDQS5}} \Box \forall x A$

    ii. $\Diamond \exists x A \vDash_{\text{CDQS5}} \exists x \Diamond A$

    iii. $\exists x \Diamond A \vDash_{\text{CDQS5}} \Diamond \exists x A$

**Challenge Questions**

1. The arguments $\forall x \Box Fx \succ \Box \forall x Fx$ and $\Diamond \exists x Fx \succ \exists x \Diamond Fx$ have no counterexamples in **CDQS5** models. In this question, your job is to find *proofs* from $\forall x \Box Fx$ to $\Box \forall x Fx$ and from $\Diamond \exists x Fx$ to $\exists x \Diamond Fx$, using the proof rules $\Box I$, $\Box E$, $\Diamond I$, $\Diamond E$, $\forall I$, $\forall E$, $\exists I$, and $\exists E$.

    You will find that given the side conditions on the rules $\Box I$ and $\Diamond E$, as they are given in chapter 9, you *cannot* find proofs that satisfy those side conditions. You will have to expand the definition of *modal formula* so that your proofs satisfy the modal side conditions. So, to answer this question, you are to do three things. (1) Clearly specify what counts as a *modal formula* in the language of first-order modal logic. (2) Explain why this choice of modal formulas results in $\Box I$ and $\Diamond E$ rules that are sound for **CDQS5** validity, by showing that if a modal formula (as you have defined it) is true at some world in a **CDQS5** model, then it is true at *all* worlds of that model. (3) Using these rules, write out full proofs for the arguments $\forall x \Box Fx \succ \Box \forall x Fx$ and $\Diamond \exists x Fx \succ \exists x \Diamond Fx$.

2. Suppose that we add the $\mathbb{A}$ operator to the language of quantified modal logic. Using $Fx$ for "$x$ is red" and $Gx$ for "$x$ is shiny," provide a formalization of Crossley and Humberstone's example

    > It is possible for everything that is *actually* red to be shiny.

    Define a **CDQS5** actuality model to be a quadruple $\langle D, W, I, g \rangle$, where $\langle D, W, I \rangle$ is a **CDQS5** model and $g \in W$. Interpret the actuality operator in these models as follows.

    - $I(\mathbb{A}A, w, v) = 1$ iff $I(A, g, v) = 1$

    Evaluate the translation of your formula in the following model, adapted from earlier in the chapter. Let $W = \{w, g\}$, let $D = \{a, b, c\}$, and interpret the predicates $F$ and $G$ as in the following tables:

    | $I(F)$ | $a$ | $b$ | $c$ |
    |--------|-----|-----|-----|
    | $w$ | 1 | 0 | 1 |
    | $g$ | 0 | 1 | 0 |

    | $I(G)$ | $a$ | $b$ | $c$ |
    |--------|-----|-----|-----|
    | $w$ | 1 | 1 | 0 |
    | $g$ | 0 | 1 | 0 |

3. In the **CDQS5** models we have defined, the interpretation of names and variables does not change from world to world. Define new models in which the interpretation and names and variables can differ from world to world, much as the interpretation of predicates can differ. What effect does this have on the logic? Can you provide any philosophical motivation for permitting the interpretation of names and variables to vary in this way? Are there strong reasons not to allow such variation?

# Glossary

**Actuality counterexample**  An ACTUALITY COUNTEREXAMPLE to the argument $X \succ A$ is an actuality model $\langle W, g, V \rangle$ and a world $w \in W$, such that $V(X, w) = 1$ and $V(A, w) = 0$.

**Actuality model**  An ACTUALITY MODEL is a triple $\langle W, g, V \rangle$ where $W$ is a set of worlds, $g \in W$, and $V$ is a valuation function from pairs of atoms and worlds to $\{0, 1\}$ such that $V(\bot, w) = 0$, for all $w \in W$.

**Assignment**  Let $M$ be a model with domain $D$. An ASSIGNMENT $v$ is a function from the variables in the language to $D$.

**Boolean counterexample**  A BOOLEAN COUNTEREXAMPLE to the argument $X \succ A$ is a Boolean model $v$ that assigns 1 to all formulas in the set $X$ and assigns 0 to $A$.

**Boolean model**  A given Boolean valuation $v$ is extended to a BOOLEAN MODEL on the whole language, a function $\mathsf{Form} \mapsto \{0, 1\}$, inductively as follows:

$$
\begin{array}{lll}
v(\bot) = 0 & & \text{always} \\
v(\neg A) = 1 & \text{iff} & v(A) = 0 \\
v(A \wedge B) = 1 & \text{iff} & v(A) = 1 \text{ and } v(B) = 1 \\
v(A \vee B) = 1 & \text{iff} & v(A) = 1 \text{ or } v(B) = 1 \\
v(A \rightarrow B) = 1 & \text{iff} & v(A) = 0 \text{ or } v(B) = 1
\end{array}
$$

**Boolean validity**  An argument from $X$ to $A$ is VALID, according to Boolean models, iff there is no Boolean model $v$ where $v(X) = 1$ and $v(A) = 0$, that is, there are no counterexamples to the argument.

**Bound variable**  If $x$ is free in $A$, then $x$ is BOUND by the displayed quantifier in $\forall x A$ and in $\exists x A$.

**CDQS5 model**  A CDQS5 MODEL is a triple $M = \langle D, W, I \rangle$ where $D$ is a nonempty set of objects (the domain), $W$ is a nonempty set of worlds, and $I$ is an interpretation function such that for each name $a$, and for all worlds $w, w' \in W$, $I(a, w) \in D$, and $I(a, w) = I(a, w')$, and for each $n$-ary predicate $F$, $I(F) : D^n \times W \mapsto \{0, 1\}$.

- $I(Ft_1 \cdots t_n, w, v) = 1$ iff $I(F)(I(t_1, v), \ldots, I(t_n, v), w) = 1$.
- $I(A \wedge B, w, v) = 1$ iff $I(A, w, v) = 1$ and $I(B, w, v) = 1$.
- $I(A \vee B, w, v) = 1$ iff $I(A, w, v) = 1$ or $I(B, w, v) = 1$.
- $I(A \rightarrow B, w, v) = 1$ iff $I(A, w, v) = 0$ or $I(B, w, v) = 1$.
- $I(\neg A, w, v) = 1$ iff $I(A, w, v) = 0$.
- $I(\bot, w, v) = 1$ never.
- $I(\Box A, w, v) = 1$ iff for every world $w' \in W$, $I(A, w', v) = 1$.
- $I(\Diamond A, w, v) = 1$ iff for some world $w' \in W$, $I(A, w', v) = 1$.
- $I(\forall x A, w, v) = 1$ iff for every $v'$ such that $v' \sim_x v$, $I(A, w, v') = 1$.
- $I(\exists x A, w, v) = 1$ iff for some $v'$ such that $v' \sim_x v$, $I(A, w, v') = 1$.

**Congruential**   An operator $\bigcirc$ in the logic L is said to be CONGRUENTIAL if and only if whenever $\vdash_L A \leftrightarrow B$, we also have $\vdash_L \bigcirc A \leftrightarrow \bigcirc B$.

**Derived rule**   A rule $R$ is a DERIVED RULE in a proof system S if, given proofs $\Pi_1, \ldots, \Pi_n$ of the premises $A_1, \ldots, A_n$ of the rule $R$, the conclusion $C$ of $R$ can be obtained by extending some of the proofs $\Pi_1, \ldots, \Pi_n$ with the primitive rules of S.

**Detour**   A DETOUR FORMULA in a proof without disjunction is a formula instance that is introduced as the conclusion of an introduction rule and then is eliminated as the major premise of an elimination rule.

**Double indexed model counterexample**   A double indexed model $\langle W, V \rangle$ is a COUNTEREXAMPLE to an argument $X \succ A$ iff there are worlds $w, v \in W$ such that $V(X, w, v) = 1$ and $V(A, w, v) = 0$.

**Double indexed model**   A DOUBLE INDEXED MODEL is given by a pair $\langle W, V \rangle$ consisting of a nonempty set $W$ of worlds and a valuation function $V$ assigning a truth value from $\{0, 1\}$ to each pair consisting of an atom and a world, where $V(\bot, w, v)$ for all $w, v \in W$. Given these data, we can define a function assigning a truth value to each formula and a *pair* of worlds, like this:

- $V(p, w, v) = 1$ iff $V(p, w) = 1$.
- $V(A \wedge B, w, v) = 1$ iff $V(A, w, v) = 1$ and $V(B, w, v) = 1$.
- $V(A \vee B, w, v) = 1$ iff $V(A, w, v) = 1$ or $V(B, w, v) = 1$.
- $V(A \rightarrow B, w, v) = 1$ iff $V(A, w, v) = 0$ or $V(B, w, v) = 1$.
- $V(\neg A, w, v) = 1$ iff $V(A, w, v) = 0$.
- $V(\bot, w, v) = 1$ never.
- $V(\Box A, w, v) = 1$ iff $V(A, u, v) = 1$ for each world $u \in W$.
- $V(\Diamond A, w, v) = 1$ iff $V(A, u, v) = 1$ for some world $u \in W$.
- $V(\mathbb{A}A, w, v) = 1$ iff $V(A, v, v) = 1$.

**Eigenvariable condition**   The $\forall$-EIGENVARIABLE CONDITION says that in an application of $\forall I$, the displayed term $b$ in the premise $A(b)$ cannot occur either in the conclusion of $\forall I$ or in any assumption in $X$ upon which the premise of the application of $\forall I$ depends.

The $\exists$-EIGENVARIABLE says that in an application of $\exists E$, the term $b$ in the discharged assumption cannot occur in $\exists x A(x)$, $B$ or in $X$.

**Equivalence counterexample**   An equivalence model $\langle W, R, V \rangle$ is a COUNTEREXAMPLE to the argument $X \succ A$ iff there is a world $w \in W$ such that $V(X, w) = 1$ and $V(A, w) = 0$.

**Equivalence model**   An EQUIVALENCE MODEL, or S5E model, is a triple $\langle W, R, V \rangle$ where $W$ is a nonempty set of worlds, $R$ is a binary relation on $W$ that is an equivalence relation, and $V$ is a function from Atom and $W$ to $\{0, 1\}$ such that $V(\bot, w) = 0$, for all $w \in W$.

**Extension**   Given that we have settled on some ordering on variables, the EXTENSION of a formula $A(x_1, \ldots, x_n)$ whose free variables are among $x_1, \ldots, x_n$, where $n \geq 2$, in a model $M$ is

$$\|A(x_1, \ldots, x_n)\| = \{\langle d_1, \ldots, d_n \rangle \in D^n : I(A(x_1, \ldots, x_n), v_{d_1, \ldots, d_n}^{x_1, \ldots, x_n}) = 1\}.$$

**First-order counterexample**   A pair of a model $M$ for a first-order language and an assignment $v$ is a COUNTEREXAMPLE to the argument $X \succ A$ iff $I(B, v) = 1$, for each $B \in X$, and $I(A, v) = 0$.

**Free for**   A term $t$ is said to be FREE FOR the variable $x$ in $A$ iff no free occurrence of $x$ in $A$ is in the scope of a quantifier that binds a variable in $t$.

**Free variable**    The set $fv(A)$ of FREE VARIABLES in a formula $A$ is defined like this:

- If $A$ is an atomic formula, $Ft_1 \cdots t_n$, then $fv(A)$ is the set of variables among the $t_i$, $1 \leq i \leq n$.
- $fv(A \wedge B) = fv(A \vee B) = fv(A \rightarrow B) = fv(A) \cup fv(B)$.
- $fv(\neg A) = fv(A)$.
- $fv(\exists x A) = fv(A) \backslash \{x\}$ and $fv(\forall x A) = fv(A) \backslash \{x\}$.

The variables in the set $fv(A)$ are said to be *free* in the formula $A$.

**Formula context**    A FORMULA CONTEXT is a formula $C(p)$ that has some occurrences of an atom $p$. A SUBSTITUTION of the formula $B$ for $p$ in that context is the formula $C(B)$, obtained by replacing all copies of $p$ in $C$ with the formula $B$.

**Formulas, predicate**    The set of FORMULAS of predicate logic is defined inductively as follows.

- If $F$ is an $n$-ary predicate and $t_1, \ldots, t_n$ are terms, then $Ft_1 \cdots t_n$ is a formula. In addition, $\perp$ is a formula, as before.
- If $A$ and $B$ are formulas, then $(A \wedge B)$, $(A \vee B)$, $(A \rightarrow B)$, and $\neg A$ are formulas.
- If $A$ is a formula and $x$ is a variable, $\forall x A$ and $\exists x A$ are formulas.
- Nothing else is a formula.

**Heyting algebra**    A HEYTING ALGEBRA is a quintuple $\langle V, \leq, \Rightarrow, 0, 1 \rangle$ such that $\langle V, \leq \rangle$ is a distributive lattice, $\Rightarrow$ is a binary operation on $V$, with $0, 1 \in V$, and the following conditions are satisfied:

- for all $x \in V$, $0 \leq x$,
- for all $x \in V$, $x \leq 1$, and
- for all $x, y, z \in V$, $x \sqcap y \leq z$ iff $x \leq y \Rightarrow z$.

**Heyting counterexample and validity**    A HEYTING COUNTEREXAMPLE $v$ to an argument $X \succ A$, with $X$ finite, is a valuation $v$ on a Heyting algebra such that $v(X) \not\leq v(A)$.
An argument $X \succ A$ is HEYTING VALID iff it has no Heyting counterexamples. When the argument $X \succ A$ is Heyting valid, we write $X \vDash_{\mathsf{H}} A$.

**Inconsistency**    We say that a set $X$ of formulas is INCONSISTENT if and only if $X \vdash_{\mathsf{I}} \perp$. That is, $X$ is inconsistent if and only if there is a proof of a contradiction from $X$.

**Instance**    If $\forall x A(x)$ is a formula and $b$ is a term free for $x$ in $A$, then $A(b)$ is an INSTANCE of the universal quantifier. Similarly, if $\exists x A(x)$ is a formula and $b$ is a term free for $x$ in $A$, then $A(b)$ is an INSTANCE of the existential quantifier.

**Intuitionistic Kripke model**    An INTUITIONISTIC KRIPKE MODEL is a triple $\langle W, R, V \rangle$ where $W$ is a nonempty set of worlds; $R$ is a reflexive, transitive binary relation on $W$; and $V$ is a function from pairs of atoms and worlds to $\{0, 1\}$, such that $V(\perp, w) = 0$ for all $w \in W$, obeying the condition that, for all atoms $p$ and all $x, y \in W$, if $V(p, x) = 1$ and $xRy$, then $V(p, y) = 1$, and the following clauses:

$$
\begin{array}{lll}
V(A \wedge B, w) = 1 & \text{iff} & V(A, w) = 1 \text{ and } V(B, w) = 1 \\
V(A \vee B, w) = 1 & \text{iff} & V(A, w) = 1 \text{ or } V(B, w) = 1 \\
V(A \rightarrow B, w) = 1 & \text{iff} & \text{for all } x \in W, \text{ if } wRx, \text{ and } V(A, x) = 1, \text{ then } V(B, x) = 1 \\
V(\neg A, w) = 1 & \text{iff} & \text{for all } x \in W, \text{ if } wRx, \text{ then } V(A, x) = 0 \\
V(\perp, w) = 0 & & \text{always}
\end{array}
$$

**Intuitionistic Kripke model counterexample**    The intuitionistic Kripke model $\langle W, R, V \rangle$ is a COUNTEREXAMPLE to the argument $X \succ A$ iff there is a world $w \in W$ such that $V(X, w) = 1$ and $V(A, w) = 0$.

**Intuitionistic logic**    Let $X$ be a set of formulas and $A$ a formula from Form. An argument $X \succ A$ is VALID according to INTUITIONISTIC LOGIC if and only if there is a proof in I from premises in the set $X$ to the conclusion $A$. If $X \succ A$ is valid in intuitionistic logic, we write $X \vdash_{\mathsf{I}} A$.

**K3 and LP counterexamples**    A trivaluation $v$ is a K3-COUNTEREXAMPLE to an argument $X \succ A$ iff $v(X) = 1$ and $v(A) \neq 1$.
A trivaluation $v$ is an LP-COUNTEREXAMPLE to an argument $X \succ A$ iff for all $B \in X$, $v(B) \neq 0$, and $v(A) = 0$.

**Lattice**

Let $\langle V, \leq \rangle$ be a partially ordered set and $\wedge$ and $\vee$ binary operations on the set $V$.

- An element $y$ is an UPPER BOUND of a subset $X \subseteq V$ iff for all $x \in X (x \leq y)$.
- An element $y$ is a LOWER BOUND of $X \subseteq V$ iff for all $x \in X (y \leq x)$.
- An element $z \in V$ is the LEAST UPPER BOUND of $X \subseteq V$ iff both
  - $z$ is an upper bound of $X$, and
  - $z \leq y$, for any upper bound $y$ of $X$.
- An element $z \in V$ is the GREATEST LOWER BOUND of $X \subseteq V$ iff both
  - $z$ is a lower bound of $X$, and
  - $y \leq z$, for any lower bound $y$ of $X$.
- A partially ordered set $\langle V, \leq \rangle$ is a LATTICE iff every pair of elements $x, y \in V$ has both
  - a least upper bound, the join $x \sqcup y$, and
  - a greatest lower bound, the meet $x \sqcap y$.
- A lattice $\langle V, \leq \rangle$ is DISTRIBUTIVE iff for all elements $x, y, z \in V$,

$$x \sqcap (y \sqcup z) = (x \sqcap y) \sqcup (x \sqcap z).$$

**Literal**

A LITERAL is an atom (not $\bot$), such as $p$, $q$, $r$, and so on, or its negation, $\neg p$, $\neg q$, $\neg r$, and so on. Nothing else is a literal.

**Modal formula**

Formulas of the form $\Box B$, $\Diamond B$, $\neg \Box B$, or $\neg \Diamond B$ are MODAL FORMULAS.

**Modal language**

The set MForm of formulas of our modal language is defined as follows. Every atomic formula in Atom is in MForm. If $A$ and $B$ are formulas in MForm, then so are the following formulas.

$$(A \wedge B) \qquad (A \vee B) \qquad (A \rightarrow B) \qquad \neg A \qquad \Box A \qquad \Diamond A$$

Nothing else is in MForm.

**Modality**

A MODALITY is a nonempty sequence of operators $\bigcirc_1, \ldots, \bigcirc_n$ where for $1 \leq i \leq n$, $\bigcirc_i$ is either $\neg, \Diamond$, or $\Box$.
A POSITIVE MODALITY is a nonempty sequence of operators $\bigcirc_1, \ldots, \bigcirc_n$ where for $1 \leq i \leq n$, $\bigcirc_i$ is either $\Diamond$ or $\Box$.

**Model**

A MODEL $M$ for a first-order language is a pair $\langle D, I \rangle$ where $D$ is a nonempty set, called the *domain*, and $I$ is a function on the language such that for each name $a$ in the language, $I(a) \in D$, and if $F$ is an $n$-ary predicate, then $I(F) : D^n \mapsto \{0, 1\}$, where $D^n$ is the set of sequences of length $n$ of objects from $D$, where $I(\bot) = 0$.

**Monotonic**

An operator $\bigcirc$ in the logic L is said to be MONOTONIC if and only if whenever $\vdash_L A \rightarrow B$, we also have $\vdash_L \bigcirc A \rightarrow \bigcirc B$.

**Necessitive**

A formula is a NECESSITIVE if and only if it has the form $\Box B$ or $\neg \Diamond B$.

**Normal proof**

A $\vee$-free proof $\Pi$ is a NORMAL PROOF iff $\Pi$ contains no detours.

**Open formula**

A formula $A$ is an OPEN FORMULA iff $fv(A) \neq \emptyset$.
A formula $A$ is a CLOSED FORMULA iff $A$ is not an open formula, that is, $fv(A) = \emptyset$.

**Partial order**

A PARTIAL ORDER, $(X, \leq)$, is a set $X$ and a binary relation $\leq$ on $X$ such that

- for all $x \in X$, $x \leq x$,
- for all $x, y \in X$, if $x \leq y$ and $y \leq x$, then $x = y$,
- and for all $x, y, z \in X$, if $x \leq y$ and $y \leq z$, then $x \leq z$.

**Possible worlds counterexample, validity**

A possible worlds model $\langle W, V \rangle$ is a COUNTEREXAMPLE to the argument $X \succ A$ iff there is a world $w \in W$ such that $V(X, w) = 1$ and $V(A, w) = 0$.
The argument $X \succ A$ is VALID according to possible worlds models iff there is no possible worlds model $\langle W, V \rangle$ with a world $w \in W$ such that $V(X, w) = 1$ and $V(A, w) = 0$.

| | |
|---|---|
| **Possible worlds model** | A POSSIBLE WORLDS MODEL is a pair $\langle W, V \rangle$, where $W$ is a nonempty set of POSSIBLE WORLDS and $V$ is a function that gives a truth value (from $\{0, 1\}$) to each choice of an atomic formula and a world, with $V(\bot, w) = 0$ for all worlds $w$. |
| **Proposition** | A PROPOSITION in a possible worlds model $\langle W, V \rangle$ is a subset of $W$. In a possible worlds model $\langle W, V \rangle$, the PROPOSITION that $A$, $\|A\|$, is $\{w \in W : V(A, w) = 1\}$. |
| **Real-world counterexample** | A double indexed model $\langle W, V \rangle$ is a REAL-WORLD COUNTEREXAMPLE to an argument $X \succ A$ iff there is a world $w \in W$ such that $V(X, w, w) = 1$ and $V(A, w, w) = 0$. |
| **Reflexive relation** | A binary relation $\leq$ on the set $W$ is REFLEXIVE iff for all $w \in W$, $w \leq w$. |
| **S4 model** | An S4 MODEL $\langle W, R, V \rangle$ is a triple consisting of a nonempty set $W$, a binary relation $R$ on $W$ that is reflexive and transitive, and a valuation function $V$ from atoms and worlds to $\{0, 1\}$ such that $V(\bot, w) = 0$, for all $w \in W$. |
| **Scope, quantifier** | The SCOPE of the initial quantifier $\forall x A$ is $A$, and the scope of the initial quantifier $\exists x A$ is $A$. |
| **Subformula property** | A proof $\Pi$ for $X \succ A$ has the SUBFORMULA PROPERTY if and only if each formula in $\Pi$ is either a subformula* of $A$ or of some formula in $X$. |
| **Substitution** | If $t$ is free for $x$ in $A$, then the SUBSTITUTION OF $t$ FOR $x$ IN A FORMULA $A$, $(A)[x/t]$, and the SUBSTITUTION OF $t$ FOR $x$ IN A TERM $u$, $u[x/t]$, are defined like this: |

- If $u$ is $x$, then $u[x/t] = t$.
- If $u$ is not $x$, then $u[x/t] = u$.
- If $A$ is $Fs_1 \cdots s_n$, then $(Fs_1 \cdots s_n)[x/t] = Fs_1[x/t] \cdots s_n[x/t]$.
- $(A \wedge B)[x/t] = A[x/t] \wedge B[x/t]$.
- $(A \vee B)[x/t] = A[x/t] \vee B[x/t]$.
- $(A \rightarrow B)[x/t] = A[x/t] \rightarrow B[x/t]$.
- $(\neg A)[x/t] = \neg(A[x/t])$.
- $(\forall x A)[x/t] = \forall x A$.
- $(\forall y A)[x/t] = \forall y((A)[x/t])$.
- $(\exists x A)[x/t] = \exists x A$.
- $(\exists y A)[x/t] = \exists y((A)[x/t])$.

| | |
|---|---|
| **Symmetric relation** | A binary relation $\leq$ is SYMMETRIC iff for all $w, v \in W$, if $w \leq v$, then $v \leq w$. |
| **Terms** | The set of TERMS of the language is defined as follows: |

- Every variable, $x, y, z, x_1, \ldots$, is a term.
- Every name, $a, b, c, a_1, \ldots$, is a term.
- Nothing else is a term.

| | |
|---|---|
| **Transitive relation** | A binary relation $\leq$ on the set $W$ is TRANSITIVE iff for all $x, y, z \in W$, if $x \leq y$ and $y \leq z$, then $x \leq z$. |
| **Tree** | A TREE $\Pi$ is a partial order, $(T, \leq)$, such that |

- there is a root, that is, an element $t$ of $T$, such that for all $x$ in $T$, $t \leq x$.
- if $x$, $y$, and $z$ are in $T$, $y \leq x$ and $z \leq x$, then either $y \leq z$ or $z \leq y$.

**Trivaluation**  A TRIVALUATION $v$ is a function from Atom to $\{0, 1, n\}$ such that $v(\bot) = 0$. The value of complex formulas in a trivaluation is determined using the truth tables below.

| $\wedge$ | 0 | $n$ | 1 | | $\vee$ | 0 | $n$ | 1 | | $\rightarrow$ | 0 | $n$ | 1 | | $\neg$ | | | $\bot$ |
|---|---|---|---|---|---|---|---|---|---|---|---|---|---|---|---|---|---|---|
| 0 | 0 | 0 | 0 | | 0 | 0 | $n$ | 1 | | 0 | 1 | 1 | 1 | | 0 | 1 | | 0 |
| $n$ | 0 | $n$ | $n$ | | $n$ | $n$ | $n$ | 1 | | $n$ | $n$ | $n$ | 1 | | $n$ | $n$ | | |
| 1 | 0 | $n$ | 1 | | 1 | 1 | 1 | 1 | | 1 | 0 | $n$ | 1 | | 1 | 0 | | |

**Variant assignment**  An assignment $u$ is an $x$-VARIANT of an assignment $v$ iff $v(y) = u(y)$, for all variables $y$ distinct from $x$. We will write $u \sim_x v$ when $u$ is an $x$-variant of $v$.

We will write $v_d^x$ for the function that assigns to $x$ the value $d$ and assigns to all other variables $y$ the same value assigned by $v$. That is:

$$v_d^x(x) = d$$

$$v_d^x(y) = v(y), \text{ if the variable } y \text{ is not } x.$$

# References

Anderson, Alan Ross, and Nuel D. Belnap. 1975. *Entailment: The Logic of Relevance and Necessity.* Vol. I. Princeton University Press.

Anderson, Alan Ross, Nuel D. Belnap, and J. Michael Dunn. 1992. *Entailment: The Logic of Relevance and Necessity.* Vol. II. Princeton University Press.

Barcan, Ruth C. 1946. "A Functional Calculus of First Order Based on Strict Implication." *Journal of Symbolic Logic* 11 (1): 1–16. https://doi.org/10.2307/2269159.

Barcan Marcus, Ruth C. 1993. *Modalities: Philosophical Essays.* Oxford University Press.

Beall, Jc, Michael Glanzberg, and David Ripley. 2018. *Formal Theories of Truth.* Oxford University Press.

Beall, Jc, and Greg Restall. 2005. *Logical Pluralism.* Oxford University Press.

Beall, Jc, Greg Restall, and Gil Sagi. 2019. "Logical Consequence." In *The Stanford Encyclopedia of Philosophy,* Spring 2019, edited by Edward N. Zalta. Metaphysics Research Lab, Stanford University. https://plato.stanford.edu/archives/spr2019/entries/logical-consequence/.

Bednarska, Kaja, and Andrzej Indrzejczak. 2015. "Hypersequent Calculi for S5: The Methods of Cut Elimination." *Logic and Logical Philosophy* 24 (3): 277–311. https://doi.org/10.12775/LLP.2015.018.

Belnap, Nuel. 1959. "The Formalization of Entailment." PhD diss., Yale University.

Belnap, Nuel. 1962. "Tonk, Plonk and Plink." *Analysis* 22:30–34.

Belnap, Nuel. 1990. "Declaratives Are Not Enough." *Philosophical Studies* 59 (1): 1–30. https://doi.org/10.1007/BF00368389.

Belnap, Nuel, Michael Perloff, and Ming Xu. 2001. *Facing the Future: Agents and Choices in Our Indeterminist World.* Oxford University Press.

Bimbó, Katalin. 2007. "Relevance Logics." In *Philosophy of Logic,* edited by Dale Jacquette, 5:723–789. Handbook of the Philosophy of Science. Elsevier.

Bimbó, Katalin. 2014. *Proof Theory: Sequent Calculi and Related Formalisms.* CRC Press.

Bimbó, Katalin, and J. Michael Dunn. 2008. *Generalized Galois Logics: Relational Semantics of Nonclassical Logical Calculi.* CSLI Publications.

Bimbó, Katalin, and J. Michael Dunn. 2013. "On the Decidability of Implicational Ticket Entailment." *Journal of Symbolic Logic* 78 (1): 214–236. https://doi.org/10.2178/jsl.7801150.

Blackburn, Patrick, Maarten de Rijke, and Yde Venema. 2002. *Modal Logic.* Cambridge Tracts in Theoretical Computer Science. Cambridge University Press.

Bochvar, D. A., and Merrie Bergmann. 1981. "On a Three-Valued Logical Calculus and Its Application to the Analysis of the Paradoxes of the Classical Extended Functional Calculus." *History and Philosophy of Logic* 2 (1–2): 87–112. https://doi.org/10.1080/01445348108837023.

Boole, George. 1854. *An Investigation of the Laws of Thought: On Which Are Founded the Mathematical Theories of Logic and Probabilities.* Walton / Maberly.

Boolos, George, John P. Burgess, and Richard C. Jeffrey. 2007. *Computability and Logic.* Cambridge University Press.

Brandom, Robert B. 2000. *Articulating Reasons: An Introduction to Inferentialism.* Harvard University Press.

Chellas, Brian F. 1980. *Modal Logic: An Introduction.* Cambridge University Press.

Cintula, Petr, Christian G. Fermüller, and Carles Noguera. 2017. "Fuzzy Logic." In *The Stanford Encyclopedia of Philosophy,* Fall 2017, edited by Edward N. Zalta. Metaphysics Research Lab, Stanford University. https://plato .stanford.edu/archives/fall2017/entries/logic-fuzzy/.

Cobreros, Pablo, Paul Egré, David Ripley, and Robert van Rooij. 2012. "Tolerant, Classical, Strict." *Journal of Philosophical Logic* 41 (2): 347–385. https://doi.org/10.1007/s10992-010-9165-z.

Cobreros, Pablo, Paul Égré, David Ripley, and Robert van Rooij. 2013. "Reaching Transparent Truth." *Mind* 122 (488): 841–866. https://doi.org/10.1093/mind/fzt110.

Coffa, J. Alberto. 1993. *The Semantic Tradition from Kant to Carnap.* Edited by Linda Wessels. Cambridge University Press.

Cook, Roy T. 2010. "Let a Thousand Flowers Bloom: A Tour of Logical Pluralism." *Philosophy Compass* 5 (6): 492–504. https://doi.org/10.1111/j.1747-9991.2010.00286.x.

Crossley, John N., and Lloyd Humberstone. 1977. "The Logic of 'Actually.'" *Reports on Mathematical Logic* 8:11–29.

Davies, Martin, and Lloyd Humberstone. 1980. "Two Notions of Necessity." *Philosophical Studies* 38 (1): 1–31. https://doi.org/10.1007/BF00354523.

Doets, Kees. 1996. *Basic Model Theory.* CSLI Publications.

Dummett, Michael. 1975. "The Philosophical Basis of Intuitionistic Logic." In *Truth and Other Enigmas,* edited by Michael Dummett, 215–247. Harvard University Press.

Dummett, Michael. 1991. *The Logical Basis of Metaphysics.* Harvard University Press.

Dummett, Michael. 2000. *Elements of Intuitionism.* 2nd ed. Oxford University Press.

Dunn, J. Michael. 1966. "The Algebra of Intensional Logics." PhD diss., University of Pittsburgh.

Dunn, J. Michael, and Gary M. Hardegree. 2001. *Algebraic Methods in Philosophical Logic.* Oxford University Press.

Dunn, J. Michael, and Greg Restall. 2002. "Relevance Logic." In *Handbook of Philosophical Logic,* 2nd ed, edited by Dov M. Gabbay and Franz Guenthner, 6:1–136. Kluwer.

Dutilh Novaes, Catarina. 2020. "Medieval Theories of Consequence." In *The Stanford Encyclopedia of Philosophy,* Fall 2020, edited by Edward N. Zalta. Metaphysics Research Lab, Stanford University. https://plato.stanford .edu/archives/fall2020/entries/consequence-medieval/.

Enderton, Herbert B. 2001. *A Mathematical Introduction to Logic.* 2nd ed. Academic Press.

Esakia, Leo. 2019. *Heyting Algebras: Duality Theory.* Springer Verlag.

Field, Hartry. 2008. *Saving Truth from Paradox.* Oxford University Press.

Fine, Kit. 1985. *Reasoning with Arbitrary Objects.* Blackwell.

Francez, Nissim. 2017. "On Harmony and Permuting Conversions." *Journal of Applied Logic* 21:14–23. https:/ /doi.org/10.1016/j.jal.2016.12.004.

Francez, Nissim, and Roy Dyckhoff. 2012. "A Note on Harmony." *Journal of Philosophical Logic* 41 (3): 613– 628. https://doi.org/10.1007/s10992-011-9208-0.

Franzén, Torkel. 2005. *Gödel's Theorem: An Incomplete Guide to Its Use and Abuse.* A. K. Peters.

French, Rohan. 2016. "Structural Reflexivity and the Paradoxes of Self-Reference." *Ergo: An Open Access Journal of Philosophy* 3. https://doi.org/10.3998/ergo.12405314.0003.005.

Fusco, Melissa. 2015. "Deontic Modality and the Semantics of Choice." *Philosophers' Imprint* 15:1–27.

Gabbay, Dov M. 1981. *Semantical Investigations in Heyting's Intuitionistic Logic.* Springer Netherlands. https:/ /doi.org/10.1007/978-94-017-2977-2.

Gabbay, Dov M., and Franz Guenthner, eds. 2001. *Handbook of Philosophical Logic.* 2nd ed. Springer Dordrecht. https://doi.org/10.1007/978-94-015-9833-0.

Garson, James W. 2013. *Modal Logic for Philosophers.* 2nd ed. Cambridge University Press.

Garson, James W. 2018. "Modal Logic." In *The Stanford Encyclopedia of Philosophy,* Fall 2018, edited by Edward N. Zalta. Metaphysics Research Lab, Stanford University. https://plato.stanford.edu/archives/fall2018 /entries/logic-modal/.

Gentzen, Gerhard. 1934. "Untersuchungen über das logische Schliessen I, II." Translated in *The Collected Papers of Gerhard Gentzen* Gentzen 1969, *Mathematische Zeitschrift* 39 (2–3): 176–210, 405–431.

Gentzen, Gerhard. 1969. *The Collected Papers of Gerhard Gentzen.* Edited by M. E. Szabo. North Holland.

Goble, Lou. 2001. *The Blackwell Guide to Philosophical Logic.* Wiley-Blackwell.

Grover, Dorothy L., Joseph L. Camp, and Nuel D. Belnap. 1975. "A Prosentential Theory of Truth." *Philosophical Studies* 27 (1): 73–125. https://doi.org/10.1007/BF01209340.

Gupta, Anil, and Nuel Belnap. 1993. *The Revision Theory of Truth.* MIT Press.

Haack, Susan. 1974. *Deviant Logic: Some Philosophical Issues.* Updated edition published as *Deviant Logic, Fuzzy Logic: Beyond the Formalism* (1996). Cambridge University Press. Cambridge University Press.

Hájek, Petr. 1998. *Metamathematics of Fuzzy Logic.* Kluwer.

Halbach, Volker. 2011. *Axiomatic Theories of Truth.* Cambridge University Press.

Halbach, Volker, and Graham E. Leigh. 2020. "Axiomatic Theories of Truth." In *The Stanford Encyclopedia of Philosophy,* Spring 2020, edited by Edward N. Zalta. Metaphysics Research Lab, Stanford University. https://plato.stanford.edu/archives/spr2020/entries/truth-axiomatic/.

Hanks, Peter. 2009. "Recent Work on Propositions." *Philosophy Compass* 4 (3): 469–486. https://doi.org/10.1111/j.1747-9991.2009.00208.x.

Hazen, Allen. 1987. "Natural Deduction and Hilbert's $\varepsilon$-Operator." *Journal of Philosophical Logic* 16 (4): 411–421. https://doi.org/10.1007/BF00431186.

Henkin, Leon. 1949. "The Completeness of the First-Order Functional Calculus." *Journal of Symbolic Logic* 14 (3): 159–166. https://doi.org/10.2307/2267044.

Hintikka, Jaakko. 1962. *Knowledge and Belief: An Introduction to the Logic of the Two Notions.* Cornell University Press.

Hodges, Wilfrid. 1997. *A Shorter Model Theory.* Cambridge University Press.

Horty, John F. 2001. *Agency and Deontic Logic.* Oxford University Press.

Hrbacek, Karel, and Thomas Jech. 1999. *Introduction to Set Theory.* Chapman/CRC.

Hughes, G. E., and M. J. Cresswell. 1996. *A New Introduction to Modal Logic.* Routledge.

Humberstone, Lloyd. 1988. "Heterogeneous Logic." *Erkenntnis* 29 (3): 395–435. https://doi.org/10.1007/BF00183072.

Humberstone, Lloyd. 2011. *The Connectives.* MIT Press.

Humberstone, Lloyd. 2016. *Philosophical Applications of Modal Logic.* College Publications.

Indrzejczak, Andrzej. 2010. *Natural Deduction, Hybrid Systems and Modal Logics.* Springer Netherlands. https://doi.org/10.1007/978-90-481-8785-0.

Jacinto, Bruno, and Stephen Read. 2017. "General-Elimination Stability." *Studia Logica* 105 (2): 361–405. https://doi.org/10.1007/s11225-016-9692-x.

Jespersen, Bjørn. 2012. "Recent Work on Structured Meaning and Propositional Unity." *Philosophy Compass* 7 (9): 620–630. https://doi.org/10.1111/j.1747-9991.2012.00509.x.

Kalish, Donald, and Richard Montague. 1964. *Logic: Techniques of Formal Reasoning.* Oxford University Press.

King, Jeffrey C. 2019. "Structured Propositions." In *The Stanford Encyclopedia of Philosophy,* Summer 2019, edited by Edward N. Zalta. Metaphysics Research Lab, Stanford University. https://plato.stanford.edu/archives/sum2019/entries/propositions-structured/.

Kripke, Saul. 1963. "Semantical Analysis of Intuitionistic Logic I." In *Formal Systems and Recursive Functions: Proceedings of the Eighth Logic Colloquium, Oxford July 1963,* edited by Michael Dummett and J. N. Crossley, 92–130. North Holland.

Kripke, Saul. 1972. *Naming and Necessity.* Harvard University Press.

Kripke, Saul. 1975. "Outline of a Theory of Truth." *Journal of Philosophy* 72:690–716.

Kürbis, Nils. 2015. "What Is Wrong with Classical Negation?" *Grazer Philosophische Studien* 92 (1): 51–86. https://doi.org/10.1163/9789004310841\_004.

Kürbis, Nils. 2019. *Proof and Falsity: A Logical Investigation.* Cambridge University Press.

Leivant, Daniel. 1979. "Assumption Classes in Natural Deduction." *Zeitschrift für Mathematische Logik und Grundlagen der Mathematik* 25 (1–2): 1–4. https://doi.org/10.1002/malq.19790250102.

Leslie, Sarah-Jane. 2007. "Generics and the Structure of the Mind." *Philosophical Perspectives* 21 (1): 375–403. https://doi.org/10.1111/j.1520-8583.2007.00138.x.

Lewis, Clarence Irving. 1918. *A Survey of Symbolic Logic.* University of California Press.

Lewis, Clarence Irving, and Cooper Langford. 1932. *Symbolic Logic.* Century Company.

MacFarlane, John. 2020. *Philosophical Logic: A Contemporary Introduction.* Routledge.

Mancosu, Paolo, Sergio Galvan, and Richard Zach. 2021. *An Introduction to Proof Theory: Normalization, Cut-Elimination, and Consistency Proofs.* Oxford University Press.

Mares, Edwin. 2016. "The Development of C. I. Lewis's Philosophy of Modal Logic." In *Logical Modalities from Aristotle to Carnap: The Story of Necessity,* edited by Max Cresswell, Edwin Mares, and Adriane Rini, 279–297. Cambridge University Press. https://doi.org/10.1017/CBO9781139939553.015.

McKinsey, J. C. C., and Alfred Tarski. 1948. "Some Theorems About the Sentential Calculi of Lewis and Heyting." *Journal of Symbolic Logic* 13 (1): 1–15. https://doi.org/10.2307/2268135.

Medeiros, Maria Da Paz N. 2006. "A New S4 Classical Modal Logic in Natural Deduction." *Journal of Symbolic Logic* 71 (3): 799–809. https://doi.org/10.2178/jsl/1154698578.

Montague, Richard. 1970. "English as a Formal Language." In *Linguaggi nella societa e nella tecnica,* edited by Bruno Visentini, 188–221. Edizioni di Communita.

Moschovakis, Joan. 2018. "Intuitionistic Logic." In *The Stanford Encyclopedia of Philosophy,* Winter 2018, edited by Edward N. Zalta. Metaphysics Research Lab, Stanford University. https://plato.stanford.edu/archives/win2018/entries/logic-intuitionistic/.

Negri, Sara. 2011. "Proof Theory for Modal Logic." *Philosophy Compass* 6 (8): 523–538. https://doi.org/10.1111/j.1747-9991.2011.00418.x.

Negri, Sara, and Jan von Plato. 2001. *Structural Proof Theory.* Cambridge University Press.

Nelson, Michael, and Edward N. Zalta. 2012. "A Defense of Contingent Logical Truths." *Philosophical Studies* 157 (1): 153–162. https://doi.org/10.1007/s11098-010-9624-y.

Omori, Hitoshi, and Heinrich Wansing. 2019. *New Essays on Belnap-Dunn Logic.* Springer Verlag.

Pacuit, Eric. 2013a. "Dynamic Epistemic Logic I: Modeling Knowledge and Belief." *Philosophy Compass* 8 (9): 798–814. https://doi.org/10.1111/phc3.12059.

Pacuit, Eric. 2013b. "Dynamic Epistemic Logic II: Logics of Information Change." *Philosophy Compass* 8 (9): 815–833. https://doi.org/10.1111/phc3.12060.

Paoli, Francesco. 2002. *Substructural Logics: A Primer.* Kluwer.

Pelletier, F., and N. Asher. 1997. "Generics and Defaults." In *Handbook of Logic and Language,* edited by J. van Benthem and A. ter Meulen, 1125–1179. MIT Press.

Pelletier, Francis. 1999. "A Brief History of Natural Deduction." *History and Philosophy of Logic* 20 (1): 1–31. https://doi.org/10.1080/014453499298165.

Plumwood, Val. 1994. *Feminism and the Mastery of Nature.* Routledge.

Poggiolesi, Francesca. 2010. *Gentzen Calculi for Modal Propositional Logic.* Springer.

Poggiolesi, Francesca, and Greg Restall. 2012. "Interpreting and Applying Proof Theories for Modal Logics." In *New Waves in Philosophical Logic,* edited by Greg Restall and Gillian Russell, 39–62. Palgrave Macmillan.

Pospesel, Howard, and David Marans. 1978. *Arguments: Deductive Logic Exercises.* 2nd ed. Prentice Hall.

Potter, Michael. 2004. *Set Theory and Its Philosophy: A Critical Introduction.* Oxford University Press.

Prawitz, Dag. 1965. *Natural Deduction: A Proof-Theoretical Study.* Almqvist and Wiksell.

Priest, Graham. 1979. "The Logic of Paradox." *Journal of Philosophical Logic* 8 (1): 219–241. https://doi.org/10.1007/BF00258428.

Priest, Graham. 2008. *An Introduction to Non-Classical Logic: From If to Is.* Cambridge University Press.

Prior, Arthur. 1961. "The Runabout Inference Ticket." *Analysis* 21:38–39.

Prior, Arthur. 1967. *Past, Present and Future.* Clarendon.

Putnam, Hillary. 1975. "The Meaning of 'Meaning.'" *Minnesota Studies in the Philosophy of Science* 7:131–193.

Raggio, Andrés. 1965. "Gentzen's Hauptsatz for the Systems NI and NK." *Logique Et Analyse* 8:91–100.

Rasiowa, Helena. 1974. *An Algebraic Approach to Non-Classical Logics.* Polish Scientific Publishers.

Rasiowa, Helena, and Roman Sikorski. 1963. *The Mathematics of Metamathematics.* Państwowe Wydawn.

Read, Stephen. 1994. *Thinking about Logic: An Introduction to the Philosophy of Logic.* Oxford University Press.

Restall, Greg. 2000. *An Introduction to Substructural Logics.* Routledge.

Restall, Greg. 2005. "Multiple Conclusions." In *Logic, Methodology and Philosophy of Science,* edited by Petr Hájek, Luis Valdés-Villanueva, and Dag Westerståhl, 189–205. College Publications.

Restall, Greg. 2006. *Logic: An Introduction.* Routledge.

Restall, Greg. 2012. "A Cut-Free Sequent System for Two-Dimensional Modal Logic, and Why It Matters." *Annals of Pure and Applied Logic* 163 (11): 1611–1623. https://doi.org/10.1016/j.apal.2011.12.012.

Rini, A. A., and M. J. Cresswell. 2012. *The World-Time Parallel: Tense and Modality in Logic and Metaphysics.* Cambridge University Press.

Ripley, David. 2013a. "Paradoxes and Failures of Cut." *Australasian Journal of Philosophy* 91 (1): 139–164. https://doi.org/10.1080/00048402.2011.630010.

Ripley, David. 2013b. "Revising Up: Strengthening Classical Logic in the Face of Paradox." *Philosophers' Imprint* 13:1–13.

Routley, R., and V. Routley. 1972. "The Semantics of First Degree Entailment." *Noûs* 6 (4): 335–359. https://doi.org/10.2307/2214309.

Routley, Richard. 2018. *Exploring Meinong's Jungle and Beyond: The Sylvan Jungle.* Edited by Maureen Eckert. Vol. 1. Springer Verlag.

Routley, Richard, Val Plumwood, Robert K. Meyer, and Ross T. Brady. 1982. *Relevant Logics and Their Rivals.* Vol. 1. Ridgeview.

Russell, Bertrand. 1905. "On Denoting." *Mind* 14:479–493. http://dx.doi.org/10.1093/mind/fzi873.

Russell, Gillian. 2019. "Logical Pluralism." In *The Stanford Encyclopedia of Philosophy,* Summer 2019, edited by Edward N. Zalta. Metaphysics Research Lab, Stanford University. https://plato.stanford.edu/archives/sum2019/entries/logical-pluralism/.

Schroeder-Heister, Peter. 2018. "Proof-Theoretic Semantics." In *The Stanford Encyclopedia of Philosophy,* Spring 2018, edited by Edward N. Zalta. Metaphysics Research Lab, Stanford University. https://plato.stanford.edu/archives/spr2018/entries/proof-theoretic-semantics/.

Schroeter, Laura. 2003. "Gruesome Diagonals." *Philosophers' Imprint* 3 (3): 1–23. http://www.philosophersimprint.org/images/3521354.0003.003.pdf.

Schroeter, Laura. 2017. "Two-Dimensional Semantics." In *The Stanford Encyclopedia of Philosophy,* Summer 2017, edited by Edward N. Zalta. Metaphysics Research Lab, Stanford University. https://plato.stanford.edu/archives/sum2017/entries/two-dimensional-semantics/.

Seldin, Jonathan P. 1989. "Normalization and Excluded Middle I." *Studia Logica* 48 (2): 193–217. https://doi.org/10.1007/bf02770512.

Shapiro, Stewart. 1991. *Foundations without Foundationalism: A Case for Second-Order Logic.* Oxford University Press.

Shoesmith, D. J., and T. J. Smiley. 1978. *Multiple-Conclusion Logic.* Cambridge University Press.

Sider, Theodore. 2010. *Logic for Philosophy.* Oxford University Press.

Slaney, J. K. 1989. "RWX Is Not Curry Paraconsistent." In *Paraconsistent Logic: Essays on the Inconsistent,* edited by Graham Priest, Richard Sylvan, and Jean Norman, 472–480. Philosophia Verlag.

Smith, Nicholas J. J. 2008. *Vagueness and Degrees of Truth.* Oxford University Press.

Smith, Peter. 2013. *An Introduction to Gödel's Theorems.* 2nd ed. Cambridge University Press.

Smullyan, Raymond M. 1968. *First-Order Logic.* Reprinted by Dover Press, 1995. Springer Verlag.

Smullyan, Raymond M. 1992. *Gödel's Incompleteness Theorems.* Oxford University Press.

Stalnaker, Robert. 1978. "Assertion." In *Context and Content,* 78–95. Oxford University Press.

Steinberger, Florian. 2011. "What Harmony Could and Could Not Be." *Australasian Journal of Philosophy* 89 (4): 617–639. https://doi.org/10.1080/00048402.2010.528781.

Sterken, Rachel Katharine. 2017. "The Meaning of Generics." *Philosophy Compass* 12 (8): e12431. https://doi.org/10.1111/phc3.12431.

Tarski, Alfred. 1944. "The Semantic Conception of Truth: And the Foundations of Semantics." *Philosophy and Phenomenological Research* 4:341–376.

Tennant, Neil. 2017. *Core Logic.* Oxford University Press.

Tent, Katrin, and Martin Ziegler. 2012. *A Course in Model Theory*. Cambridge University Press. https://doi.org/10.1017/CBO9781139015417.

Tranchini, Luca. 2018. "Stabilizing Quantum Disjunction." *Journal of Philosophical Logic* 47 (6): 1029–1047. https://doi.org/10.1007/s10992-018-9460-7.

Uckelman, Sara L. 2021. "What Problem Did Ladd-Franklin (Think She) Solve(d)?" *Notre Dame Journal of Formal Logic* 62 (3): 527–552. https://doi.org/10.1215/00294527-2021-0026.

van Benthem, Johan. 1991. *The Logic of Time: A Model-Theoretic Investigation into the Varieties of Temporal Ontology and Temporal Discourse*. Kluwer Academic.

van Benthem, Johan. 1996. *Exploring Logical Dynamics*. CSLI Publications.

van Benthem, Johan. 2010. *Modal Logic for Open Minds*. CSLI Publications.

van Dalen, Dirk. 2002. "Intuitionistic Logic." In *Handbook of Philosophical Logic,* edited by Dov M. Gabbay and F. Guenthner, 1–114. Springer Netherlands. https://doi.org/10.1007/978-94-017-0458-8_1.

van Ditmarsch, Hans, Wiebe van der Hoek, and Barteld Kooi. 2007. *Dynamic Epistemic Logic*. Synthese Library. Springer.

von Plato, Jan. 2013. *Elements of Logical Reasoning*. Cambridge University Press.

von Plato, Jan, and Gerhard Gentzen. 2008. "Gentzen's Proof of Normalization for Natural Deduction." *Bulletin of Symbolic Logic* 14 (2): 240–257. https://doi.org/10.2178/bsl/1208442829.

Walsh, Sean, and Tim Button. 2018. *Philosophy and Model Theory*. Oxford University Press.

Wansing, Heinrich. 1998. *Displaying Modal Logic*. Kluwer.

Williamson, Timothy. 2013. *Modal Logic as Metaphysics*. Oxford University Press.

Zach, Richard. 2019. *Boxes and Diamonds: An Open Introduction to Modal Logic*. Open Logic Project. https://bd.openlogicproject.org.

Zach, Richard. 2020. *Incompleteness and Computability*. Open Logic Project. https://ic.openlogicproject.org.

Zardini, Elia. 2011. "Truth without Contra(di)ction." *Review of Symbolic Logic* 4 (4): 498–535. https://doi.org/10.1017/s1755020311000177.

# Symbol Index

# Subject Index